Handbook of Integration

Jones and Bartlett Books in Mathematics

Baum, R.J., *Philosophy and Mathematics*
ISBN 0-86720-514-2

Eisenbud, D., and Huneke, C., *Free Resolutions in Commutative Algebra and Algebraic Geometry*
ISBN 0-86720-285-8

Epstein, D.B.A., *et al.*, *Word Processing in Groups*
ISBN 0-86720-241-6

Epstein, D.B.A., and Gunn, C., *Supplement to Not Knot*
ISBN 0-86720-297-1

Geometry Center, University of Minnesota, *Not Knot* (VHS video)
ISBN 0-86720-240-8

Gleason, A., *Fundamentals of Abstract Analysis*
ISBN 0-86720-209-2

Hansen, V.L., *Geometry in Nature*
ISBN 0-86720-238-6

Harpaz, A., *Relativity Theory: Concepts and Basic Principles*
ISBN 0-86720-220-3

Loomis, L.H., and Sternberg, S., *Advanced Calculus*
ISBN 0-86720-122-2

Morgan, F., *Riemannian Geometry: A Beginner's Guide*
ISBN 0-86720-242-4

Protter, M.H., and Protter, P.E., *Calculus, Fourth Edition*
ISBN 0-86720-093-6

Redheffer, R., *Differential Equations: Theory and Applications*
ISBN 0-86720-200-9

Redheffer, R., *Introduction to Differential Equations*
ISBN 0-86720-289-0

Ruskai, M.B., *et al.*, *Wavelets and Their Applications*
ISBN 0-86720-225-4

Serre, J.-P., *Topics in Galois Theory*
ISBN 0-86720-210-6

Zwillinger, D., *Handbook of Integration*
ISBN 0-86720-293-9

Handbook of Integration

Daniel Zwillinger

Department of Mathematical Sciences
Rensselaer Polytechnic Institute
Troy, New York

Jones and Bartlett Publishers
Boston London

Editorial, Sales, and Customer Service Offices
Jones and Bartlett Publishers
One Exeter Plaza
Boston, MA 02116

Jones and Bartlett Publishers International
P.O. Box 1498
London W6 7RS
England

This book was typeset by the author using TEX.

Figures 57.1 and 57.2 originally appeared on pages 79 and 80 in R. Piessens, E. de Doncker-Kapenga, C.W. Überhuber, and D.K. Kahaner, *Quadpack*, Springer-Verlag, 1983. Reprinted courtesy of Springer-Verlag.

Library of Congress Cataloging-in-Publication Data

Zwillinger, Daniel, 1957–
Handbook of integration / Daniel Zwillinger.
p. cm.
Includes bibiliographical references and index.
ISBN 0-86720-293-9
1. Numerical integration. I. Title.
QA299.3.Z85 1992
515′.43—dc20 92-14050
CIP

Printed in the United States of America

96 95 94 93 92 10 9 8 7 6 5 4 3 2 1

Table of Contents

I Applications of Integration

II Concepts and Definitions

III Exact Analytical Methods

IV Approximate Analytical Methods

V Numerical Methods: Concepts

VI Numerical Methods: Techniques

Preface

This book was begun when I was a graduate student in applied mathematics at the California Institute of Technology. Being able to integrate functions easily is a skill that is presumed at the graduate level. Yet, some integrals can only be simplified by using clever manipulations. I found it useful to create a list of manipulation techniques. Each technique on this list had a brief description of how the method was used and to what types of integrals it applied. As I learned more techniques they were added to the list. This book is a direct outgrowth of that list.

In performing mathematical analysis, analytic evaluation of integrals is often required. Other times, an approximate integration may be more informative than a representation of the exact answer. (The exact representation could, for example, be in the form of an infinite series.) Lastly, a numerical approximation to an integral may be all that is required in some applications.

This book is therefore divided into five sections:

- *Applications of Integration* which shows how integration is used in differential equations, geometry, probability and performing summations;
- *Concepts and Definitions* which defines several different types of integrals and operations on them;
- *Exact Techniques* which indicates several ways in which integrals may be evaluated exactly;
- *Approximate Techniques* which indicates several ways in which integrals may be evaluated approximately; and
- *Numerical Techniques* which indicates several ways in which integrals may be evaluated numerically.

This handbook has been designed as a reference book. Many of the techniques in this book are standard in an advanced course in mathematical methods. Each technique is accompanied by several current references;

these allow each topic to be studied in more detail. This book should be useful to students and also to practicing engineers or scientists who must evaluate integrals on an occasional basis.

Had this book been available when I was a graduate student, it would have saved me much time. It has saved me time in evaluating integrals that arose from my own work in industry (the Jet Propulsion Laboratory, Sandia Laboratories, EXXON Research and Engineering, and the MITRE Corporation).

Unfortunately, there may still be some errors in the text; I would greatly appreciate receiving notice of any such errors. Please send these comments care of Jones and Bartlett.

No book is created in a vacuum, and this one is no exception. Thanks are extended to Harry Dym, David K. Kahaner, Jay Ramanthan, Doug Reinelt, and Michael Strauss for reviewing the manuscript. Their help has been instrumental in clarifying the text. Lastly, this book would have not been possible without the enthusiasm of my editor, Alice Peters.

Boston, MA 1992 Daniel Zwillinger

Introduction

This book is a compilation of the most important and widely applicable methods for evaluating and approximating integrals. As a reference book, it provides convenient access to these methods and contains examples showing their use.

The book is divided into five parts. The first part lists several applications of integration. The second part contains definitions and concepts and has some useful transformations of integrals. This section of the book defines many different types of integrals, indicates what Feynman diagrams are, and describes many useful transformations.

The third part of the book is a collection of exact analytical evaluation techniques for integrals. For nearly every technique the following are given:

- the types of integrals to which the method is applicable
- the idea behind the method
- the procedure for carrying out the method
- at least one simple example of the method
- notes for more advanced users
- references to the literature for more discussion or examples.

The material for each method has deliberately been kept short to simplify use. Proofs have been intentionally omitted.

It is hoped that, by working through the simple example(s) given, the method will be understood. Enough insight should be gained from working the example(s) to apply the method to other integrals. References are given for each method so that the principle may be studied in more detail, or more examples seen. Note that not all of the references listed at the end of a section may be referred to in the text.

The author has found that computer languages that perform symbolic manipulations (such as Macsyma) are very useful when performing the calculations necessary to analyze integrals. Examples of several symbolic manipulation computer languages are given.

Not all integrals can be evaluated analytically in terms of elementary functions; sometimes an approximate evaluation will have to do. Other times, an approximate evaluation may be *more* useful than an exact evaluation. For instance, an exact evaluation in terms of a slowly converging infinite series may be laborious to approximate numerically. The same integral may have a simple approximation that indicates some characteristic behavior or easily allows a numerical value to be obtained.

The fourth part of this book deals with approximate analytical solution techniques. For the methods in this part of the book, the format is similar to that used for the exact solution techniques. We classify a method as an approximate method if it gives some information about the value of an integral but will not specify the value of the integral at all values of the independent variable(s) appearing in the integral. The methods in this section describe, for example, the method of stationary phase and the method of steepest descent.

When an exact or an approximate solution technique cannot be found, it may be necessary to find the solution numerically. Other times, a numerical solution may convey more information than an exact or approximate analytical solution. The fifth part of this book deals with the most important methods for obtaining numerical approximations to integrals. From a vast literature of techniques available for numerically approximating integrals, this book has only tried to illustrate some of the more important techniques. At the beginning of the fifth section is a brief introduction to the concepts and terms used in numerical methods.

This book is not designed to be read at one sitting. Rather, it should be consulted as needed. This book contains many references to other books. While some books cover only one or two topics well, some books cover all their topics well. The following books are recommended as a first source for detailed understanding of the integration techniques they cover: Each is broad in scope and easy to read.

References

[1] C. M. Bender and S. A. Orszag, *Advanced Mathematical Methods for Scientists and Engineers*, McGraw–Hill, New York, 1978.

[2] P. J. Davis and P. Rabinowitz, *Methods of Numerical Integration*, Second Edition, Academic Press, Orlando, Florida, 1984.

[3] W. Squire, *Integration for Engineers and Scientists*, American Elsevier Publishing Company, New York, 1970.

How to Use This Book

This book has been designed to be easy to use when evaluating integrals, whether exactly, approximately, or numerically. This introductory section outlines how this book may be used to analyze a given integral.

First, determine if the integral has been studied in the literature. A list of many integrals may be found in the "Look Up Technique" section beginning on page 170. If the integral you wish to analyze is contained in one of the lists in that section, then see the indicated reference. This technique is the single most useful technique in this book.

Special Forms

[1] If the integral has a special form, then it may be evaluated in closed form without too much difficulty. If the integral has the form:

(A) $\int^x R(x)\,dx$, where $R(x)$ is a rational function then the integral can be evaluated in terms of logarithms and arc-tangents (see page 183).

(B) $\int^x P(x,\sqrt{R})\log Q(x,\sqrt{R})\,dx$, where $P(\,,\,)$ and $Q(\,,\,)$ are rational functions and $R = A^2 + Bx + Cx^2$, then the integral can be evaluated in terms of dilogarithms (see page 145).

(C) $\int^x R(x,\sqrt{T(x)})\,dx$ where $R(\,,\,)$ is a rational function of its arguments and $T(x)$ is a third of fourth order polynomial, then the integral can be evaluated in terms of elliptic functions (see page 147).

(D) $\int_0^{2\pi} f(\cos\theta,\sin\theta)\,d\theta$, then the integral may be re-formulated as a contour integral (see page 129).

[2] If the integral is a contour integral, see page 129.

[3] If the integral is a path integral, see page 86.

[4] If the integral is a principal-value integral (i.e., the integral sign looks like $⨍$), then see page 92.

[5] If the integral is a finite-part integral (i.e., the integral sign looks like $\fint$), then see page 73.

[6] If the integral is a loop integral (i.e, the integral sign looks like $\oint$), or if the integration is over a closed curve in the complex plane, then see pages 129 or 164.

[7] If the integral appears to be divergent, then the integral might need to be interpreted as a principal-value integral (see page 92) or as a finite-part integral (see page 73).

Looking for an Exact Evaluation

[1] If you have access to a symbolic manipulation computer language (such as Maple, Macsyma, or Derive), then see page 117.

[2] For a given integral, if one integration technique does not work, try another. Most integrals that can be analytically evaluated can be evaluated by more than one technique. For example, the integral

$$I = \int_0^\infty \frac{\sin x}{x}\, dx$$

is shown to converge on page 66. Then I is evaluated (using different methods) on pages 118, 133, 144, 145, and 185.

Looking for an Approximate Evaluation

[1] If λ is large, C is an integration contour, and the integral has the form:

(A) $\int_C e^{\lambda f(x)} g(x)\, dx$, then the method of steepest descents may be used (see page 229).

(B) $\int_C e^{\lambda f(x)} g(x)\, dx$, where $f(x)$ is a real function, then Laplace's method may be used (see page 221).

(C) $\int_C e^{i\lambda f(x)} g(x)\, dx$, where $f(x)$ is a real function, then the method of stationary phase may be used (see page 226).

There is a collection of other special forms on page 181.

[2] Interval analysis techniques, whether implemented analytically or numerically, permit exact upper and lower bounds to be determined for an integral (see page 218).

Looking for a Numerical Evaluation

[1] It is often easiest to use commercial software packages when looking for a numerical solution (see page 254). The type of routine needed may be determined from the taxonomy section (see page 258). The taxonomy classification may then be used as the entry in the table of software starting on page 260.

[2] If a low accuracy solution is acceptable, then a Monte Carlo solution technique may be used, see page 304.

[3] If the integral in question has a very high dimension, then Monte Carlo methods may be the only usable technique, see page 304.

[4] If a parallel computer is available to you, then see page 315.

[5] If the integrand is periodic, then lattice rules may be appropriate. See page 300.

[6] References for quadrature rules involving specific geometric regions, or for integrands with a specific functional form, may be found on page 337.

[7] Examples of some one-dimensional and two-dimensional quadratures rules may be found on page 340.

[8] A listing of some integrals that have been tabulated in the literature may be found on page 348.

Other Things to Consider

[1] Is fractional integration involved? See page 75.

[2] Is a proof that the integral cannot be evaluated in terms of elementary functions desired? See page 77.

[3] Does the equation involve a large or small parameter? See the asymptotic methods described on pages 195 and 199.

I

Applications of Integration

1. Differential Equations: Integral Representations

Applicable to Linear differential equations.

Idea

Sometimes the solution of a linear ordinary differential equation can be written as a contour integral.

Procedure

Let $L_z[\cdot]$ be a linear differential operator with respect to z, and suppose that the ordinary differential equation we wish to solve has the form

$$L_z[u(z)] = 0. \tag{1.1}$$

We look for a solution of (1.1) in the form of an integral

$$u(z) = \int_C K(z, \xi) v(\xi) \, d\xi, \tag{1.2}$$

for some function $v(\xi)$ and some contour C in the complex ξ plane. The function $K(z,\xi)$ is called the *kernel.* Some common kernels are:

$$\begin{aligned} &\text{Euler kernel:} && K(z,\xi) = (z-\xi)^n \\ &\text{Laplace kernel:} && K(z,\xi) = e^{\xi z} \\ &\text{Mellin kernel:} && K(z,\xi) = z^{\xi} \end{aligned}$$

We combine (1.2) and (1.1) for

$$\int_C L_z[K(z,\xi)]v(\xi)\,d\xi = 0. \tag{1.3}$$

Now we must (conceptually) find a linear differential operator $A_\xi[\cdot]$, operating with respect to ξ, such that $L_z[K(z,\xi)] = A_\xi[K(z,\xi)]$. After $A_\xi[\cdot]$ has been found, then (1.3) can be rewritten as

$$\int_C A_\xi[K(z,\xi)]v(\xi)\,d\xi = 0. \tag{1.4}$$

Now we integrate (1.4) by parts. The resulting expression will be a differential equation to be solved for $v(\xi)$ and some boundary terms. The boundary terms will determine the contour C. Knowing both $v(\xi)$ and C, the solution to (1.1) is given by the integral in (1.2).

Example

Consider Airy's differential equation

$$L_z[u] = u''(z) - zu(z) = 0. \tag{1.5}$$

We assume that the solution of (1.5) has the form

$$u(z) = \int_C e^{z\xi}v(\xi)\,d\xi, \tag{1.6}$$

for some $v(\xi)$ and some contour C. Substituting (1.6) into (1.5) we find

$$\int_C \xi^2 v(\xi)e^{z\xi}\,d\xi - z\int_C v(\xi)e^{z\xi}\,d\xi = 0. \tag{1.7}$$

The second term in (1.7) can be integrated by parts to obtain

$$\int_C \xi^2 v(\xi)e^{z\xi}\,d\xi - \left[v(\xi)e^{z\xi}\right|_C + \int_C v'(\xi)e^{z\xi}\,d\xi = 0,$$

or

$$-\left[v(\xi)e^{z\xi}\right|_C + \int_C e^{z\xi}\left[\xi^2 v(\xi) + v'(\xi)\right]\,d\xi = 0. \tag{1.8}$$

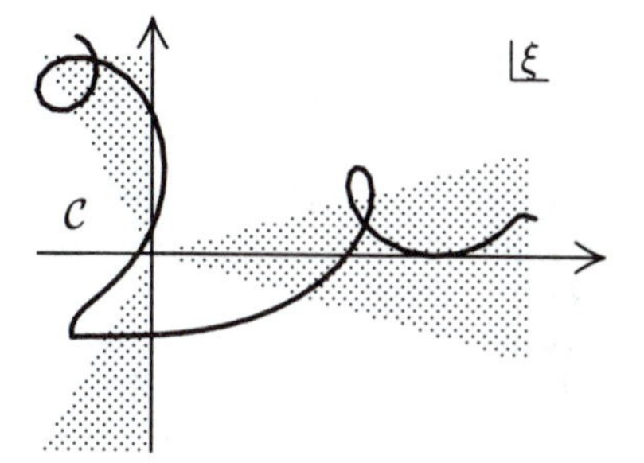

Figure 1. A solution to (1.5) is determined by any contour $\mathcal{C}$ that starts and ends in the shaded regions. All of the shaded regions extend to infinity. One possible contour is shown.

We choose

$$A_\xi[v] = \xi^2 v(\xi) + v'(\xi) = 0, \tag{1.9}$$

and the boundary conditions

$$\left[v(\xi) e^{z\xi} \right|_{\mathcal{C}} = 0. \tag{1.10}$$

With these choices, equation (1.8) is satisfied. From (1.9) we can solve for $v(\xi)$:

$$v(\xi) = \exp\left(-\frac{\xi^3}{3} \right). \tag{1.11}$$

Using (1.11) in (1.10) we must choose the contour $\mathcal{C}$ so that

$$\left[v(\xi) e^{z\xi} \right|_{\mathcal{C}} = \left[\exp\left(z\xi - \frac{\xi^3}{3} \right) \right|_{\mathcal{C}} = 0, \tag{1.12}$$

for all real values of z. The only restriction that (1.12) places on $\mathcal{C}$ is that the contour start and end in one of the shaded regions shown in Figure 1. Finally, the solution to (1.5) can be written as the integral

$$u(z) = \int_{\mathcal{C}} e^{(\xi z - \xi^2/3)}\, d\xi. \tag{1.13}$$

Asymptotic methods can be applied to (1.13) to determine information about $u(z)$.

Notes

[1] This method is also known as Laplace's method.

[2] Loop integrals are contour integrals in which the path of integration is given by a loop. For example, the integral $\int_{-\infty}^{(0+)}$ indicates an integral that starts at negative infinity, loops around the origin once, in the clockwise sense, and then returns to negative infinity.

Using the methods is this section, the fundamental solutions to Legendre's differential equation, $(1-x^2)y'' - 2xy' + n(n+1)y = 0$, can be written in the form of loop integrals:

$$\begin{aligned} P_\nu(z) &= \frac{1}{2\pi i} \oint^{(1+,z+)} \frac{(\zeta^2-1)^\nu}{2^\nu (z-\zeta)^{\nu+1}}\, d\zeta \\ Q_\nu(z) &= \frac{1}{4i \sin \nu\pi} \oint^{(1+,-1+)} \frac{(\zeta^2-1)^\nu}{2^\nu (z-\zeta)^{\nu+1}}\, d\zeta \end{aligned} \tag{1.14.a–b}$$

where the integration contour in (1.14.a) is a closed curve with positive direction passing through the ζ-plane, avoiding the half-line $(-\infty,-1)$, and admitting 1 and z as inner points of the domain it bounds. The contour in (1.14.b) is a closed ∞-shaped curve encircling the point 1 once in the negative direction and the point -1 once in the positive direction. Equation (1.14.a) is known as Schläfli's integral representation.

[3] Since there are three regions in Figure 1, there are three different contours that can start and end in one of these regions; each corresponds to a solution of (1.5). The functions $\mathrm{Ai}(x)$ and $\mathrm{Bi}(x)$, appropriately scaled, are obtained by two of these three choices for the contour (see page 171). The third solution is a linear combination of the functions $\mathrm{Ai}(x)$ and $\mathrm{Bi}(x)$.

[4] Sometimes a double integral may be required to find an integral representation. In this case, a solution of the form $u(z) = \iint K(z;s,t)w(s,t)\, ds\, dt$ is proposed. Details may be found in Ince [8], page 197. As an example, the equation

$$(x^2-1)\frac{d^2y}{dx^2} + (a+b+1)x\frac{dy}{dx} + aby = 0$$

has the two linearly independent solutions

$$y_\pm(x) = \int_0^\infty \int_0^\infty \exp\left[\pm xst - \frac{1}{2}(s^2+t^2)\right] s^{a-1} t^{b-1}\, ds\, dt.$$

[5] Poisson's integral formula is an integral representation of the solution to Laplace's differential equation. If $u(r,\theta)$ satisfies Laplace's equation $\nabla^2 u = u_{rr} + r^{-1}u_r + r^{-2}u_{\theta\theta} = 0$ for $0 < r < R$, and $u(R,\theta) = f(\theta)$ for $0 \le \theta < 2\pi$, then $u(r,\theta)$ for $0 < r < R$ is given by

$$u(r,\theta) = \frac{1}{2\pi} \int_0^{2\pi} \frac{R^2 - r^2}{R^2 - 2Rr\cos(\theta-\phi) + r^2} f(\phi)\, d\phi. \tag{1.15}$$

This is known as the Poisson formula for a circle. Integral solutions for Laplace's equation are also known when the geometry is a sphere, a half-plane, a half-space, or an annulus. See Zwillinger [10] for details.

[6] Pfaffian differential equations, which are equations of the form

$$\left[x^n F\left(x\frac{d}{dx}\right) + G\left(x\frac{d}{dx}\right)\right] y = 0,$$

can also be solved by this method. See Zwillinger [10].

[7] An application of this method to partial differential equations may be found in Bateman [2], pages 268–275.

[8] The Mellin–Barnes integral representation for an ordinary differential equation has the form

$$u(z) = \int_C v(\xi) z^\xi \left[\frac{\prod_{j=1}^{m} \Gamma(b_j - \xi) \prod_{j=1}^{n} \Gamma(1 - a_j + \xi)}{\prod_{j=m+1}^{q} \Gamma(1 - b_j + \xi) \prod_{j=n+1}^{r} \Gamma(a_j - \xi)}\right] d\xi.$$

In this representation, only the contour C and the constants $\{a_i, b_j, m, n, q, r\}$ are to be determined (see Babister [1] for details).

[9] The ordinary integral $\int_{x_0}^{x} A(t)\,dt$ is a construction that solves the initial-value problem: $\mathbf{y}'(x) = A(x)$ with $\mathbf{y}(x_0) = O$ (here O is the matrix of all zeros). The product integral is an analogous construction that solves the initial-value problem: $\mathbf{y}'(x) = A(x)\mathbf{y}(x)$ with $\mathbf{y}(x_0) = I$ (here I is the identity matrix). See Dollard and Friedman [6] for details.

[10] Given a linear differential equation (ordinary or partial): $L[u] = f(\mathbf{x})$, the Green's function $G(\mathbf{x}, \mathbf{z})$ satisfies $L[G] = \delta(\mathbf{x} - \mathbf{z})$, and a few technical conditions. (Here, δ represents the usual delta function.) The solution to the original equation can then be written as the integral $u(\mathbf{x}) = \int f(\mathbf{z}) G(\mathbf{x}, \mathbf{z})\, d\mathbf{z}$. See Zwillinger [10] for details.

References

[1] A. W. Babister, *Transcendental Functions Satisfying Nonhomogeneous Linear Differential Equations*, The MacMillan Company, New York, 1967, pages 24–26.

[2] H. Bateman, *Differential Equations*, Longmans, Green and Co., 1926, Chapter 10, pages 260–264.

[3] R. G. Buschman, "Simple contiguous function relations for functions defined by Mellin–Barnes integrals," *Indian J. Math.*, **32**, No. 1, 1990, pages 25–32.

[4] G. F. Carrier, M. Krook, and C. E. Pearson, *Functions of a Complex Variable*, McGraw–Hill Book Company, New York, 1966, pages 231–239.

[5] B. Davies, *Integral Transforms and Their Applications — Second Edition*, Springer–Verlag, New York, 1985, pages 342–367.

[6] J. D. Dollard and C. N. Friedman, *Product Integration with Applications to Differential Equations*, Addison–Wesley Publishing Co., Reading, MA, 1979.

[7] R. A. Gustafson, "Some Q-Beta and Mellin-Barnes Integrals with Many Parameters Associated to the Classical Groups," *SIAM J. Math. Anal.*, **23**, No. 2, March 1992, pages 525–551.

[8] E. L. Ince, *Ordinary Differential Equations*, Dover Publications, Inc., New York, 1964, pages 186–203 and 438–468.

[9] F. W. J. Olver, *Asymptotics and Special Functions*, Academic Press, New York, 1974.

[10] D. Zwillinger, *Handbook of Differential Equations*, Academic Press, New York, Second Edition, 1992.

2. Differential Equations: Integral Transforms

Applicable to Linear differential equations.

Idea

In order to solve a linear differential equation, it is sometimes easier to transform the equation to some "space," solve the equation in that "space," and then transform the solution back.

Procedure

Given a linear differential equation, multiply the equation by a kernel and integrate over a specified region (see Table 2.1 and Table 2.2 for a listing of common kernels and limits of integration). Use integration by parts to obtain an equation for the transform of the dependent variable.

You will have used the "correct" transform (i.e., you have chosen the correct kernel and limits) if the boundary conditions given with the original equation have been utilized. Now solve the equation for the transform of the dependent variable. From this, obtain the solution by multiplying by the inverse kernel and performing another integration. Table 2.1 and Table 2.2 also list the inverse kernel.

Example

Suppose we have the boundary value problem for $y = y(x)$

$$\begin{gathered} y_{xx} + y = 1, \\ y(0) = 0, \quad y(1) = 0. \end{gathered} \tag{2.1.a–c}$$

Since the solution vanishes at both of the endpoints, we suspect that a finite sine transform might be a useful transform to try. Define the finite sine transform of $y(x)$ to be $z(\xi)$, so that

$$z(\xi) = \int_0^1 y(x) \sin \xi x \, dx. \tag{2.2}$$

(See "finite sine transform–2" in Table 2.1). Now multiply equation (2.1.a) by $\sin \xi x$ and integrate with respect to x from 0 to 1. This results in

$$\int_0^1 y_{xx}(x) \sin \xi x \, dx + \int_0^1 y(x) \sin \xi x \, dx = \int_0^1 \sin \xi x \, dx. \tag{2.3}$$

If we integrate the first term in (2.3) by parts, twice, we obtain

$$\begin{aligned} \int_0^1 y_{xx}(x) \sin \xi x \, dx = y_x(x) \sin \xi x \Big|_{x=0}^{x=1} &- \xi y(x) \cos \xi x \Big|_{x=0}^{x=1} \\ &- \xi^2 \int_0^1 y(x) \sin \xi x \, dx. \end{aligned} \tag{2.4}$$

Since we will only use $\xi = 0, \pi, 2\pi, \ldots$ (see Table 2.1), the first term on the right-hand side of (2.4) is identically zero. Because of the boundary conditions in (2.1.b-c), the second term on the right-hand side of (2.4) also vanishes. (Since we have used the given boundary conditions to simplify certain terms appearing in the transformed equation, we suspect we have used an appropriate transform. If we had taken a finite cosine transform, instead of the one that we did, the boundary terms from the integration by parts would not have vanished.)

Using (2.4), simplified, in (2.3) results in

$$-\xi^2 \int_0^1 y(x) \sin \xi x \, dx + \int_0^1 y(x) \sin \xi x \, dx = \frac{1 - \cos \xi}{\xi}.$$

Using the definition of $z(\xi)$ (from (2.2)) this becomes

$$-\xi^2 z(\xi) + z(\xi) = \frac{1 - \cos \xi}{\xi},$$

or

$$z(\xi) = \frac{1 - \cos \xi}{(1 - \xi^2)\xi}.$$

Now that we have found an explicit formula for the transformed function, we can use the summation formula (inverse transform) in Table 2.1 to determine that

$$\begin{aligned}
y(x) &= \sum_{\xi = 0, \pi, 2\pi, \ldots} 2z(\xi) \sin \xi x, \\
&= \sum_{\xi = 0, \pi, 2\pi, \ldots} 2 \frac{1 - \cos \xi}{(1 - \xi^2)\xi} \sin \xi x, \\
&= \sum_{k=0}^{\infty} 2 \frac{1 - (-1)^k}{(1 - \pi^2 k^2)\pi k} \sin k\pi x, \\
&= \sum_{k = 1, 3, 5, \ldots} \frac{4 \sin k\pi x}{(1 - \pi^2 k^2)\pi k},
\end{aligned} \tag{2.5}$$

where we have defined $k = \xi / \pi$.

The exact solution of (2.1) is $y(x) = 1 - \cos x + \dfrac{\cos 1 - 1}{\sin 1} \sin x$. If this solution is expanded in a finite Fourier series, we obtain the representation in (2.5).

Table 2.1 Different transform pairs of the form

$$v(\xi_k) = \int_\alpha^\beta u(x) K(x, \xi_k)\, dx, \qquad u(x) = \sum_{\xi_k} H(x, \xi_k) v(\xi_k).$$

Finite cosine transform – 1, (see Miles [17], page 86) here l and h are arbitrary, and the $\{\xi_k\}$ satisfy $\xi_k \tan \xi_k l = h$.

$$v(\xi_k) = \int_0^1 u(x) \cos(x\xi_k)\, dx, \quad u(x) = \sum_{\xi_k} \frac{(2 - \delta_{\xi_k 0})(\xi_k^2 + h^2) \cos(\xi_k x)}{h + l(\xi_k^2 + h^2)} v(\xi_k).$$

Finite cosine transform – 2, (see Butkov [3], page 161) this is the last transform with $h = 0$, $l = 1$, so that $\xi_k = 0, \pi, 2\pi, \ldots$.

$$v(\xi_k) = \int_0^1 u(x) \cos(x\xi_k)\, dx, \quad u(x) = \sum_{\xi_k} (2 - \delta_{\xi_k 0}) \cos(\xi_k x)\, v(\xi_k).$$

Finite sine transform – 1, (see Miles [17], page 86) here l and h are arbitrary, and the $\{\xi_k\}$ satisfy $\xi_k \cot(\xi_k l) = -h$.

$$v(\xi_k) = \int_0^1 u(x) \sin(x\xi_k)\, dx, \quad u(x) = \sum_{\xi_k} 2 \frac{(\xi_k^2 + h^2) \sin(\xi_k x)}{h + l(\xi_k^2 + h^2)} v(\xi_k).$$

Finite sine transform – 2, (see Butkov [3], page 161) this is the last transform with $h = 0$, $l = 1$, so that $\xi_k = 0, \pi, 2\pi, \ldots$.

$$v(\xi_k) = \int_0^1 u(x) \sin(x\xi_k)\, dx, \quad u(x) = \sum_{\xi_k} 2 \sin(\xi_k x)\, v(\xi_k).$$

Finite Hankel transform – 1, (see Tranter [24], page 88) here n is arbitrary and the $\{\xi_k\}$ are positive and satisfy $J_n(\xi_k) = 0$.

$$v(\xi_k) = \int_0^1 u(x) x J_n(x\xi_k)\, dx, \quad u(x) = \sum_{\xi_k} 2 \frac{J_n(x\xi_k)}{J_{m+1}^2(\xi_k)} v(\xi_k).$$

Finite Hankel transform – 2, (see Miles [17], page 86) here n and h are arbitrary and the $\{\xi_k\}$ are positive and satisfy $\xi_k J_n'(a\xi_k) + h J_n(a\xi_k) = 0$.

$$v(\xi_k) = \int_0^a u(x) x J_n(x\xi_k)\, dx, \quad u(x) = \sum_{\xi_k} \frac{2\xi_k^2 J_n(x\xi_k)}{\{(h^2 + \xi_k^2) a^2 - m^2\} J_n^2(a\xi_k)} v(\xi_k).$$

Finite Hankel transform – 3, (see Miles [17], page 86) here $b > a$, the $\{\xi_k\}$ are positive and satisfy $Y_n(a\xi_k)J_n(b\xi_k) = J_n(a\xi_k)Y_n(b\xi_k)$, and $Z_n(x\xi_k) := Y_n(a\xi_k)J_n(x\xi_k) - J_n(a\xi_k)Y_n(x\xi_k)$.

$$v(\xi_k) = \int_a^b u(x)xZ_n(x\xi_k)\,dx, \quad u(x) = \sum_{\xi_k} \frac{\pi^2}{2}\frac{\xi_k^2 J_n^2(b\xi_k)Z_n(x\xi_k)}{J_n^2(a\xi_k) - J_n^2(b\xi_k)} v(\xi_k).$$

Legendre transform, (see Miles [17], page 86) here $\xi_k = 0, 1, 2, \ldots$.

$$v(\xi_k) = \int_{-1}^1 u(x)P_{\xi_k}(x)\,dx, \quad u(x) = \sum_{\xi_k} \frac{2\xi_k + 1}{2} P_{\xi_k}(x)v(\xi_k).$$

Table 2.2 Different integral transform pairs of the form

$$v(\xi) = \int_\alpha^\beta K(x,\xi)u(x)\,dx, \qquad u(x) = \int_a^b H(x,\xi)v(\xi)\,d\xi.$$

Fourier transform, (see Butkov [16], Chapter 7)

$$v(\xi) = \frac{1}{\sqrt{2\pi}}\int_{-\infty}^\infty e^{ix\xi}\,u(x)\,dx, \quad u(x) = \frac{1}{\sqrt{2\pi}}\int_{-\infty}^\infty e^{-ix\xi}\,v(\xi)\,d\xi.$$

Fourier cosine transform, (see Butkov [16], page 274)

$$v(\xi) = \sqrt{\frac{2}{\pi}}\int_0^\infty \cos(x\xi)\,u(x)\,dx, \quad u(x) = \sqrt{\frac{2}{\pi}}\int_0^\infty \cos(x\xi)\,v(\xi)\,d\xi.$$

Fourier sine transform, (see Butkov [16], page 274)

$$v(\xi) = \sqrt{\frac{2}{\pi}}\int_0^\infty \sin(x\xi)\,u(x)\,dx, \quad u(x) = \sqrt{\frac{2}{\pi}}\int_0^\infty \sin(x\xi)\,v(\xi)\,d\xi.$$

Hankel transform, (see Sneddon [20], Chapter 5)

$$v(\xi) = \int_0^\infty u(x)xJ_\nu(x\xi)\,dx, \quad u(x) = \int_0^\infty \xi J_\nu(x\xi)v(\xi)\,d\xi.$$

Hartley transform, (see Bracewell [5])

$$v(\xi) = \frac{1}{\sqrt{2\pi}} \int_{-\infty}^{\infty} (\cos x\xi + \sin x\xi)\, u(x)\, dx,$$
$$u(x) = \frac{1}{\sqrt{2\pi}} \int_{-\infty}^{\infty} (\cos x\xi + \sin x\xi)\, v(\xi)\, d\xi.$$

Hilbert transform, (see Sneddon [20], pages 233–238)

$$v(\xi) = \fint_{-\infty}^{\infty} u(x) \frac{1}{\pi(x-\xi)}\, dx, \quad u(x) = \fint_{-\infty}^{\infty} \frac{1}{\pi(\xi - x)} v(\xi)\, d\xi.$$

***K*–transform**, (see Bateman [3])

$$v(\xi) = \int_0^{\infty} K_\nu(x\xi)\sqrt{\xi x}\, u(x)\, dx, \quad u(x) = \frac{1}{\pi i} \int_{\sigma - i\infty}^{\sigma + i\infty} I_\nu(x\xi)\sqrt{\xi x}\, v(\xi)\, d\xi.$$

Kontorovich–Lebedev transform, (see Sneddon [20], Chapter 6)

$$v(\xi) = \int_0^{\infty} \frac{K_{i\xi}(x)}{x} u(x)\, dx, \quad u(x) = \frac{2}{\pi^2} \int_0^{\infty} \xi \sinh(\pi\xi) K_{i\xi}(x)\, v(\xi)\, d\xi.$$

Kontorovich–Lebedev transform (alternative form), (see Jones [14])

$$v(\xi) = \int_0^{\infty} H_\xi^{(2)}(x)\, u(x)\, dx, \quad u(x) = -\frac{1}{2x} \int_{-i\infty}^{i\infty} \xi J_\xi(x)\, v(\xi)\, d\xi.$$

Laplace transform, (see Sneddon [20], Chapter 3)

$$v(\xi) = \int_0^{\infty} e^{-x\xi}\, u(x)\, dx, \quad u(x) = \frac{1}{2\pi i} \int_{\sigma - i\infty}^{\sigma + i\infty} e^{x\xi}\, v(\xi)\, d\xi.$$

Mehler–Fock transform of order m, (see Sneddon [20], Chapter 7)

$$v(\xi) = \int_0^{\infty} \sinh(x) P^m_{i\xi - 1/2}(\cosh x)\, u(x)\, dx,$$
$$u(x) = \int_0^{\infty} \xi \tanh(\pi\xi) P^m_{i\xi - 1/2}(\cosh x)\, v(\xi)\, d\xi.$$

Mellin transform, (see Sneddon [20], Chapter 4)

$$v(\xi) = \int_0^{\infty} x^{\xi - 1}\, u(x)\, dx, \quad u(x) = \frac{1}{2\pi i} \int_{\sigma - i\infty}^{\sigma + i\infty} x^{-\xi}\, v(\xi)\, d\xi.$$

Weber formula, (see Titchmarsh [23], page 75)

$$v(\xi) = \int_a^\infty \sqrt{x}\left[J_\nu(x\xi)Y_\nu(a\xi) - Y_\nu(x\xi)J_\nu(a\xi)\right] u(x)\,dx,$$

$$u(x) = \sqrt{x}\int_0^\infty \frac{J_\nu(x\xi)Y_\nu(a\xi) - Y_\nu(x\xi)J_\nu(a\xi)}{J_\nu^2(a\xi) + Y_\nu^2(a\xi)}\, v(\xi)\,d\xi.$$

Weierstrass transform, (see Hirschman and Widder [11], Chapter 8)

$$v(\xi) = \frac{1}{\sqrt{4\pi}}\int_{-\infty}^{\infty} e^{(\xi-x)^2/4}\, u(x)\,dx, \quad u(x) = \frac{1}{\sqrt{4\pi}}\lim_{T\to\infty}\int_{-T}^{T} e^{(x-i\xi)^2/4} v(i\xi)\,d\xi.$$

Notes

[1] There are many tables of transforms available (see Bateman [3] or Magnus, Oberhettinger, and Soni [16]). It is generally easier to look up a transform than to compute it.

[2] Transform techniques may also be used with systems of linear equations.

[3] Transforms may also be evaluated numerically. There are many results on how to compute the more popular transforms numerically, like the Laplace transform. See, for example, Strain [22].

[4] The finite Hankel transforms are useful for differential equations that contain the operator $L_H[u]$ and the Legendre transform is useful for differential equations that contain the operator $L_L[u]$, where

$$L_H[u] = u_{rr} + \frac{u_r}{r} - \frac{n^2}{r^2}u \qquad \text{and} \qquad L_L[u] = \frac{\partial}{\partial r}\left((1-r^2)\frac{\partial u}{\partial r}\right).$$

For example, the Legendre transform of $L_L[u]$ is simply $-\xi_k(\xi_k+1)v(\xi_k)$.

[5] Integral transforms are generally created for solving a specific differential equation with a specific class of boundary conditions. The Mathieu integral transform (see Inayat-Hussain [12]) has been constructed for the two-dimensional Helmholtz equation in elliptic-cylinder coordinates.

[6] Integral transforms can also be constructed by integrating the Green's function for a Sturm–Liouville eigenvalue problem. See Zwillinger [25] for details.

[7] Note that many of the transforms in Table 2.1 and Table 2.2 do not have a standard form. In the Fourier transform, for example, the two $\sqrt{2\pi}$ terms might not be symmetrically placed as we have shown them. Also, a small variation of the K-transform is known as the Meijer transform (see Ditkin and Prudnikov [8], page 75).

[8] If a function $f(x, y)$ has radial symmetry, then a Fourier transform in both x and y is equivalent to a Hankel transform of $f(r) = f(x, y)$, where $r^2 = x^2 + y^2$. See Sneddon [20], pages 79–83.

[9] Two transform pairs that are continuous in one variable and discrete in the other variable, on an infinite interval, are the Hermite transform

$$u(x) = \sum_{n=0}^{\infty} v_n H_n(x) e^{-x^2/2}, \qquad v_n = \frac{1}{(2^n)!\sqrt{\pi}} \int_{-\infty}^{\infty} u(x) H_n(x) e^{-x^2/2}\, dx,$$

where $H_n(x)$ is the n-th Hermite polynomial, and the Laguerre transform

$$u(x) = \sum_{n=0}^{\infty} v_n L_n^{\alpha}(x) \frac{n!}{\Gamma(n+\alpha+1)}, \qquad v_n = \int_0^{\infty} u(x) L_n^{\alpha}(x) x^{\alpha} e^{-x}\, dx,$$

where $L_n^{\alpha}(x)$ is the Laguerre polynomial of degree n, and $\alpha \geq 0$. See Haimo [10] for details.

[10] Classically, the Fourier transform of a function only exists if the function being transformed decays quickly enough at $\pm\infty$. The Fourier transform can be extended, though, to handle generalized functions. For example, the Fourier transform of the n-th derivative of the delta function is given by $\mathcal{F}\left[\delta^{(n)}(t)\right] = (i\omega)^n$.

Another way to approach the Fourier transform of functions that do not decay quickly enough at either ∞ or $-\infty$ is to use the *one-sided Fourier transforms*. See Chester [6] for details.

[11] Many of the transforms listed generalize naturally to n dimensions. For example, in n dimensions we have:

(A) Fourier transform:
$$\begin{cases} v(\boldsymbol{\xi}) = (2\pi)^{-n/2} \displaystyle\int_{\mathbf{R}^n} e^{i\boldsymbol{\xi}\cdot\mathbf{x}} u(\mathbf{x})\, d\mathbf{x}, \\ u(\mathbf{x}) = (2\pi)^{-n/2} \displaystyle\int_{\mathbf{R}^n} e^{-i\boldsymbol{\xi}\cdot\mathbf{x}} v(\boldsymbol{\xi})\, d\boldsymbol{\xi}. \end{cases}$$

(B) Hilbert transform (see Bitsadze [4]):

$$\frac{\partial f}{\partial x_i} = \frac{\Gamma(n/2)}{\pi^{n/2}} \int_{R^{n-1}} \frac{y_i - x_i}{|\mathbf{y}-\mathbf{x}|^n} \phi(y)\, d\mathbf{y}, \quad i = 1, 2, \ldots, n-1,$$

$$\phi(y) = -\frac{\Gamma(n/2)}{\pi^{n/2}} \int_{R^{n-1}} \frac{(\mathbf{y}-\mathbf{x})\cdot\nabla f}{|\mathbf{y}-\mathbf{x}|^n}\, d\mathbf{y}.$$

[12] Apelblat [2] has found that repeated use of integral transforms can lead to the simplification of some infinite integrals. For example, let $F_s(y)$ denote the Fourier sine transform of the function $f(x)$, $F_s(y) = \int_0^{\infty} f(x) \sin yx\, dx$. Taking the Laplace transform of this results in

$$\begin{aligned} G(s) &= \int_0^{\infty} e^{-sy} \left\{ \int_0^{\infty} f(x) \sin yx\, dx \right\} dy \\ &= \int_0^{\infty} \frac{x f(x)}{s^2 + x^2}\, dx, \end{aligned} \tag{2.6}$$

where the order of integration has been changed and then the inner integral evaluated. For some functions $f(x)$ it may be possible to find the corresponding $F_s(y)$ and $G(s)$ using comprehensive tables of integral transforms.

Equating this to the expression in (2.6) may result in a definite integral hard to evaluate in other ways.

As a simple example of the technique, consider using $f(x) = \dfrac{1}{x(a^2+x^2)}$. With this we can find $F_s(y) = \frac{\pi}{2a^2}\left(1 - e^{-ay}\right)$. Taking the Laplace transform of this, and equating the result to (2.6), we have found the integral

$$\int_0^\infty \frac{dx}{(a^2+x^2)(s^2+x^2)} = \frac{\pi}{2as(a+s)}.$$

[13] Carson's integral is the integral transformation $\Omega(p) = p\int_0^\infty e^{-pt} f(t)\,dt$. See Iyanaga and Kawada [8].

[14] A transform pair that is continuous in each variable, on a finite interval, is the finite Hilbert transform

$$v(\xi) = \frac{1}{\pi}\int_{-1}^{1} \frac{u(x)}{x-\xi}\,dx, \quad u(x) = \frac{1}{\sqrt{1-x^2}}\left[C - \frac{1}{\pi}\int_{-1}^{1} \frac{\sqrt{1-\xi^2}}{\xi - x} v(\xi)\,d\xi\right],$$

where C is an arbitrary constant, and the integrals are principal value integrals. See Sneddon [20], page 467, for details.

[15] Note that, for the Hilbert transform, the integrals in Table 2.2 are principal value integrals.

References

[1] M. Abramowitz and I. A. Stegun, *Handbook of Mathematical Functions*, National Bureau of Standards, Washington, DC, 1964, pages 1019–1030.

[2] A. Apelblat, "Repeating Use of Integral Transforms—A New Method for Evaluation of Some Infinite Integrals," *IMA J. Appl. Mathematics*, **27**, 1981, pages 481–496.

[3] Staff of the Bateman Manuscript Project, A. Erdélyi (ed.), *Tables of Integral Transforms*, in 3 volumes, McGraw–Hill Book Company, New York, 1954.

[4] A. V. Bitsadze, "The Multidimensional Hilbert Transform," *Soviet Math. Dokl.*, **35**, No. 2, 1987, pages 390–392.

[5] R. N. Bracewell, *The Hartley Transform*, Oxford University Press, New York, 1986.

[6] C. R. Chester, *Techniques in Partial Differential Equations*, McGraw–Hill Book Company, New York, 1970.

[7] B. Davies, *Integral Transforms and Their Applications — Second Edition*, Springer–Verlag, New York, 1985.

[8] V. A. Ditkin and A. P. Prudnikov, *Integral Transforms and Operational Calculus*, translated by D. E. Brown, English translation edited by I. N. Sneddon, Pergamon Press, New York, 1965.

[9] H.-J. Glaeske, "Operational Properties of a Generalized Hermite Transformation," *Aequationes Mathematicae*, **32**, 1987, pages 155–170.

[10] D. T. Haimo, "The Dual Weierstrass–Laguerre Transform," *Trans. AMS*, **290**, No. 2, August 1985, pages 597–613.

[11] I. I. Hirschman and D. V. Widder, *The Convolution Transform*, Princeton University Press, Princeton, NJ, 1955.

[12] A. A. Inayat-Hussain, "Mathieu Integral Transforms," *J. Math. Physics*, **32**, No. 3, March 1991, pages 669–675.

[13] S. Iyanaga and Y. Kawada, *Encyclopedic Dictionary of Mathematics*, MIT Press, Cambridge, MA, 1980.

[14] D. S. Jones, "The Kontorovich–Lebedev Transform," *J. Inst. Maths. Applics*, **26**, 1980, pages 133–141.

[15] O. I. Marichev, *Handbook of Integral Transforms of Higher Transcendental Functions: Theory and Algorithmic Tables*, translated by L. W. Longdon, Halstead Press, John Wiley & Sons, New York, 1983.

[16] W. Magnus, F. Oberhettinger, and R. P. Soni, *Formulas and Theorems for the Special Functions of Mathematical Physics*, Springer–Verlag, New York, 1966.

[17] J. W. Miles, *Integral Transforms in Applied Mathematics*, Cambridge University Press, 1971.

[18] C. Nasim, "The Mehler–Fock Transform of General Order and Arbitrary Index and Its Inversion," *Int. J. Math. & Math. Sci.*, **7**, No. 1, 1984, pages 171–180.

[19] F. Oberhettinger and T. P. Higgins, *Tables of Lebedev, Mehler, and Generalized Mehler Transforms*, Mathematical Note No. 246, Boeing Scientific Research Laboratories, October 1961.

[20] I. N. Sneddon, *The Use of Integral Transforms*, McGraw–Hill Book Company, New York, 1972.

[21] I. Stakgold, *Green's Functions and Boundary Value Problems*, John Wiley & Sons, New York, 1979.

[22] J. Strain, "A Fast Laplace Transform Based on Laguerre Functions," *Math. of Comp.*, **58**, No. 197, January 1992, pages 275–283.

[23] E. C. Titchmarsh, *Eigenfunction Expansions Associated with Second–Order Differential Equations*, Clarendon Press, Oxford, 1946.

[24] C. J. Tranter, *Integral Transforms in Mathematical Physics*, Methuen & Co. Ltd., London, 1966.

[25] D. Zwillinger, *Handbook of Differential Equations*, Academic Press, New York, Second Edition, 1992.

3. Extremal Problems

Applicable to Finding a function that maximizes (or minimizes) an integral.

Yields

A differential equation for the critical function.

Procedure

Given the functional

$$J[u] = \iint_R L(\mathbf{x}, \partial_{x_j}) u(\mathbf{x}) \, d\mathbf{x}, \tag{3.1}$$

where the operator $L(\,)$ is a linear or nonlinear function of its arguments, how can $u(\mathbf{x})$ be determined so that $J[u]$ is critical (i.e., either a maximum or a minimum)?

The variational principle that is most often used is $\delta J = 0$, which states that the integral $J[u]$ should be stationary with respect to small changes in $u(\mathbf{x})$. If we let $h(\mathbf{x})$ be a "small," continuously differentiable function, then we can form

$$J[u+h] - J[u] = \iint_R \left[L(\mathbf{x}, \partial_{x_j})(u(\mathbf{x}) + h(\mathbf{x})) - L(\mathbf{x}, \partial_{x_j}) u(\mathbf{x}) \right] d\mathbf{x}. \tag{3.2}$$

By integration by parts, (3.2) can often be written as

$$J[u+h] - J[u] = \iint_R N(\mathbf{x}, \partial_{x_j}) u(\mathbf{x}) \, d\mathbf{x} + O(||h||^2)$$

plus some boundary terms. The variational principle requires that $\delta J := J[u+h] - J[u]$ vanishes to leading order, or that

$$N(\mathbf{x}, \partial_{x_j}) u(\mathbf{x}) = 0. \tag{3.3}$$

Equation (3.3) is called the first variation of (3.1), or the *Euler–Lagrange equation* corresponding to (3.1). (This is also called the *Euler equation.*) A functional in the form of (3.1) determines an Euler–Lagrange equation. Conversely, given an Euler–Lagrange equation, a corresponding functional can sometimes be obtained.

Many approximate and numerical techniques for differential equations utilize the functional associated with a given system of Euler–Lagrange equations. For example, both the Rayleigh–Ritz method and the finite element method create (in principle) integrals that are then analyzed (see Zwillinger [5]).

The following collection of examples assume that the dependent variable in the given differential equation has *natural boundary conditions.* If the dependent variable did not have these specific boundary conditions, then the boundary terms that were discarded in going from (3.2) to (3.3) would have to be satisfied in addition to the Euler–Lagrange equation.

Example 1

Suppose that we have the functional $J[y] = \int_R F(x, y, y', y'')\, dx$. Forming (3.2) we find

$$\begin{aligned} J[u+h] - J[u] &= \int_R \Big[F(x, u+h, u'+h', u''+h'') - F(x, u, u', u'') \Big]\, dx \\ &= \int_R \Big[\Big(F(\bullet) + F_y(\bullet)h + F_{y'}(\bullet)h' + F_{y''}(\bullet)h'' + O\left(h^2\right) \Big) \\ &\qquad - F(\bullet) \Big]\, dx \\ &= \int_R \Big[F_y(\bullet)h + F_{y'}(\bullet)h' + F_{y''}(\bullet)h'' \Big]\, dx + O\left(h^2\right), \end{aligned} \tag{3.4}$$

where the bullet stands for (x, u, u', u''), and the $O\left(h^2\right)$ terms should really be written as $O\left(h^2, (h')^2, (h'')^2\right)$. Now integration by parts can be used to find

$$\begin{aligned} \int_R F_{y'}(\bullet)h'\, dx &= F_{y'}(\bullet)h \Big|_{\partial R} - \int_R h \frac{d}{dx}\left(F_{y'}\right) \\ \int_R F_{y''}(\bullet)h''\, dx &= F_{y''}(\bullet)h' \Big|_{\partial R} - \int_R h' \frac{d}{dx}\left(F_{y''}\right) \\ &= F_{y''}(\bullet)h' \Big|_{\partial R} - \frac{d}{dx}\left(F_{y'}(\bullet)\right)h \Big|_{\partial R} + \int_R h \frac{d^2}{dx^2}\left(F_{y''}\right). \end{aligned}$$

If we take the natural boundary conditions $F_{y'}(\bullet)\Big|_{\partial R} = 0$ and $F_{y''}(\bullet)\Big|_{\partial R} = 0$, then (3.4) becomes

$$\begin{aligned} J[u+h] - J[u] &= \int_R \left[F_y(\bullet)h - \frac{d}{dx}F_{y'}(\bullet)h + \frac{d^2}{dx^2}F_{y''}(\bullet)h \right] dx + O\left(h^2\right) \\ &= \int_R h \left[F_y(\bullet) - \frac{d}{dx}F_{y'}(\bullet) + \frac{d^2}{dx^2}F_{y''}(\bullet) \right] dx + O\left(h^2\right). \end{aligned} \tag{3.5}$$

If (3.5) is to vanish for all "small" functions h, then it must be that

$$F_y(\bullet) - \frac{d}{dx}F_{y'}(\bullet) + \frac{d^2}{dx^2}F_{y''}(\bullet) = 0.$$

This is the Euler–Lagrange equation associated with the original functional.

Example 2

The Euler–Lagrange equation for the functional

$$J[y] = \int_R F\left(x, y, y', \ldots, y^{(n)}\right)\, dx,$$

where $y = y(x)$ is

$$\frac{\partial F}{\partial y} - \frac{d}{dx}\left(\frac{\partial F}{\partial y'}\right) + \frac{d^2}{dx^2}\left(\frac{\partial F}{\partial y''}\right) - \cdots + (-1)^n \frac{d^n}{dx^n}\left(\frac{\partial F}{\partial y^{(n)}}\right) = 0.$$

For this equation the natural boundary conditions are given by

$$\begin{aligned} &y(x_0) = y_0, \quad y'(x_0) = y_0', \quad \ldots, \quad y^{(n-1)}(x_0) = y_0^{(n-1)}, \\ &y(x_1) = y_1, \quad y'(x_1) = y_1', \quad \ldots, \quad y^{(n-1)}(x_1) = y_1^{(n-1)}. \end{aligned}$$

Example 3

The Euler–Lagrange equation for the functional

$$J[u] = \iint\limits_R F(x, y, u, u_x, u_y, u_{xx}, u_{xy}, u_{yy})\, dx\, dy, \tag{3.6}$$

where $u = u(x, y)$ is

$$\begin{aligned} \frac{\partial F}{\partial u} - \frac{\partial}{\partial x}\left(\frac{\partial F}{\partial u_x}\right) - \frac{\partial}{\partial y}\left(\frac{\partial F}{\partial u_y}\right) + \frac{\partial^2}{\partial x^2}\left(\frac{\partial F}{\partial u_{xx}}\right) \\ + \frac{\partial^2}{\partial x \partial y}\left(\frac{\partial F}{\partial u_{xy}}\right) + \frac{\partial^2}{\partial y^2}\left(\frac{\partial F}{\partial u_{yy}}\right) = 0. \end{aligned}$$

Example 4

The Euler–Lagrange equation for the functional

$$J[u] = \iint\limits_R \left[a\left(\frac{\partial u}{\partial x}\right)^2 + b\left(\frac{\partial u}{\partial x}\right)^2 + cu^2 + 2fu\right] dx\, dy,$$

which is a special case of (3.6), is: $\dfrac{\partial}{\partial x}\left(a\dfrac{\partial u}{\partial x}\right) + \dfrac{\partial}{\partial y}\left(b\dfrac{\partial u}{\partial y}\right) - cu = f.$

Example 5

For the $2m$-th order ordinary differential equation (in formally self-adjoint form)

$$\sum_{k=0}^{m}(-1)^k \frac{d^k}{dx^k}\left(p_k(x)\frac{d^k u}{dx^k}\right) = f(x),$$
$$u(a) = u'(a) = \cdots = u^{(m-1)}(a) = 0,$$
$$u(b) = u'(b) = \cdots = u^{(m-1)}(b) = 0,$$

a corresponding functional is

$$J[u] = \int_a^b \left(\sum_{k=0}^{m} p_k(x)\left(\frac{d^k u}{dx^k}\right)^2 - 2f(x)u(x)\right) dx.$$

Notes

[1] Note that two different functionals can yield the same set of Euler–Lagrange equations. For example, $\delta \int J\,dx = \delta \int (J + y + xy')\,dx$. The reason that $\delta \int (y + xy')\,dx = 0$ is because the integrand is an exact differential ($\int (y + xy')\,dx = \int d(xy)$). Hence, this integral is path independent; its value is determined by the boundary conditions.

The Euler–Lagrange equations for the two functionals $\int\int u_{xx}u_{yy}\,dx\,dy$ and $\int\int (u_{xy})^2\,dx\,dy$ are also the same.

[2] Even if the boundary conditions given with a differential equation are not natural, a variational principle may sometimes be found. Consider

$$J[u] = \int_{x_1}^{x_2} F(x,u,u')\,dx - g_1(x,u)\Big|_{x=x_1} + g_2(x,u)\Big|_{x=x_2},$$

where $g_1(x,u)$ and $g_2(x,u)$ are unspecified functions. The necessary conditions for u to minimize $J[u]$ are (see Mitchell and Wait [3])

$$\frac{\partial F}{\partial u} - \frac{d}{dx}\frac{\partial F}{\partial u'} = 0,$$
$$\left.\frac{\partial F}{\partial u'} + \frac{\partial g_1}{\partial u}\right|_{x=x_1} = 0, \qquad \left.\frac{\partial F}{\partial u'} + \frac{\partial g_2}{\partial u}\right|_{x=x_2} = 0.$$

If g_1 and g_2 are identically zero, then we recover the natural boundary conditions. However, we may choose g_1 and g_2 to suit other boundary conditions. For example, the problem

$$u'' + f(x) = 0,$$
$$u' + \alpha u\Big|_{x=x_1} = 0, \qquad u' + \beta u\Big|_{x=x_2} = 0,$$

corresponds to the functional

$$J[u] = \int_{x_1}^{x_2}\left[\frac{1}{2}\left(u'\right)^2 - f(x)u\right] dx + \frac{\beta u^2}{2}\Big|_{x=x_2} - \frac{\alpha u^2}{2}\Big|_{x=x_1}.$$

[3] This technique can be used in higher dimensions. For example, consider the functional

$$J[u] = \iint_R F(x, y, u, u_x, u_y, u_{xx}, u_{xy}, u_{yy})\, dx\, dy + \int_{\partial R} G(x, y, u, u_\sigma, u_{\sigma\sigma}, u_n)\, d\sigma,$$

where $\partial/\partial\sigma$ and $\partial/\partial n$ are partial differential operators in the directions of the tangent and normal to the curve ∂R. Necessary conditions for $J[u]$ to have a minimum are the Euler–Lagrange equations (given in (3.6)) together with the boundary conditions:

$$\begin{gathered}
\left[\frac{\partial F}{\partial u_x} - \frac{\partial}{\partial x}\frac{\partial F}{\partial u_{xx}}\right] y_\sigma - \left[\frac{\partial F}{\partial u_y} - \frac{\partial}{\partial y}\frac{\partial F}{\partial u_{yy}}\right] x_\sigma \\
- \left[\frac{\partial}{\partial\sigma}\left(\frac{\partial F}{\partial u_{xx}} - \frac{\partial F}{\partial u_{yy}}\right)\right] x_\sigma y_\sigma + \frac{1}{2}\left[\frac{\partial}{\partial\sigma}\frac{\partial F}{\partial u_{xy}}\left(x_\sigma^2 - y_\sigma^2\right)\right] \\
+\frac{1}{2}\left[\left(\frac{\partial}{\partial x}\frac{\partial F}{\partial u_{xy}}\right) x_\sigma - \left(\frac{\partial}{\partial y}\frac{\partial F}{\partial u_{xy}}\right) y_\sigma\right] \\
+G_u - \frac{\partial}{\partial\sigma}\frac{\partial G}{\partial u_\sigma} + \frac{\partial^2}{\partial\sigma^2}\frac{\partial G}{\partial u_{\sigma\sigma}} = 0, \\
\frac{\partial G}{\partial u_n} + \frac{\partial F}{\partial u_{xx}} y_\sigma^2 + \frac{\partial F}{\partial u_{yy}} x_\sigma^2 + \frac{\partial F}{\partial u_{xy}} x_\sigma y_\sigma = 0,
\end{gathered}$$

where $x_\sigma = dx/d\sigma$ and $y_\sigma = dy/d\sigma$. See Mitchell and Wait [3] for details.

References

[1] E. Butkov, *Mathematical Physics*, Addison–Wesley Publishing Co., Reading, MA, 1968, pages 573–588.

[2] L. V. Kantorovich and V. I. Krylov, *Approximate Methods of Higher Analysis*, Interscience Publishers, New York, 1958, Chapter 4, pages 241–357.

[3] A. R. Mitchell and R. Wait, *The Finite Element Method in Differential Equations*, Wiley, New York, 1977, pages 27–31.

[4] H. Rund, *The Hamilton–Jacobi Theory in the Calculus of Variations*, D. Van Nostrand Company, Inc., New York, 1966.

[5] D. Zwillinger, *Handbook of Differential Equations*, Academic Press, New York, Second Edition, 1992.

4. Function Representation

Idea

Certain integrals can be used to represent functions.

Procedure

This section contains several different representational theorems. Each has found many important applications in the literature.

Bochner–Martinelli Representation

> Let f be a holomorphic function in a domain $D \subset C^n$, with piecewise smooth boundary ∂D, and let f be continuous in its closure $\overline{D}$. Then we have the representation

$$\frac{(n-1)!}{(2\pi i)^n} \int_{\partial D} \frac{f(\boldsymbol{\zeta})}{|\boldsymbol{\zeta} - \mathbf{z}|^{2n}} \sum_{j=1}^{n} \left(\overline{\zeta}_j - \overline{z}_j\right) d\widehat{\zeta}_i = \begin{cases} f(z), & \text{if } z \in D, \\ 0, & \text{if } z \notin D, \end{cases}$$

> where $d\widehat{\zeta}_i = d\overline{\zeta}_1 \wedge d\zeta_1 \wedge \cdots \wedge [d\overline{\zeta}_i] \wedge \cdots \wedge d\overline{\zeta}_n \wedge d\zeta_n$, and $[d\overline{\zeta}_i]$ means that the term $d\overline{\zeta}_i$ is to be omitted.

For $n = 1$, this is identical to the Cauchy representation. Another way to write this result is as follows:

> Let $H(G)$ be the ring of holomorphic functions in G. Let G_j be a domain in the z_j-plane with piecewise smooth boundary C_j. If $f \in H(G)$ (where $G := \prod_{j=1}^{n} G_j$) is continuous on $\overline{G}$, then

$$\frac{1}{(2\pi i)^n} \int_{C_1 \times \ldots \times C_n} \frac{f(\boldsymbol{\zeta})}{(\zeta_1 - z_1) \ldots (\zeta_n - z_n)} d\zeta_1 \wedge \ldots \wedge d\zeta_n = \begin{cases} f(z), & \text{for } z \in G, \\ 0, & \text{for } z \notin G. \end{cases}$$

For details see Krantz [5] or Iyanaga and Kawada [6] (page 101).

Cauchy Representation

Cauchy's integral formula states that if a domain D is bounded by a finite union of simple closed curves Γ, and f is analytic within D and across Γ, then

$$f(\xi) = \frac{1}{2\pi i} \int_{\Gamma} \frac{f(z)}{z - \xi} \, dz,$$

for $\xi \in D$. (See page 129 for several applications of this formula.)

If D is the disk $|z| < R$, then Cauchy's theorem becomes Poisson's integral formula

$$f(z) = \frac{1}{2\pi} \int_0^{2\pi} f(Re^{i\phi}) \frac{R^2 - r^2}{R^2 + r^2 - 2Rr\cos(\theta - \phi)} \, d\phi.$$

There is an analogous formula, called Villat's integral formula, when D is an annulus. See Iyanaga and Kawada [6], page 636.

There are also extensions of this formula when $f(z)$ is not analytic. In terms of the differential operator $\partial_{\bar{z}} = \frac{1}{2}(\partial_x + \partial_y)$, the Cauchy–Green formula is

$$f(\xi) = \frac{1}{2\pi i} \int_\Gamma \frac{f(z)}{z-\xi}\, dz - \frac{1}{\pi} \iint_D \frac{\partial f}{\partial \bar{z}} \frac{1}{z-\xi}\, dx\, dy,$$

for $\xi \in D$. This formula is valid whenever f is smooth enough for the derivative $\partial_{\bar{z}}$ to make sense. If f is analytic, then the Cauchy–Riemann equations hold; these equations are equivalent to $\partial_{\bar{z}} = 0$. See Khavinson [7] for details.

Green's Representation Theorems

- *Three dimensions:* If ϕ and $\nabla^2\phi$ are defined within a volume V bounded by a simple closed surface S, p is an interior point of V, and n represents the outward unit normal, then

$$\phi(p) = -\frac{1}{4\pi} \int_V \frac{\nabla^2\phi}{r}\, dV + \frac{1}{4\pi} \int_S \frac{1}{r} \frac{\partial \phi}{\partial n}\, dS - \frac{1}{4\pi} \int_S \phi \frac{\partial}{\partial n} \left(\frac{1}{r}\right) dS. \tag{4.1}$$

  Note that if ϕ is harmonic (i.e., $\nabla^2\phi = 0$), then the right-hand side of (4.1) simplifies.
- *Two dimensions:* If ϕ and $\nabla^2\phi$ are defined within a planar region S bounded by a simple closed curve C, p is an interior point of S, and n_q represents the outward unit normal at the point q, then

$$\begin{aligned} \phi(p) = \frac{1}{2\pi} \int_S \nabla^2\phi(q) \log|p-q|\, dS &+ \frac{1}{2\pi} \int_C \phi(q) \frac{\partial}{\partial n_q} \log|p-q|\, dq \\ &- \frac{1}{2\pi} \int_C \log|p-q| \frac{\partial}{\partial n_q} \phi(q)\, dq. \end{aligned}$$

- *n dimensions:* If ϕ and its second derivatives are defined within a region Ω in $\mathbf{R}^n$ bounded by the surface Σ, and n_q represents the outward unit normal at the point q, then for points p not on the surface Σ we have (if $n > 3$)

$$\begin{aligned} \phi(p) = &-\frac{1}{(n-2)\sigma_n} \int_\Omega \frac{\nabla^2\phi(q)}{|p-q|^{n-2}}\, d\Omega_q \\ &+ \frac{1}{(n-2)\sigma_n} \int_\Sigma \left(\frac{1}{|p-q|^{n-2}} \frac{\partial \phi(q)}{\partial n_q} - \phi(q) \frac{\partial}{\partial n_q} \frac{1}{|p-q|^{n-2}} \right) d\Sigma_q \end{aligned}$$

  where $\sigma_n = 2\pi^{n/2}/\Gamma(n/2)$ is the area of a unit sphere in $\mathbf{R}^n$.

See Gradshteyn and Ryzhik [2], 10.717, pages 1089–1090 for details.

Herglotz's Integral Representation

The Herglotz integral representation is based on Poisson's integral representation. It states:

> Let $f(z)$ be holomorphic in $|z| < R$ with positive real part. Then
> $$f(z) = \int_0^{2\pi} \frac{Re^{i\phi} + z}{Re^{i\phi} - z}\, d\rho(\phi),$$
> for $|z| < R$, where $\rho(\phi)$ is a monotonic increasing real-valued function with total variation unity. This function is determined uniquely, up to an additive constant, by $f(z)$.

See Iyanaga and Kawada [6], page 161. Another statement of this integral representation is (see Hazewinkel [3], page 124):

> Let $f(z)$ be regular in the unit disk $D = \{z \mid |z| < 1\}$, and assume that it has a positive real part (i.e., $\operatorname{Re} f(z) < 0$), then $f(z)$ can be represented as
> $$f(z) = \int \frac{\xi + z}{\xi - z}\, d\mu(\xi) + ic,$$
> where the imaginary part of c is zero. Here μ is a positive measure concentrated on the circle $\{\xi \mid |\xi| = 1\}$.

Parametric Representation of a Univalent Function

From Hazewinkel [3], page 124, we have:

> Let $f(z)$ be analytic in the unit disk $D = \{z \mid |z| < 1\}$, and assume that $\operatorname{Im} f(x) = 0$ for $-1 < x < 1$ and $\operatorname{Im} f(z) \operatorname{Im} z > 0$ for $\operatorname{Im} z \neq 0$. Then $f(z)$ can be represented as
> $$f(z) = \int \frac{z\, d\mu(\xi)}{(z - \xi)(z - \overline{\xi})},$$
> where μ is a measure concentrated on the circle $\{\xi \mid |\xi| = 1\}$ and normalized by $||\mu|| = \int d\mu(\xi) = 1$.

Pompeiu Formula

From Henrici [4] we have the following theorem:

> **Theorem:** Let R be a region bounded by a system Γ of regular closed curves such that points in R have winding number 1 with respect to Γ. If f is a complex-valued function that is real-differentiable in a region containing $R \cup \Gamma$, then for any point $z \in R$ there holds
> $$f(z) = \frac{1}{2\pi i} \int_\Gamma \frac{f(z)}{t - z}\, dt - \frac{1}{\pi} \iint_R \frac{\partial f(t)}{\partial \overline{z}} \frac{1}{t - z}\, dx\, dy,$$
> where $t = x + iy$.

Note that if $f(z)$ is analytic, then this reduces to Cauchy's formula.

Solutions to the Biharmonic Equation

Some function representations require that the function have some specific properties. For example, if u is biharmonic in a bounded region R (that is, $\nabla^4 u = 0$), and if $f = \nabla^2 u$, then $u(z)$ may be written as (for $z \in R$):

$$u(z) = \frac{1}{2\pi} \iint\limits_R f(t) \log \frac{1}{|t-z|} \, dx \, dy + v(z)$$

where $t = x + iy$ and v is harmonic in R (that is, v satisfies Laplace's equation $\nabla^2 v = 0$). See Henrici [4].

Notes

[1] Schläfli's integral representation is an integral representation of the Legendre function of the second kind.

[2] See also the section on integrals used to represent the solutions of differential equations (page 1).

[3] If a domain D is simply connected, and the vector field $\mathbf{V}$ tends sufficiently rapidly to zero near the boundary of D and at infinity, then we have Helmholtz's theorem: $\mathbf{V} = \nabla\phi + \nabla \times \mathbf{A}$, where

$$\phi = -\iiint\limits_D \frac{\nabla \cdot \mathbf{V}}{4\pi r} \, dV \qquad \text{and} \qquad \mathbf{A} = \iiint\limits_D \frac{\nabla \times \mathbf{V}}{4\pi r} \, dV.$$

[4] If we define the one-form

$$\omega(\boldsymbol{\zeta}, \mathbf{z}) = \frac{(n-1)!}{(2\pi i)^n} \frac{1}{|\boldsymbol{\zeta} - \mathbf{z}|^{2n}} \sum_{j=1}^{n} \left(\overline{\zeta}_j - \overline{z}_j\right) \widehat{d\zeta_i}$$

then a generalization of the Bochner–Martinelli representation, which is analogous to the Cauchy–Green formula is given by (see Hazewinkel [3], page 404):

> If the function f is continuously differentiable in the closure of the domain $D \subset \mathbf{C}^n$ with piecewise-smooth boundary ∂D, then, for any point $z \in D$,

$$f(z) = \int_{\partial D} f(\boldsymbol{\zeta}) \omega(\boldsymbol{\zeta}, \mathbf{z}) - \int_D \overline{\partial} f(\boldsymbol{\zeta}) \wedge \omega(\boldsymbol{\zeta}, \mathbf{z}).$$

References

[1] I. A. Ayzenberg and A. P. Yuzhakov, *Integral Representations and Residues in Multidimensional Complex Analysis*, Translations of Mathematical Monographs, Volume 58, Amer. Math. Soc., Providence, Rhode Island, 1983.

[2] I. S. Gradshteyn and I. M. Ryzhik, *Tables of Integrals, Series, and Products*, Academic Press, New York, 1980.

[3] M. Hazewinkel (managing ed.), *Encyclopaedia of Mathematics*, Kluwer Academic Publishers, Dordrecht, The Netherlands, 1988.

[4] P. Henrici, *Applied and Computational Complex Analysis*, Volume 3, John Wiley & Sons, New York, 1986, pages 290, 302.

[5] S. Krantz, *Function Theory of Several Complex Variables*, John Wiley & Sons, New York, 1982.

[6] S. Iyanaga and Y. Kawada, *Encyclopedic Dictionary of Mathematics*, MIT Press, Cambridge, MA, 1980.

[7] D. Khavinson, "The Cauchy–Green Formula and Its Application to Problems in Rational Approximation on Sets with a Finite Perimeter in the Complex Plane," *J. Funct. Anal.*, **64**, 1985, pages 112–123.

[8] R. M. Range, *Holomorphic Functions and Integral Representations in Several Variables*, Springer–Verlag, New York, 1986.

5. Geometric Applications

Idea

Integrals and integration have many uses in geometry.

Length

If a two-dimensional curve is parameterized by $\mathbf{x}(t) = (x(t), y(t))$ for $a \leq t \leq b$, then the length of the curve is given by

$$L = \int_a^b \sqrt{\left(\frac{dx}{dt}\right)^2 + \left(\frac{dy}{dt}\right)^2}\, dt.$$

For a curve defined by $y = y(x)$, for $a \leq x \leq b$, this simplifies to

$$L = \int_a^b \sqrt{1 + \left(\frac{dy}{dx}\right)^2}\, dx.$$

A curve in three-dimensional space $\{x(t), y(t), z(t)\}$, for $a \leq t \leq b$, has length

$$L = \int_a^b \sqrt{\left(\frac{dx}{dt}\right)^2 + \left(\frac{dy}{dt}\right)^2 + \left(\frac{dz}{dt}\right)^2}\, dt.$$

Area

If a surface is described by

$$z = f(x, y), \qquad \text{for } (x, y) \text{ in the region } R_{xy},$$

then the area of the surface, S, is given by

$$S = \iint_{R_{xy}} \sqrt{1 + \left(\frac{\partial z}{\partial x}\right)^2 + \left(\frac{\partial z}{\partial y}\right)^2}\, dx\, dy.$$

If, instead, the surface is described parametrically by $\mathbf{x} = (x, y, z)$ with $x = x(u, v)$, $y = y(u, v)$, $z = z(u, v)$, for (u, v) in the region R_{uv}, then the area of the surface, S, is given by

$$S = \iint_{R_{uv}} |\mathbf{x}_u \times \mathbf{x}_v|\, du\, dv = \iint_{R_{uv}} \sqrt{EG - F^2}\, du\, dv,$$

where

$$\begin{aligned} E &= \mathbf{x}_u \cdot \mathbf{x}_u = x_u^2 + y_u^2 + z_u^2, \\ F &= \mathbf{x}_u \cdot \mathbf{x}_v = x_u x_v + y_u y_v + z_u z_v, \\ G &= \mathbf{x}_v \cdot \mathbf{x}_v = x_v^2 + y_v^2 + z_v^2. \end{aligned} \tag{5.1}$$

General Coordinate Systems

In a three-dimensional orthogonal coordinate system, let $\{\mathbf{a}_i\}$ denote the unit vectors in each of the three coordinate directions, and let $\{u_i\}$ denote distance along each of these axes. The coordinate system may be designated by the *metric coefficients* $\{g_{11}, g_{22}, g_{33}\}$, defined by

$$g_{ii} = \left(\frac{\partial x_1}{\partial u_i}\right)^2 + \left(\frac{\partial x_2}{\partial u_i}\right)^2 + \left(\frac{\partial x_3}{\partial u_i}\right)^2, \tag{5.2}$$

where $\{x_1, x_2, x_3\}$ represent rectangular coordinates. Then an element of area on the $u_1 u_2$ surface (i.e., u_3 is held constant) is given by $dS_{12} = \left[\sqrt{g_{11}}du_1\right]\left[\sqrt{g_{22}}du_2\right]$. Metric coefficients for some common orthogonal coordinate systems may be found on page 113. Moon and Spencer [2] list the metric coefficients for 43 different orthogonal coordinate systems. (These consist of 11 general systems, 21 cylindrical systems, and 11 rotational systems.)

Operations for orthogonal coordinate systems are sometimes written in terms of $\{h_i\}$ functions, instead of the $\{g_{ii}\}$ terms. Here, $h_i = \sqrt{g_{ii}}$, so that, for example, $dS_{12} = [h_1 du_1]\,[h_2 du_2]$.

Volume

Using the metric coefficients defined in (5.2), we define $g = g_{11}g_{22}g_{33}$. An element of volume is then given by

$$dV = \sqrt{g_{11}g_{22}g_{33}}\, du_1\, du_2\, du_3 = \sqrt{g}\, du_1\, du_2\, du_3.$$

Moments of Inertia

For a bounded set S with positive area A and a density function $\rho(x, y)$, we have the following definitions:

$\int_s \int \rho(x, y)\, dA = M$ = total mass
$\int_s \int \rho(x, y) x\, dA = M_y$ = first moment with respect to the x-axis
$\int_s \int \rho(x, y) y\, dA = M_x$ = first moment with respect to the y-axis
$\int_s \int \rho(x, y) x^2\, dA = I_y$ = second moment with respect to the y-axis
$\int_s \int \rho(x, y) y^2\, dA = I_x$ = second moment with respect to the x-axis
$\int_s \int \rho(x, y)(x^2 + y^2)\, dA = I_0$ = polar second moment with respect to the origin.

Example 1

Consider a helix defined by $\mathbf{x}(t) = (a\cos t, a\sin t, bt)$ for t in the range $[0, 2\pi]$. The length of this curve is

$$L = \int_0^{2\pi} \sqrt{a^2 \sin^2 t + a^2 \cos^2 t + b^2}\, dt = \int_0^{2\pi} \sqrt{a^2 + b^2}\, dt = 2\pi\sqrt{a^2 + b^2}.$$

Example 2

Consider a torus defined by $\mathbf{x} = ((b + a\sin\phi)\cos\theta,\ (b + a\sin\phi)\sin\theta,\ a\cos\phi)$, where $0 \le \theta \le 2\pi$ and $0 \le \phi \le 2\pi$. From (5.1), we can compute $E = \mathbf{x}_\theta \cdot \mathbf{x}_\theta = (b + a\sin\phi)^2$, $F = \mathbf{x}_\theta \cdot \mathbf{x}_\phi = 0$, and $G = \mathbf{x}_\phi \cdot \mathbf{x}_\phi = a^2$. Therefore, the surface area of the torus is

$$S = \int_0^{2\pi} \int_0^{2\pi} \sqrt{EG - F^2}\, d\theta\, d\phi = \int_0^{2\pi} \int_0^{2\pi} a(b + a\sin\phi)\, d\theta\, d\phi = 4\pi^2 ab.$$

Example 3

In cylindrical coordinates we have $\{x_1 = r\cos\phi,\ x_2 = r\sin\phi,\ x_3 = z\}$ so that $\{h_r = 1,\ h_\theta = r,\ h_z = 1\}$. Consider a cylinder of radius R and height H. This cylinder has three possible areas we can determine:

$$S_{\theta z} = \int_0^H \int_0^{2\pi} dS_{\theta z} = \int_0^H \int_0^{2\pi} h_\theta h_z\, d\theta\, dz = 2\pi RH,$$

$$S_{\theta r} = \int_0^R \int_0^{2\pi} dS_{\theta r} = \int_0^R \int_0^{2\pi} h_\theta h_r\, d\theta\, dr = \pi R^2,$$

$$S_{rz} = \int_0^H \int_0^R dS_{rz} = \int_0^H \int_0^R h_r h_z\, dr\, dz = RH.$$

We can identify each of these: $S_{\theta z}$ is the area of the outside of the cylinder, $S_{\theta r}$ is the area of an end of the cylinder, and S_{rz} is the area of a radial slice (that is, a vertical cross-section from the center of the cylinder).

We can also compute the volume of this cylinder to be

$$V = \int_0^H \int_0^R \int_0^{2\pi} h_\theta h_r h_z \, d\theta \, dr \, dz = \int_0^H \int_0^R \int_0^{2\pi} r \, d\theta \, dr \, dz = \pi R^2 H.$$

Notes

[1] If C is a simple closed curve, positively oriented, that is piecewise continuous, then the line integrals $\oint_C x \, dy$ and $-\oint_C y \, dx$ both have the same value, which is equal to the area enclosed by C. This is an application of Green's theorem, see page 164.

[2] The quadratic form (see (5.1)) $I = d\mathbf{x} \cdot d\mathbf{x} = E \, du^2 + 2F \, du \, dv + G \, dv^2$ is called the *first fundamental form* of $\mathbf{x} = \mathbf{x}(u, v)$. The length of a curve described by $\mathbf{x}(u(t), v(t))$, for t in the range $[a, b]$ is

$$L = \int_a^b \sqrt{E \left(\frac{du}{dt}\right)^2 + 2F \left(\frac{du}{dt}\right)\left(\frac{dv}{dt}\right) + G \left(\frac{dv}{dt}\right)^2} \, dt.$$

[3] The Gauss–Bonnet formula relates the exterior angles of an object with the curvature of an object (see Lipschutz [2]):

> Let C be a curvilinear polygon of class C^2 on a patch of a surface of class greater than or equal to 3. We presume that C has a positive orientation and that its interior on the patch is simple connected. Then
>
> $$\int_C \kappa_g \, ds + \iint_R K \, dS = 2\pi - \sum_i \theta_i,$$
>
> where κ_g is the geodesic curvature along C, K is the Gaussian curvature, R is the union of C and its interior, and the $\{\theta_i\}$ are the exterior angles on C.

For example, consider a geodesic triangle formed from three geodesics. Along a geodesic we have $\kappa_g = 0$, so that $\sum_i \theta_i = 2\pi - \iint_R K \, ds$. For a planar surface $K = 0$. Hence, we have found that the sum of the exterior angles in a planar triangle is 2π. (This is equivalent to the usual conclusion that the sum of the interior angles of a planar triangle is π.)

For a sphere of radius a, we have $K = 1/a^2$. Therefore, the sum of the exterior angles on a spherical triangle of area A is $2\pi - A/a^2$.

References

[1] W. Kaplan, *Advanced Calculus*, Addison–Wesley Publishing Co., Reading, MA, 1952.

[2] M. M. Lipschutz, *Differential Geometry*, McGraw–Hill Book Company, New York, 1969. Schaum Outline Series.

[3] P. Moon and D. E. Spencer, *Field Theory Handbook*, Springer–Verlag, New York, 1961.

6. MIT Integration Bee

Every year at the Massachusetts Institute of Technology (MIT) there is an "Integration Bee" open to undergraduates. This consists of an hour-long written exam, with the highest scorers going on to a verbal exam run like a Spelling Bee. It is claimed that completion of first semester calculus is adequate to evaluate all of the integrals.

In 1991 the written exam was given on January 15 and consisted of the following forty integrals that had to be evaluated:

(1) $\int e^{1991x}\,dx$

(2) $\int (\sin x - \cos x)^2\,dx$

(3) $\int \log x\,dx$

(4) $\int \frac{dx}{\pi x}$

(5) $\int e^{\sin^2 x} e^{\cos^2 x}\,dx$

(6) $\int \frac{dx}{x \log x}$

(7) $\int \frac{x}{x^4+1}\,dx$

(8) $\int \frac{x+1}{\sqrt[3]{x^2+2x+2}}\,dx$

(9) $\int x e^x \sin x\,dx$

(10) $\int e^{(e^x+x)}\,dx$

(11) $\int \frac{dx}{\sec x + \tan x \sin x}$

(12) $\int \frac{e^{5x}+e^{7x}}{e^x+e^{-x}}\,dx$

(13) $\int \sqrt{-1+\frac{2}{1+3x}}\,dx$

(14) $\int \sinh x - \cosh x\,dx$

(15) $\int \frac{\sin x\, e^{\sec x}}{\cos^2 x}\,dx$

(16) $\int \frac{x^2+1}{x^4-x^2+1}\,dx$

(17) $\int \frac{dx}{\pi x^2 + \tan^{-1} x + x^2 \tan^{-1} x + \pi}$

(18) $\int \sec^3 x\,dx$

(19) $\int \frac{dx}{x^2-10x+26}$

(20) $\int \frac{dx}{x^2-11x-26}$

(21) $\int \frac{dx}{12+13\cos x}$

(22) $\int \frac{x^3+1}{x+1}\,dx$

(23) $\int \frac{\left(1-4x^4\right)^{-1/2}}{(4x)^{-1}}\,dx$

(24) $\int e^{1991}\,dx$

(25) $\int (\log x + 1)x^x \, dx$

(26) $\int (\cos 2x)(\sin 6x) \, dx$

(27) $\int \frac{dx}{\sqrt{x}\,(1+\sqrt{x})}$

(28) $\int e^{1/x} x^{-3} \, dx$

(29) $\int \sqrt{\csc x - \sin x} \, dx$

(30) $\int \frac{x^2+1}{x^3-x} \, dx$

(31) $\int 42^x \, dx$

(32) $\int x^5 e^x \, dx$

(33) $\int x e^{x^2} \, dx$

(34) $\int \frac{dx}{(x^2+1)^2}$

(35) $\int \frac{dx}{e^x + e^{-x}}$

(36) $\int \tan x \log|\sec x| \, dx$

(37) $\int \cos(\sin x) \cos x \, dx$

(38) $\int \frac{dx}{x^2 - 9}$

(39) $\int \frac{\pi \, dx}{\sqrt{16 - e^2}}$

(40) $\int \sqrt{\tan x} \, dx.$

On January 22, the top 11 scorers on the written exam participated in the Integration Bee. (These people had obtained between 26 and 35 correct answers to the above written exam.) The first few rounds were run with a fixed time in which to simplify a specific integral. The integrals, and the time allowed for each, were:

- 1 minute for $\int \sin^{-1} x \, dx$
- 2 minutes for $\int \frac{x^2 - 2x + 2}{x^2 + 1} \, dx$,
- 2 minutes for $\int \frac{\sin^2 x \cos^2 x}{1 + \cos 2x} \, dx$,
- 2 minutes for $\int \sqrt{x + x^2\sqrt{x}} \, dx$ (since five people in a row did not obtain the correct answer, this integral was discarded and the people who could not integrate it were not penalized),
- 2 minutes for $\int \cos 4x \cos 2x \, dx$,
- 2 minutes for $\int \frac{\sqrt{x^3 - 1}}{x} \, dx$.

After these integrals, there were four finalists. The ranking of the finalists was achieved by four rounds of competitive integration (a pair was given the same integral; whoever obtained the correct answer first was the winner of that round). The integrals to be evaluated were

$$\int \frac{e^x(x-2)}{x^3} \, dx \qquad \int \frac{\cot x}{\log(\sin x)} \, dx$$

$$\int x \sec^2 x \, dx \qquad \int x \sec x \, (x \tan x + 2) \, dx.$$

The 1991 title of "Grand Integrator of MIT" was awarded to Chris Teixeira. The second and third place winners were Belle Yseng and Trac Tran.

7. Probability

Idea

This section describes how integration is used in probability theory.

Procedure

If $p(x)$ represents the density function of the random variable X then the expectation of the function $g(X)$ is given by

$$E[g(X)] = \int p(x)g(x)\,dx,$$

where the range of integration is specified by the density function. Expectations of certain functions have special names and notations. For example,

$$\begin{aligned}
\text{mean of } X &= \mu = \mu_1 = E[X], \\
\text{variance of } X &= \operatorname{Var}(X) = \sigma^2 = \mu_2' = E[(X-\mu)^2], \\
n\text{-th moment of } X &= \mu_n = E[X^n], \\
n\text{-th central moment of } X &= \mu_n' = E[(X-\mu)^n], \\
\text{characteristic function of } X &= \phi(t) = E\left[e^{itX}\right], \\
\text{generating function of } X &= \psi(s) = E\left[s^X\right].
\end{aligned}$$

The random variable X, with density function $f(x)$, has the *distribution function* $F(x) = \int_{-\infty}^{x} f(t)\,dt$. The probability that $X < x$ is then given by $F(x)$.

Notes

[1] The mean is sometimes called the "average." The skewness is defined to be to be μ_3/σ^3, and the excess is defined to be $\mu_4/\sigma^4 - 3$.

[2] If the random variable X has the density function $f(x)$, then the *entropy* of X is defined to be (see McEliece [3])

$$H(X) = -E[\log f(X)] = \int_0^\infty f(x)\log f(x)\,dx.$$

[3] If the random variable X_n (for $n = 1, 2, \ldots$) has the distribution Φ_n (for $n = 1, 2, \ldots$), respectively, and if

$$\lim_{n\to\infty} \int_{-\infty}^{\infty} f(x)\,d\Phi_n(x) = \int_{-\infty}^{\infty} f(x)\,d\Phi_\infty(x)$$

for every continuous function f with compact support, then the sequence $\{\Phi_n\}$ is said to *converge in distribution* to X_∞.

[4] For a continuous parameter random variable $\{X(t)\}$, we can also define

$$\begin{aligned} \text{mean of } X &= \mu(t) = E[X], \\ \text{variance of } X &= \mathrm{Var}(t) = E[(X(t) - \mu(t))^2], \\ \text{covariance of } X &= \mathrm{Cov}(s,t) = E[(X(t) - \mu(t))(X(s) - \mu(s))], \end{aligned}$$

References

[1] W. Feller, *An Introduction to Probability Theory and Its Applications*, John Wiley & Sons, New York, 1968.

[2] C. W. Helstrom and J. A. Ritcey, "Evaluation of the Noncentral F-Distribution by Numerical Contour Integration," *SIAM J. Sci. Stat. Comput.*, **6**, No. 3, 1985, pages 505–514.

[3] R. J. McEliece, *The Theory of Information and Coding*, Addison–Wesley Publishing Co., Reading, MA, 1977.

8. Summations: Combinatorial

Applicable to Evaluation of combinatorial sums.

Procedure

A combinatorial sum may sometimes be written as a summation over contour integrals. Interchanging the order of integration (when permitted), allows a different integral to be evaluated. Evaluating this new integral will then yield the desired sum.

Finding the contour integral representation of the terms in the summation may be aided by Table 8.

Example 1

Consider the sum

$$S_n(m) = \sum_{k=0}^{n} (-4)^k \binom{k}{m} \binom{n+k}{2k},$$

where m is an integer. By use of Table 8 we make the identification

$$S_n(m) = \sum_{k=0}^{n} (-4)^k \left(\frac{1}{2\pi i} \int_{|x|=\rho_1} \frac{(1+x)^k}{x^{m+1}}\, dx \right) \left(\frac{1}{2\pi i} \int_{|y|=\rho_2} \frac{(1+y)^{n+k}}{y^{2k+1}}\, dy \right).$$

If we choose ρ_1 and ρ_2 appropriately (i.e., in this case we require $|y|^2 > 4(1+x)(1+y)$ in the integration), then we may move the summation inside the integrals and evaluate the sum on k to obtain

$$S_n(m) = \frac{1}{(2\pi i)^2} \int_{|x|=\rho_1, |y|=\rho_2} \frac{(1+y)^n y}{x^{m+1}(y^2 + 4(1+y)(1+x))}\, dx\, dy.$$

Table 8. Representations of combinatorial objects as contour integrals. Here, $\operatorname*{res}_x F(x)$ denotes the sum of the residues of $F(x)$ at all poles within some region centered about the origin. That is: $\operatorname*{res}_x F(x) = \frac{1}{2\pi i}\int_\Gamma F(x)\,dx$.

Binomial Coefficients (where $0 < \rho < 1$):

$$\binom{m}{n} = \operatorname*{res}_x (1+x)^m x^{-n-1} = \frac{1}{2\pi i}\int_{|x|=\rho}(1+x)^m x^{-n-1}\,dx.$$

Multinomial Coefficients (where $\Gamma(\rho) = \left\{\mathbf{x} = (x_1, \ldots, x_k) \mid |\mathbf{x}| = \rho_i,\ 0 < \rho_i < 1,\ i = 1, \ldots, k\right\}$):

$$\begin{aligned}\binom{m}{n_1, n_2, \ldots, n_k} &= \operatorname*{res}_{x_1, x_2, \ldots, x_k} (1 + x_1 + \cdots + x_k)^m x_1^{-n_1-1} x_2^{-n_2-1} \cdots \\ &\qquad \times x_k^{-n_k-1} \\ &= \frac{1}{(2\pi i)^k}\int_{\Gamma(\rho)} (1 + x_1 + \cdots + x_k)^m x_1^{-n_1-1} x_2^{-n_2-1} \cdots \\ &\qquad \times x_k^{-n_k-1}\,dx_1 \cdots dx_k.\end{aligned}$$

Bernoulli numbers:	$B_n = n! \operatorname*{res}_x (e^x - 1)^{-1} x^{-n}.$
Euler numbers:	$E_n = n! \operatorname*{res}_x \cosh^{-1}(x) x^{-n-1}.$
Power terms	$\frac{m^n}{n!} = \operatorname*{res}_x \left(e^{mx} x^{-n-1}\right).$

Since m is an integer, the integral with respect to x may be evaluated by the residue theorem to obtain

$$S_n(m) = 4^m \frac{(-1)^m}{2\pi i}\int_{|y|=\rho_2} \frac{(1+y)^{n+m}}{(y+2)^{2m+2}}\,dy.$$

Evaluating this last integral, by another application of the residue theorem, we obtain our final form for the summation

$$S_n(m) = (-1)^n 4^m \binom{n+m}{2m} \frac{2n+1}{2m+1}.$$

This is one of the so-called Moriety identities.

Example 2

Consider the summation

$$R(m,n,p) := \sum_{k=0}^{\min(m,n)} \binom{n}{k}\binom{m}{k}\binom{m+n+p-k}{m+n}, \tag{8.1}$$

where m, n, and p are non-negative integers. Using Table 8, it is easy to show that

$$R(m,n,p) := \sum_{k=0}^{\infty} \frac{1}{(2\pi i)^3} \left(\int_{|x_1|=\rho_1} \frac{(1+x_1)^m}{x_1^{k+1}}\, dx_1 \right) \left(\int_{|x_2|=\rho_2} \frac{(1+x_2)^n}{x_2^{k+1}}\, dx_2 \right) \times \left(\int_{|x_3|=\rho_3} \frac{(1+x_3)^{m+n+p-k}}{x_3^{m+n+1}}\, dx_3 \right)$$

The reason that the k summation can be extended to include large values of k is because there are no contributions from these values. By defining $S_x = \{\mathbf{x} = (x_1, x_2, x_3) \mid |x_1| = |x_2| = 2, |x_3| = \frac{1}{2}\}$, this integral can be written as

$$R(m,n,p) = \frac{1}{(2\pi i)^3} \int_{S_x} (1+x_3)^{m+n+p}(1+x_1)^m(1+x_2)^n x_3^{-m-n-1} \times \left(\sum_{k=0}^{\infty} (1+x_3)^{-k}(x_1x_2)^{-k-1} \right) d\mathbf{x}.$$

If we introduce the new variables t_1 and t_2 and define the curve $S_t = \{\mathbf{t} = (t_1, t_2) \mid |t_1| = |t_2| = \frac{1}{10}\}$, then this last three-dimensional integral and summation may be written as the following five-dimensional contour integral:

$$R(m,n,p) = \frac{1}{(2\pi i)^5} \int_{S_t \times S_x} \frac{(1+x_3)^{p+1} x_3}{f(t_1,x_1) f(t_2,x_2)(x_1x_2(1+x_3)-1)t_1^{m+1}t_2^{n+1}}\, d\mathbf{x}\, d\mathbf{t},$$

where $f(a,b) := x_3 - a(1+b)(1+x_3)$. If this five-dimensional integral is evaluated with respect to x_1, x_2, and x_3, in that order, then we obtain

$$R(m,n,p) = \frac{1}{(2\pi i)^2} \int_{S_t} (1-t_1)^{-p-1}(1-t_2)^{-p-1} t_1^{-n-1} t_1^{-m-1}\, d\mathbf{t}.$$

Using Table 8, this two-dimensional integral is equal to

$$R(m,n,p) = \binom{m+p}{m}\binom{n+p}{n}. \tag{8.2}$$

The final result, equations (8.1) and (8.2), can be evaluated for different choices of the parameters to obtain, for instance,

$$R(n,n,2n) = \sum_{k=0}^{n} \binom{n}{k}^2 \binom{4n-k}{2n} = \sum_{k=0}^{n} \binom{n}{k}^2 \binom{3n+k}{2n} = \binom{3n}{n}^2$$

and

$$R(m,m,n) = \sum_{k=0}^{n} \binom{m}{k}^2 \binom{n+2m-k}{2m} = \binom{m+n}{n}^2.$$

Notes

[1] Both of the examples in this section are from Egorychev [2], pages 52 and 169.

[2] In the paper by Gillis *et al.* [3] the following representation of the Legendre polynomials is used to evaluate the integral $\int_{-1}^{1} P_{n_1}(x) \cdots P_{n_k}(x)\,dx$, where $n_1, \ldots, n_k$ are non-negative integers:

$$P_n(x) = 2^{-n} \sum_{k \leq n/2} (-1)^k \binom{n}{k} \binom{2n-2k}{n} x^{n-2k}.$$

[3] Bressoud [1] uses a combinatorial approach to evaluate integrals of the form $\int_{-\pi/2}^{\pi/2} \cdots \int_{-\pi/2}^{\pi/2} \prod_{\sigma \in S} \sin^{2k(\sigma)} \sigma \, d\alpha_1 \ldots d\alpha_1$, where S is a set of nonlinear sums of elements of the $\{\alpha_i\}$, and k is an integer-valued function.

References

[1] D. M. Bressoud, "Definite Integral Evaluation by Enumeration, Partial Results in the MacDonald Conjectures," *Combinatoire énumérative*, Lecture Notes in Mathematics #1234, Springer–Verlag, New York, 1986, pages 48–57.

[2] G. P. Egorychev, *Integral Representation and the Computation of Combinatorial Sums*, Translations of Mathematical Monographs, **59**, Amer. Math. Soc., Providence, Rhode Island, 1984.

[3] J. Gillis, J. Jedwab, and D. Zeilberger, "A Combinatorial Interpretation of the Integral of the Product of Legendre Polynomials," *SIAM J. Math. Anal.*, **19**, No. 6, November 1988, pages 1455–1461.

[4] K. Mimachi, "A Proof of Ramanujan's Identity by Use of Loop Integrals," *SIAM J. Math. Anal.*, **19**, No. 6, November 1988, pages 1490–1493.

9. Summations: Other

Idea

Some summations can be determined by simple manipulations of integrals.

Procedure

One technique for evaluating infinite sums is by use of the Watson transform (see page 44). Under suitable convergence and analyticity constraints, we have:

> **Theorem**: If $g(z)$ is analytic in a domain D with a Jordan contour $\mathcal{C}$, then
>
> $$\frac{1}{2\pi i}\int_{\mathcal{C}} g(z)\cot \pi z\, dz = \sum g(n)$$
>
> for those integers n that are within $\mathcal{C}$.

Alternately (see Iyanaga and Kawada [1], page 1164): If an analytic function $f(z)$ is holomorphic except at poles a_n $(n = 1, 2, \ldots, k)$ in a domain bounded by the simple closed curve $\mathcal{C}$ and containing the points $z = m$ (for $m = 1, 2, \ldots, N$), then

$$\sum_{m=1}^{N} f(m) = \frac{1}{2\pi i}\int_{\mathcal{C}} \pi(\cot \pi z) f(z)\, dz - \sum_{n=1}^{k} \operatorname{Res}\left[\pi(\cot \pi z) f(z)\right]\Big|_{z=a_n}.$$

When the left-hand side is replaced by $\sum_{m=1}^{N}(-1)^m f(m)$, then the $\cot \pi z$ must be replaced by $\operatorname{cosec} \pi z$.

Another technique that can be used to evaluate summations is the Euler–Maclaurin summation formula. From Wong [3] (page 36) we find

> If $f(t)$ is defined on $0 \le t < \infty$, and if $f^{(2n)}(t)$ is absolutely integrable on $(0, \infty)$ then, for $n = 1, 2, \ldots$
>
> $$\begin{aligned} f(0) + \ldots + f(n) &= \int_0^{\infty} f(x)\, dx + \tfrac{1}{2}\left[f(0) + f(n)\right] \\ &\quad + \sum_{s=1}^{m-1} \frac{B_{2s}}{(2s)!}\left[f^{(2s-1)}(n) - f^{(2s-1)}(0)\right] + R_M(n), \end{aligned} \tag{9.1}$$
>
> where the remainder is $R_m(n) = \displaystyle\int_0^n \frac{B_{2m} - B_{2m}(x - \lfloor x \rfloor)}{(2m)!} f^{(2m)}(x)\, dx.$
>
> The remainder can be bounded by
>
> $$|R_m(n)| \le \left(2 - 2^{1-2m}\right)\frac{|B_{2m}|}{(2m)!}\int_0^n \left|f^{(2m)}(x)\right|\, dx.$$

In this theorem, the Bernoulli polynomials $\{B_s(x)\}$ are defined by the generating function

$$\frac{te^{xt}}{e^t - 1} = \sum_{s=0}^{\infty} B_s(x)\frac{t^s}{s!}. \tag{9.2}$$

The Bernoulli numbers $\{B_s\}$ are given by $B_s = B_s(0)$ and a generating function for them can be obtained from (9.2), by setting $x = 0$.

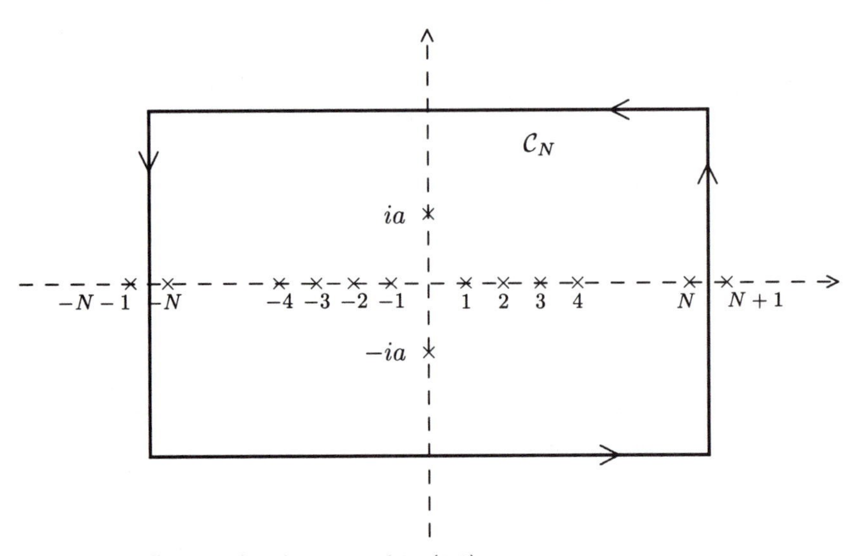

Figure 9.1 Contour for the integral in (9.3).

Example 1

As an example of the Watson transform, consider the sum $S = \sum_{n=1}^{\infty} 1/(n^2 + a^2)$. We define the integral

$$I = \frac{1}{2\pi i} \int_{\mathcal{C}_N} \frac{\pi(\cot \pi z)}{z^2 + a^2}\, dz \tag{9.3}$$

where $\mathcal{C}_N$ is the contour shown in Figure 9.1. Note that the vertical and horizontal sides to $\mathcal{C}_N$ are at the values $-N - \frac{1}{2}$ and $N + \frac{1}{2}$.

The contour integral in (9.3) can be evaluated by using Cauchy's theorem (see page 129). The poles within the contour are at $z = \pm ia$, $0, \pm 1, \pm 2, \dots, \pm N$. The residues at $\pm ia$ are $\pi \cot(\pm i\pi a)/(\pm 2ia)$, and the residue at $z = n$ (for $n = 0, \pm 1, \dots, \pm N$) is $1/(n^2 + a^2)$. Hence,

$$I = \sum_{n=-N}^{N} \frac{1}{n^2 + a^2} + \left[\frac{\pi \cot(i\pi a)}{2ia} + \frac{\pi \cot(-i\pi a)}{-2ia} \right]. \tag{9.4}$$

As $N \to \infty$, it is easy to show from (9.3) that $I \to 0$. Indeed, since the cotangent function is bounded, we have $I = O\left(N^{-2}\right) O(N) \to 0$ as $N \to \infty$. Taking the limit as $N \to \infty$, and combining (9.3) and (9.4), we find

$$\sum_{n=-\infty}^{\infty} \frac{1}{n^2 + a^2} + \left[\frac{\pi \cot(i\pi a)}{2ia} + \frac{\pi \cot(-i\pi a)}{-2ia} \right] = 0$$

or

$$\sum_{n=1}^{\infty} \frac{1}{n^2 + a^2} = \frac{\pi}{2a} \coth \pi a - \frac{1}{2a^2}. \tag{9.5}$$

If the limit $a \to 0$ is taken in this formula, then we obtain the well-known result (see also Example 3): $\sum_{n=1}^{\infty} n^{-2} = \pi^2/6$.

Example 2

As an example of the Euler–Maclaurin summation formula, consider the harmonic numbers, defined by $H_n = \frac{1}{1} + \frac{1}{2} + \ldots + \frac{1}{n}$. Using (9.1) we find:

$$H_n = \log n + \frac{1}{2}\left(1 + \frac{1}{n}\right) + \sum_{s=1}^{m-1} \frac{B_{2s}}{(2s)!}\left[-\frac{(2s-1)!}{n^{2s}} + (2s-1)!\right] + R_m(n-1), \tag{9.6}$$

where

$$R_m(n-1) = \int_0^{n-1} \frac{B_{2m} - B_{2m}(x - \lfloor x \rfloor)}{(1+x)^{2m+1}}\, dx.$$

Taking the limit of $n \to \infty$ in (9.6) results in an expression for Euler's constant γ:

$$\gamma = \lim_{n\to\infty}\left(\sum_{k=1}^{n} \frac{1}{k} - \log n\right) = \frac{1}{2} + \sum_{s=1}^{m-1} \frac{B_{2s}}{2s} + E_m,$$

where the error term is given by $E_m = \displaystyle\int_0^{\infty} \frac{B_{2m} - B_{2m}(x - \lfloor x \rfloor)}{(1+x)^{2m+1}}\, dx.$

Example 3

As an example of a different technique, consider the evaluation of the zeta function at an argument of two: $\zeta(2) := \sum_{n=1}^{\infty} n^{-2}$. We have

$$\begin{aligned}
\zeta(2) &= \sum_{n=0}^{\infty} \frac{1}{(2n+1)^2} + \sum_{n=1}^{\infty} \frac{1}{(2n)^2} \\
&= \sum_{n=0}^{\infty} \frac{1}{(2n+1)^2} + \frac{1}{4}\zeta(2) \\
&= \frac{4}{3} \sum_{n=0}^{\infty} \frac{1}{(2n+1)^2}.
\end{aligned}$$

Since $\int_0^1 x^{2n}\,dx = 1/(2n+1)$, we can write $\left(\int_0^1 y^{2n}\,dy\right)\left(\int_0^1 x^{2n}\,dx\right) = 1/(2n+1)^2$. Therefore, we have

$$\begin{aligned}
\zeta(2) &= \frac{4}{3}\sum_{n=0}^{\infty}\frac{1}{(2n+1)^2} \\
&= \frac{4}{3}\sum_{n=0}^{\infty}\int_0^1\int_0^1 (xy)^{2n}\,dx\,dy \\
&= \frac{4}{3}\int_0^1\int_0^1 \sum_{n=0}^{\infty}(xy)^{2n}\,dx\,dy \\
&= \frac{4}{3}\int_0^1\int_0^1 \frac{dx\,dy}{1-x^2y^2}.
\end{aligned}$$

Now make the change of variables from $\{x, y\}$ to $\{u, v\}$ via $x = \sin u/\cos v$, $y = \sin v/\cos u$. The Jacobian of the transformation is given by

$$\begin{aligned}
J = \left|\frac{\partial(x,y)}{\partial(u,v)}\right| &= \begin{vmatrix} x_u & y_u \\ x_v & y_v \end{vmatrix} \\
&= \begin{vmatrix} \dfrac{\cos u}{\cos v} & \dfrac{\sin u \sin v}{\cos^2 v} \\ \dfrac{\sin v \sin u}{\cos^2 u} & \dfrac{\cos v}{\cos u} \end{vmatrix} \\
&= 1 - \frac{\sin^2 u \sin^2 v}{\cos^2 u \cos^2 v} \\
&= 1 - x^2y^2.
\end{aligned}$$

Continuing the calculation of the zeta function, we find

$$\begin{aligned}
\zeta(2) &= \frac{4}{3}\int_0^1\int_0^1 \frac{dx\,dy}{1-x^2y^2} \\
&= \frac{4}{3}\int_0^1\int_0^1 \frac{J}{1-x^2y^2}\,du\,dv \\
&= \frac{4}{3}\int_{x=0}^{x=1}\int_{y=0}^{y=1} du\,dv.
\end{aligned} \tag{9.7}$$

The region of integration in the (u, v) plane becomes the triangle with vertices at $(u = 0, v = 0)$, $(u = 0, v = \pi/2)$, and $(u = \pi/2, u = 0)$ (see Figure 9.2). Since this triangle has area $\pi^2/8$ we finally determine

$$\zeta(2) = \frac{4}{3}\frac{\pi^2}{8} = \frac{\pi^2}{6}.$$

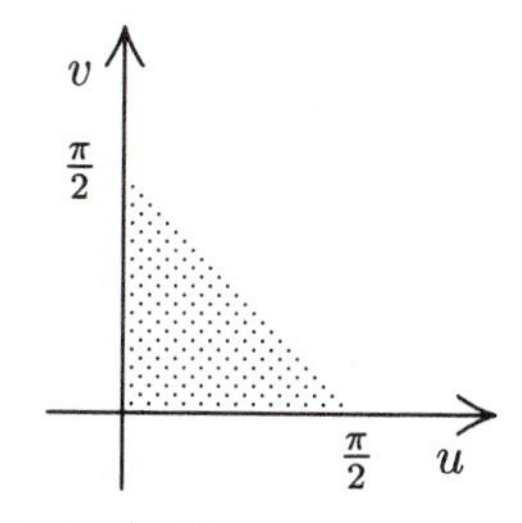

Figure 9.2 Integration region in (9.7).

Notes

[1] Under some continuity and convergence assumptions, the Poisson summation formula states (see Iyanaga and Kawada [1], page 924)

$$\sum_{n=-\infty}^{\infty} f(n) = \sum_{n=-\infty}^{\infty} \int_{-\infty}^{\infty} e^{2\pi i n t} f(t)\, dt.$$

(This formula can also be extended to functions of several variables.)

For example, if we take $f(t) = e^{-\pi t^2 x}$ (for some fixed $x > 0$), then we obtain

$$\begin{aligned} \sum_{n=-\infty}^{\infty} e^{-\pi n^2 x} &= \sum_{n=-\infty}^{\infty} \int_{-\infty}^{\infty} e^{2\pi i n t - \pi t^2 x}\, dt \\ &= \sum_{n=-\infty}^{\infty} \frac{1}{\sqrt{x}} \exp\left(-\frac{\pi n^2}{x}\right). \end{aligned} \tag{9.8}$$

For small values of x, the sum on the right-hand side of (9.8) converges much more quickly than the sum on the left-hand side. See also Smith [2]. We note in passing that the equation in (9.8) represents the following functional relationship of theta functions: $\theta(x) = \frac{1}{\sqrt{x}} \theta\left(\frac{1}{x}\right)$.

[2] For another example similar to Example 1, the summation

$$\sum_{n=1}^{\infty} \frac{\cos nt}{n^2 + a^2} = \frac{\pi \cosh a(\pi - t)}{a \sinh \pi a} - \frac{1}{2a^2}$$

can be derived from the integral $frac12\pi i \int_C \frac{\pi \cos z(\pi - t)}{(z^2 + a^2) \sin \pi z}\, dz$.

References

[1] S. Iyanaga and Y. Kawada, *Encyclopedic Dictionary of Mathematics*, MIT Press, Cambridge, MA, 1980.

[2] P. J. Smith, "A New Technique for Calculating Fourier Integrals Based on the Poisson Summation Formula," *J. Statist. Comput. Simulation*, **33**, No. 3, 1989, pages 135–147.

[3] R. Wong, *Asymptotic Approximation of Integrals*, Academic Press, New York, 1989.

10. Zeros of Functions

Applicable to Functions with zeros that we would like to characterize.

Idea

By evaluating certain integrals, information about the location of zeros of functions can be obtained.

Procedure

There are several theorems that can be used to determine the location of zeros of functions. We illustrate two such theorems.

A standard theorem from complex analysis states (see Levinson and Redheffer [6], Theorem 6.1):

> **Theorem**: Let $f(z)$ be a meromorphic function in a simply connected domain D containing a Jordan contour $\mathcal{C}$. Suppose f has no zeros or poles on $\mathcal{C}$. Let N be the number of zeros and P the number of poles of f in $\mathcal{C}$, where a multiple zero or pole is counted according to its multiplicity. Then
>
> $$\frac{1}{2\pi i}\int_{\mathcal{C}} \frac{f'(z)}{f(z)}\, dz = N - P. \tag{10.1}$$

The *principle of the argument* is the name given to the statement:

$$N - P = \frac{1}{2\pi} \arg f(z)\bigg|_{\mathcal{C}}, \tag{10.2}$$

which is just a reformulation of (10.1). (Note that "arg" denotes the argument or phase of the following function.) The quantity $N - P$ is also known as the *index* of f relative to the contour $\mathcal{C}$. See Example 1 for an application of (10.2).

Another useful theorem is (see Bharucha-Reid and Sambandham [1], Lemma 4.9):

> **Theorem**: If $f(t) \in C^1$ for $a < t < b$ and $f(t)$ has a finite number of points in $a < t < b$ with $f'(t) = 0$, then the number of zeros of $f(t)$ in the interval $a < t < b$ is given by
>
> $$n(a, b; f) = \frac{1}{2\pi}\int_{-\infty}^{\infty} d\zeta \int_a^b \cos[\zeta f(t)]|f'(t)|\, dt,$$
>
> where multiple zeros have been counted once.

See Example 2 for an application of this theorem.

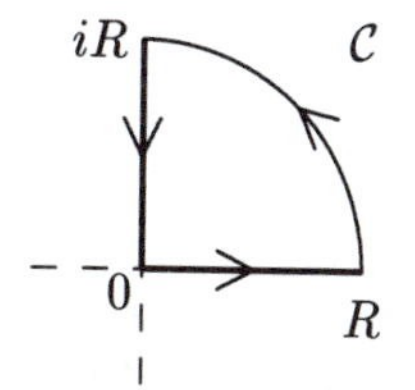

Figure 10. The contour used in Example 1.

Example 1

Consider the polynomial $z^3 - z^2 + 2$. How many roots does it have in the first quadrant? We will use the principle of the argument with $f(z) = z^3 - z^2 + 2$ and the contour in Figure 10.1 as $R \to \infty$.

To use (10.2), we must determine how the argument of $f(z)$ changes on the three components of the contour $\mathcal{C}$:

- The horizontal component ($y = 0$ and $0 \le x \le R$): We observe that $f(0) > 0$ and $f(\infty) > 0$. Since there is only one inflection point of $f(z)$ on this component (at $z = \frac{2}{3}$), we conclude that $f(z)$ is always positive on this segment. Hence, there is no change in $\arg f$ on this segment:
$$\arg f(x)\Big|_{x=0}^{x=R} = 0. \tag{10.3}$$
- The curved component $z = Re^{i\theta}$ with $0 \le \theta \le \frac{\pi}{2}$: On this component, $f(z) = R^3 e^{3i\theta}(1 + w)$ where $|w| < 2/R$ for large values of R. Hence, $\arg f\left(Re^{i\theta}\right) = 3\theta + \arg(1 + w)$. Therefore,
$$\arg f\left(Re^{i\theta}\right)\Big|_{\theta=0}^{\theta=\pi/2} = \frac{3\pi}{2} + \varepsilon_1, \tag{10.4}$$
where $\varepsilon_1 \to 0$ as $R \to \infty$.
- The vertical component ($x = 0$ and $0 \le y \le R$): We observe that $f(iy) = (-y^2 + 2) + i(-y^3)$. As y decreases from R to 0, $\operatorname{Re} f(iy)$ changes its sign at $y = \sqrt{2}$ from negative to positive, while $\operatorname{Im} f(iy)$ remains negative. Hence, as y decreases from R to 0, $f(iy)$ starts in the third quadrant and ends in the fourth quadrant. Therefore,
$$\arg f(iy)\Big|_{y=R}^{y=0} = 2\pi - \frac{3\pi}{2} + \varepsilon_2 = \frac{\pi}{2} + \varepsilon_2, \tag{10.5}$$
where $\varepsilon_2 \to 0$ as $R \to \infty$.

Combining the results in (10.3), (10.4), and (10.5) we find that
$$\arg f(z)|_{\mathcal{C}} = 0 + \left(\frac{3\pi}{2} + \varepsilon_1\right) + \left(\frac{\pi}{2} + \varepsilon_2\right).$$
As $R \to \infty$ we conclude that $\arg f(z)\big|_{\mathcal{C}} = 2\pi$. From (10.2) we conclude, therefore, that there is exactly one root of $f(z)$ in the first quadrant.

Example 2

In this example we answer the question: "What is the average number of real roots of the polynomial $f(x;\mathbf{a}) := a_0 + a_1 x + \ldots + a_{n-1}x^{n-1}$ when the $\{a_i\}$ are chosen randomly?" The analysis here is from Kac [5]; see also Bharucha-Reid and Sambandham [1].

For definiteness, we presume that the $\{a_i\}$ lie on the surface of the n-dimensional sphere of radius unity, $S_n(1)$ (i.e., the $\{a_i\}$ satisfy $|\mathbf{a}|^2 = \sum_{k=0}^{n-1} a_k^2 = 1$). Define $N_n(\mathbf{a})$ to be the number of real roots of $f(x;\mathbf{a})$. A simple scaling argument shows that $N_n(\alpha\mathbf{a}) = N_n(\mathbf{a})$ for any non-zero constant α; this will be needed later.

Define M_n to be the average number of real roots of the polynomial $f(x;\mathbf{a})$ as $\mathbf{a}$ varies over $S_n(1)$. That is,

$$M_n := \frac{1}{|S_n(1)|}\int_{S_n(1)} N_n(\mathbf{a})\,d\sigma.$$

(Here $d\sigma$ represents the surface element on a sphere.) It is not difficult to show that M_n can also be represented in the form

$$M_n = \frac{1}{(2\pi)^{n/2}}\int_{-\infty}^{\infty}\cdots\int_{-\infty}^{\infty} e^{-\frac{1}{2}|\mathbf{a}|^2} N_n(\mathbf{a})\,d\mathbf{a}$$

(where $d\mathbf{a} = da_0\,da_1\,\ldots\,da_{n-1}$), since this last integral can be rewritten as follows:

$$\begin{aligned}
M_n &= \frac{1}{(2\pi)^{n/2}}\int_0^{\infty} e^{-r^2/2}\left(\int_{S_n(r)} N_n(\mathbf{a})\,d\sigma_r\right)dr \\
&= \frac{1}{(2\pi)^{n/2}}\int_0^{\infty} e^{-r^2/2}\left(r^{n-1}\int_{S_n(1)} N_n(\mathbf{a})\,d\sigma\right)dr \\
&= \frac{1}{|S_n(1)|}\left(\int_{S_n(1)} N_n(\mathbf{a})\,d\sigma\right)
\end{aligned}$$

If we we use the notation $()^{(1)}$ to denote the number (or fraction) of real roots in the interval $(-1,1)$ and $()^{(2)}$ to denote the number (or fraction) of real roots not in the interval $(-1,1)$, then $N_n^{(1)}(\mathbf{a}) = N_n^{(2)}(\mathbf{a})$ because $\sum_{k=0}^{n-1} a_k x^k = \left(\sum_{k=0}^{n-1} a_{n-1-k}x^{-k}\right)x^{n-1}$. (That is, for every polynomial having x as a root, there is a corresponding polynomial with x^{-1} as a root.) This implies that $M_n^{(1)}(\mathbf{a}) = M_n^{(2)}(\mathbf{a})$ and $M_n(\mathbf{a}) = 2M_n^{(1)}(\mathbf{a})$.

Using the second theorem stated in the procedure, with $a = -1$ and $b = 1$, we find that

$$M_n^{(1)} = \frac{1}{(2\pi)^{n/2}}\int_{-\infty}^{\infty}\cdots\int_{-\infty}^{\infty} e^{-|\mathbf{a}|^2/2}\,d\mathbf{a}\,\frac{1}{2\pi}\int_{-\infty}^{\infty} d\zeta\int_{-1}^{1}\cos[\zeta f(t)]|f'(t)|\,dt. \tag{10.6}$$

(Note that we write $f(t)$ for the function $f(t;\mathbf{a})$.) Interchanging the order of integration, and recognizing that the absolute value function has the simple integral representation $|y| = \frac{1}{\pi}\int_{-\infty}^{\infty} \frac{1-\cos\eta y}{\eta^2}\, d\eta$, allows (10.6) to be written as

$$M_n^{(1)} = \frac{1}{2\pi}\int_{-1}^{1} dt \int_{-\infty}^{\infty} d\zeta\, R(\zeta, t),$$

where

$$R(\zeta,t) = \frac{1}{\pi}\int_{-\infty}^{\infty} \Big[S(t,\zeta,0) - S(t,\zeta,\eta)\Big]\frac{d\eta}{\eta^2}$$

and

$$S(t,\zeta,\eta) = \frac{1}{(2\pi)^{n/2}}\int_{-\infty}^{\infty}\cdots\int_{-\infty}^{\infty} e^{-|\mathbf{a}|^2/2}\, d\mathbf{a}\cos[\zeta f(t)]\cos[\eta f'(t)].$$

Writing the cosine function in complex exponential form allows the **a**-integrals to be evaluated (recall that $f(t) = f(t;\mathbf{a})$) to determine that

$$S(t,\zeta,\eta) = \frac{1}{2}\left[\exp\left(-\frac{1}{2}\sum_{k=0}^{n-1}\left(\zeta t^k + k\eta t^{k-1}\right)^2\right) + \exp\left(-\frac{1}{2}\sum_{k=0}^{n-1}\left(\zeta t^k - k\eta t^{k-1}\right)^2\right)\right].$$

Hence,

$$R(\zeta,t) = \frac{1}{\pi}\int_{-\infty}^{\infty}\left[e^{-(\zeta^2/2)\sum_{k=0}^{n-1} t^{2k}} - e^{-(1/2)\sum_{k=0}^{n-1}(\zeta t^k + k\eta t^{k-1})^2}\right]\frac{d\eta}{\eta^2}.$$

If we define the functions $A_n(t) = \sum_{k=0}^{n-1} t^k$, $B_n(t) = \sum_{k=0}^{n-1} kt^{2k-1}$, and $C_n(t) = \sum_{k=0}^{n-1} k^2 t^{2k}$, then we can finally find

$$M_n^{(1)} = \frac{1}{\pi}\int_{-1}^{1}\frac{\sqrt{A_n(t)C_n(t) - B_n^2(t)}}{A_n(t)}\, dt.$$

Our final answer is therefore

$$M_n = 2M_n^{(1)} = \frac{1}{\pi}\int_{-1}^{1}\frac{\sqrt{1-\left(\dfrac{nt^{n-1}(1-t^2)}{1-t^{2n}}\right)^2}}{1-t^2}\, dt. \qquad (10.7)$$

An asymptotic analysis then reveals that $M_n \approx \frac{2}{\pi}\log n$ as $n \to \infty$.

Notes

[1] The same result in (10.7) is obtained in three different cases:
  (A) The $\{a_i\}$ are chosen to be uniformly distributed on the interval $(-1, 1)$;
  (B) The $\{a_i\}$ are chosen to be equal to $+1$ and -1 with equal probability;
  (C) The $\{a_i\}$ are chosen to be uniformly distributed on the unit ball (as shown in Example 2).

[2] If $\mathcal{C}^*$ is the image of $\mathcal{C}$ under f in the first theorem in the Procedure, then $N - P$ turns out to be equal to the number of times $\mathcal{C}^*$ winds around the origin; i.e., it is the winding number of $\mathcal{C}^*$ with respect to the origin.

[3] In example 1, the polynomial $f(z) = z^3 - z^2 + 2$ has the roots $z = -1$ and $z = 1 \pm i$.

[4] If $g(z)$ is analytic in D, the zeros of $f(z)$ are simple and occur at the points $\{a_i\}$, the poles of $f(z)$ are simple and occur at the points $\{b_i\}$, then the result in (10.1) may be extended to

$$\frac{1}{2\pi i} \int_{\mathcal{C}} \frac{f'(z)}{f(z)} g(z)\, dz = \sum g(a_i) - \sum g(b_i). \tag{10.8}$$

If we choose $f(z) = \sin \pi z$, then (10.8) becomes

$$\frac{1}{2\pi i} \int_{\mathcal{C}} g(z) \cot \pi z \, dz = \sum g(n),$$

for those integers n that are within the contour $\mathcal{C}$. This formula is very useful for evaluating infinite sums; it is known as the Watson transform. An example of its usage may be found in the section beginning on page 34.

[5] Rouché's theorem compares the number of zeros of two related functions. It states (see Levinson and Redheffer [6], Theorem 6.2):

> Let $f(z)$ and $g(z)$ be analytic in a simple connected domain D containing a Jordan contour C. Let $|f(z)| > |g(z)|$ on C. Then $f(z)$ and $f(z) + g(z)$ have the same number of zeros inside C.

This theorem can be used to prove that a polynomial of degree n has n roots. For the polynomial $h(z) = \sum_{i=0}^{n} a_i z^i$ of degree n, choose $f(z) = a_0 z^n$ and $g(z) = h(z) - f(z)$.

References

[1] A. T. Bharucha-Reid and M. Sambandham, *Random Polynomials*, Academic Press, New York, 1986, Chapter 4, pages 49–102.

[2] M. P. Carpentier, "Computation of the Index of an Analytic Function," in O. Keast and G. Fairweather (eds.), *Numerical Integration: Recent Developments, Software and Applications*, Reidel, Dordrecht, The Netherlands, 1987, pages 83–90.

[3] M. P. Carpentier and A. F. Dos Santos, "Solution of Equations Involving Analytic Functions," *J. Comput. Physics*, **45**, 1982, pages 210–220.

[4] N. I. Ioakimidis, "Quadrature Methods for the Determination of Zeros of Transcendental Functions—a Review," in P. Keast and G. Fairweather (eds.), *Numerical Integration: Recent Developments, Software and Applications*, Reidel, Dordrecht, The Netherlands, 1987, pages 61–82.
[5] M. Kac, "On the Average Number of Real Roots of a Random Algebraic Equation," *Bull. Amer. Math. Soc.*, **49**, pages 314–320, 1943.
[6] N. Levinson and R. M. Redheffer, *Complex Variables*, Holden–Day, Inc., San Francisco, 1970, Section 4.6, pages 216–223.

11. Miscellaneous Applications

Idea

This section describes other uses of integration.

Physics

Let $\mathbf{F}$ be the force on an object in three-dimensional space. The work done in moving an object from point $\mathbf{a}$ to point $\mathbf{b}$ is defined by the line integral

$$W = \int_{\mathbf{a}}^{\mathbf{b}} \mathbf{F} \cdot d\mathbf{s},$$

where $\mathbf{s}$ is an element of the path traversed from $\mathbf{a}$ to $\mathbf{b}$. In a conservative force field, the force can be written as the gradient of a scalar potential field: $\mathbf{F} = \nabla P$. In this case, the amount of work performed is independent of the path and is given by $W = P(\mathbf{b}) - P(\mathbf{a})$.

For example, since gravity is a conservative force field, $\mathbf{g} = (0, 0, g) = \nabla(g\mathbf{z})$, the work performed in moving an object from the location $\mathbf{a} = (a_x, a_y, a_z)$ to the location $\mathbf{b} = (b_x, b_y, b_z)$ is just $W = g(b_z - a_z)$.

Mechanics

The momentum of a rigid body K is defined by $\mathbf{Q} = \int (d\mathbf{r}/dt)\, dm$ where dm is the mass of the volume element at a point $\mathbf{r}$ and $(d\mathbf{r}/dt)$ is its velocity. See Iyanaga and Kawada [6], page 454.

The angular momentum of a rigid body, about an arbitrary point $\mathbf{r}_0$, is defined by $\mathbf{H} = \int (\mathbf{r} - \mathbf{r}_0) \times (d\mathbf{r}/dt)\, dm$. See Iyanaga and Kawada [6], page 455.

Mechanics

If $\nabla^2 u = -1$ in G, and $u = 0$ along ∂G, then the *torsional rigidity* of the domain G is defined to be $P = 4 \iint_G u\, dA$. For a disk of radius R, we have $P = \pi R^4/2$. See Hersch [4].

Notes

[1] Using an integral representation of derivatives, Calio *et al.* [2] demonstrate how quadrature formulas can be used to differentiate analytic functions.

[2] Let Q be a bounded two-dimensional domain with a partly smooth curve of boundary. Aizenberg [1] contains an integral representation for the difference between the number of lattice points of Q and its volume. Then a similar result for a three-dimensional domain is given.

[3] Ioakimidis [5] uses contour integrals to find the location of branch points.

References

[1] L. A. Aizenberg, "Application of the Multidimensional Logarithmic Residue to Number Theory. An Integral Formula for the Difference Between the Number of Lattice Points in a Domain and its Volume," *Ann. Polon. Math.*, **46**, 1985, pages 395–401.

[2] F. Calio, M. Frontini, and G. V. Milovanović, "Numerical Differentiation of Analytic Functions Using Quadratures on the Semicircle," *Comp. & Maths. with Appls.*, **22**, No. 10, 1991, pages 99–106.

[3] H. M. Haitjema, "Evaluating Solid Angles Using Contour Integrals," *Appl. Math. Modelling* **11**, No. 1, 1987, pages 69–71.

[4] J. Hersch, "Isoperimetric Monotonicity: Some Properties and Conjectures (Connections Between Isoperimetric Inequalities)," *SIAM Review*, **30**, No. 4, December 1988, pages 551–577.

[5] N. I. Ioakimidis, "Locating Branch Points of Sectionally Analytic Functions by Using Contour Integrals and Numerical Integration Rules," *Int. J. Comp. Math.*, **41**, 1992, pages 215-222.

[6] S. Iyanaga and Y. Kawada, *Encyclopedic Dictionary of Mathematics*, MIT Press, Cambridge, MA, 1980.

[7] I. Vardi, "Integrals, an Introduction to Analytic Number Theory," *Amer. Math. Monthly*, **95**, No. 4, 1988, pages 308–315.

II

Concepts and Definitions

12. Definitions

Asymptotic Expansion Let $f(x)$ be continuous in a region R and let $\{\phi_n(x)\}$ be an asymptotic sequence as $x \to x_0$. Then the formal series $\sum_{n=0}^{\infty} a_n \phi_n(x)$ is said to be an *infinite asymptotic expansion* of $f(x)$, as $x \to x_0$, with respect to $\{\phi_n(x)\}$ if the equivalent sets of conditions

$$f(x) = \sum_{n=0}^{m} a_n \phi_n(x) + O(\phi_{m+1}(x)), \quad \text{as } x \to x_0, \tag{12.1}$$

$$\lim_{x \to x_0} \left[\frac{f(x) - \sum_{n=0}^{m} a_n \phi_n(x)}{\phi_m(x)} \right] = 0, \tag{12.2}$$

for each $m = 0, 1, 2, \ldots$ are satisfied. If, instead, (12.1) only holds for $m = 0, 1, 2, \ldots, N-1$, then

$$f(x) \sim \sum_{n=0}^{N-1} a_n \phi_n(x), \quad \text{as } x \to x_0$$

is said to be an *asymptotic expansion of f to N terms* with respect to the asymptotic sequence $\{\phi_n(x)\}$. Note that $O(\,)$ is defined on page 351.

Asymptotic Sequence The sequence of functions $\{\phi_n(x)\}$, $n = 0, 1, 2, \ldots$ is called an asymptotic sequence as $x \to x_0$ in some region R if, for every n, $\phi_n(x)$ is defined and continuous in R and $\phi_{n+1}(x) = o(\phi_n(x))$, as $x \to x_0$.

Auxiliary Asymptotic Sequence Let $\{\phi_n(x)\}$ be an asymptotic sequence as $x \to x_0$. Then the formal series $\sum_{n=0}^{\infty} a_n f_n(x)$ is said to be an asymptotic expansion of $f(x)$, with respect to the *auxiliary asymptotic sequence* $\{\phi_n(x)\}$ if

$$f(x) = \sum_{n=0}^{N} f_n(x) + o(\phi_N(x)), \text{as } x \to x_0,$$

for each $m = 0, 1, 2, \ldots$. This is denoted $f(x) \sim \sum_{n=0}^{\infty} f_n(x)$ when the auxiliary asymptotic sequence is understood.

Borel Field of Sets Given a system of sets M, the Borel field of sets is the smallest system of sets containing M and closed with respect to the operations of countable union and taking complements.

Bromwich Integral The Bromwich integral is a contour integral that has the shape of a semicircular region, extending to infinity, with the flat portion vertical in the complex domain.

Holomorphic Function A function is holomorphic in a domain if it is holomorphic at every point in the domain. A function is holomorphic at a point if it satisfies the Cauchy–Riemann equations at that point.

Inner Product An inner product is a binary operation on a vector space which produces a scalar. The inner product must also satisfy the following properties (where **a**, **b**, and **c** are vectors, z is a scalar, and an overbar indicates complex conjugation):

[1] $(\mathbf{a}, \mathbf{b}) = \overline{(\mathbf{b}, \mathbf{a})}$
[2] $(\mathbf{a}, \mathbf{b} + \mathbf{c}) = (\mathbf{a}, \mathbf{b}) + (\mathbf{a}, \mathbf{c})$
[3] $(z\mathbf{a}, \mathbf{b}) = z(\mathbf{a}, \mathbf{b})$
[4] $(\mathbf{a}, \mathbf{a}) > 0$, unless $\mathbf{a} = \mathbf{0}$, when $(\mathbf{0}, \mathbf{0}) = 0$

Integrand In the integral $\int_a^b f(x)\,dx$, the integrand is $f(x)$.

Jacobian The Jacobian, or the Jacobian determinant, is denoted by $\partial(F, G, H, \ldots)/\partial(u, v, w, \ldots)$ and is defined to be the determinant

$$\frac{\partial(F, G, H, \ldots)}{\partial(u, v, w, \ldots)} = \begin{vmatrix} F_u & F_v & F_w & \cdots \\ G_u & G_v & G_w & \cdots \\ H_u & H_v & H_w & \cdots \\ \vdots & \vdots & \vdots & \ddots \end{vmatrix}.$$

When used in a change of variable computation (see page 109), the absolute value of the Jacobian is used.

L_p Functions A measurable function $f(x)$ is said to belong to L_p if $\int_0^\infty |f(x)|^p\,dx$ is finite.

Lebesgue Measurable Set A set of $\mathbf{R}^p$ is called a Lebesgue measurable set (or simply a measurable set) if it belongs to the smallest σ-algebra containing the Borel sets and the sets of measure zero of $\mathbf{R}^p$.

Leibniz's Rule Leibniz's rule states that

$$\frac{d}{dt}\left(\int_{f(t)}^{g(t)} h(t,\xi)\,d\xi\right) = g'(t)h(t,g(t)) - f'(t)h(t,f(t)) + \int_{f(t)}^{g(t)} \frac{\partial h}{\partial t}(t,\xi)\,d\xi.$$

Linear Independence Given the smooth functions $\{y_1, y_2, \ldots, y_n\}$, the Wronskian is the determinant

$$\begin{vmatrix} y_1 & y_2 & \cdots & y_n \\ y_1' & y_2' & \cdots & y_n' \\ \vdots & \vdots & \ddots & \vdots \\ y_1^{(n-1)} & y_2^{(n-1)} & \cdots & y_n^{(n-1)} \end{vmatrix}.$$

If the Wronskian does not vanish in an interval, then the functions are linearly independent.

Lower Limit, Upper Limit In the integral $\int_a^b f(x)\,dx$, the *lower limit* is the value a, the *upper limit* is the value b.

Measure A (positive) measure on a σ-algebra A is a mapping μ of A into $[0,\infty]$ such that if E is a disjoint union of a sequence of sets $E_n \in A$, then $\mu(E) = \sum \mu(E_n)$.

Meromorphic Function A meromorphic function is analytic, except possibly for the presence of poles.

Norms If f is a measurable function on $\mathbf{R}^n$ then we define the L_p norm of f (for $0 < p < \infty$) by $||f||_p = \left(\int_{R^n} |f(\mathbf{x})|^p\,d\mathbf{x}\right)^{1/p}$.

Orthogonal Two functions $f(x)$ and $g(x)$ are said to be orthogonal with respect to a weighting function $w(x)$ if the inner product vanishes, i.e., $(f(x), g(x)) := \int f(x)w(x)\bar{g}(x)\,dx = 0$ over some appropriate range of integration. Here, an overbar indicates the complex conjugate.

Pole An isolated singularity of $f(z)$ at α is said to be a pole if $f(z) = g(z)/(z-\alpha)^m$, where $m \geq 1$ is an integer, $g(z)$ is analytic in a neighborhood if α and $g(\alpha) \neq 0$. The integer m is called the order of the pole.

Sigma-algebra A family A of subsets of a set X is called a σ-algebra if the empty set is in A, and if A is closed under complementation and countable union.

Set of Measure Zero A set E in $\mathbf{R}^p$ is called a set of measure zero (or a negligible set) if there exists a Borel set A such that $E \subset A$ and $\nu(A) = 0$.

Variations Let $f(x)$ be a real bounded function defined on $[a, b]$. Given the subdivision $a = x_0 < x_1 < \ldots < x_n = b$, denote the sum of positive (negative) differences $f(x_i) - f(x_{i-1})$ by P $(-N)$. The suprema of P, N, and $P + N$, for all possible subdivisions of $[a, b]$, are called the *positive variation*, the *negative variation*, and the *total variation* of $f(x)$ on $[a, b]$. If any one of these three values is finite, then they are all finite. In this case, $f(x)$ is said to be of *bounded variation.*

The continuous function $x \sin \frac{1}{x}$ is not of bounded variation, while the discontinuous function $\operatorname{sgn} x$ is of bounded variation.

If $g(x)$ is an increasing function on $[a, b]$ then the total variation of $g(x)$ on $[a, b]$, written $\operatorname{var}_{[a,b]} g$, is given by $g(b) - g(a)$. Hence, by writing an arbitrary continuous function $f(x)$ as the difference of two increasing functions $f(x) = f_1(x) - f_2(x)$ we find

$$\operatorname{var}_{[a,b]} f = \operatorname{var}_{[a,b]} f_1 + \operatorname{var}_{[a,b]} f_1 = (f_1(b) - f_1(a)) + (f_2(b) - f_2(a)).$$

One way to form the decomposition $f(x) = f_1(x) - f_2(x)$ is by

$$f_1(x) = \int_a^x \left[f'(t)\right]^+ \, dt, \qquad f_2(x) = \int_a^x \left[f'(t)\right]^- \, dt.$$

(Note that the notation $[]^-$ and $[]^+$ is defined on page 352.)

Bounded Variation A function $g(x)$ is of bounded variation in $[a, b]$ if and only if there exists a number M such that $\sum_{i=1}^m |g(x_i) - g(x_{i-1})| < M$ for all partitions $a = x_0 < x_1 < x_2 < \cdots < x_m = b$ of the interval. Alternately, $g(x)$ is of bounded variation if and only if it can be written in the form $g(x) = g_1(x) - g_2(x)$ where the functions $g_1(x)$ and $g_2(x)$ are bounded and nondecreasing in $[a, b]$.

Weyl's Integral Formula Let G be a compact connected semisimple Lie group and H a Cartan subgroup of G. If μ, β, and λ are all normalized to be of total measure 1, then

$$\int_G f(g) \, d\mu(g) = \frac{1}{w} \int_H \int_{G/H} f\left(ghg^{-1}\right) J(h) \, d\lambda(g^*) \, d\beta(h)$$

for every continuous function f on G, where w is the order of the Weyl group of G. Here J is given by

$$J(\exp X) = \left| \prod_{\alpha \in P} \left(e^{\alpha(X)/2} - e^{-\alpha(X)/2}\right) \right|^2,$$

where P is the set of all positive roots α of G with respect to H and X is an arbitrary element of the Lie algebra of H.

13. Integral Definitions

Idea

There are many different types of integrals of interest. These integrals include the following:

Abelian (see below)
contour (see page 129)
fractional (see page 75)
improper (see below)
Lebesgue (see below)
loop (see page 4)
Riemann (see below)
Stratonovich (see page 186)
Cauchy (see page 92)
Feynman (see page 70)
Henstock (see below)
Ito (see page 186)
line (see page 164)
path (see page 86)
stochastic (see below)
surface (see page 24)

Properties of Integrals

Lebesgue [16] defined six properties that the integral of a bounded function should have. These properties are:

[1] $\int_a^b f(x)\,dx = \int_{a+h}^{b+h} f(x-h)\,dx$;
[2] $\int_a^b f(x)\,dx + \int_b^c f(x)\,dx + \int_c^a f(x)\,dx = 0$;
[3] $\int_a^b [f_1(x) + f_2(x)]\,dx = \int_a^b f_1(x)\,dx + \int_a^b f_2(x)\,dx$;
[4] $\int_0^1 1\,dx = 1$;
[5] If $f > 0$ and $b > a$, then $\int_a^b f(x)\,dx > 0$;
[6] If $f_n(x) \le f_{n+1}(x)$ and $\lim_{n\to\infty} f_n(x) = f(x)$ for all x, then $\lim_{n\to\infty} \int_a^b f_n(x)\,dx = \int_a^b f(x)\,dx$.

Squire [22] indicates that more than sixty kinds of integrals have been developed that satisfy the above criteria, in different degrees of generality. In the following sections we describe only a few of the different types of integrals. Pesin [20] has a very comprehensive review of many types of integrals.

Abelian Integral

Suppose that we have an algebraic curve whose equation is $G(x, y) = 0$. Let $y = f(x)$ be the algebraic function satisfying this equation, and define S to be the associated Riemann surface on which y is single-valued. Define the rational function R by $R(x, y) = P(x, y)/Q(x, y)$, where both P and Q are polynomial functions. Note that R is single-valued on S and that the only singularities that R has on S are a finite number of poles.

An Abelian integral has the form $I(x, y) = \int_{x_0,y_0}^{x,y} R(x, y)\,dx$, where the path of integration is on the surface S. The value of this integral depends upon the integration path. Note that $I(x, y)$ is regular for all finite paths that avoid the poles of the integrand. There are only three kinds of Abelian integrals; an Abelian integral is of

[1] the first kind if it is regular everywhere,
[2] the second kind if its only singularities are poles,
[3] the third kind if it has logarithmic singularities.

No other types of singularities are possible for an Abelian integral.

Note that an Abelian integral can be of the first kind and not be constant; Liouville's theorem does not apply since I is defined on a Riemann surface, not the complex plane.

If the limits of integration are fixed, then all possible values of an Abelian integral can be determined by considering the combinatorial topology of S. Fixing the points A and B on S, define $J = \int_A^B R(x,y)\,dx$. If P is a specific path on S from A to B, then any other path from A to B is of the form $P + \Gamma$, where Γ is a closed path passing through A and B. Define K to be the Abelian integral associated with the path Γ: $K = \int_\Gamma R(x,y)\,dx$. Note that the value of K is not changed as Γ is continuously distorted, provided that Γ stays on S and does not cross any poles of $R(x,y)$.

As an example, elliptic integrals can be defined by: $R(x,y) = 1/y$ and $G(x,y) = (1-x^2)(1-k^2x^2) - y^2 = 0$.

Much research has been performed on the inversion of Abelian integrals. For example, Theorem 6.2 of Bliss [1] (page 170) states that:

> **Theorem**: If an Abelian integral $u = \int_{(x_0,y_0)}^{(x,y)} \eta(x,y)\,dx$ on the Riemann surface T of an irreducible algebraic equation $f(x,y) = 0$ defines a single-valued inverse function $x(u)$, $y(u)$, then the genus of the curve $f = 0$ must be either $p = 0$ or $p = 1$. In the case $p = 0$ the integral is either of the second kind with a single simple pole, or of the third kind with two simple logarithmic places and no other singularities. In the case $p = 1$ the integral is of the first kind.

For details, see Hazewinkel [8] (pages 14–16), or Lang [15].

Henstock Integral

Given the interval $[a,b]$ and a positive function $\delta : [a,b] \to \mathbf{R}$, define a partition to be given by $\{(t_i, [x_{i-1}, x_i])\}_{i=1}^n$, where the intervals $[x_{i-1}, x_i]$ are non-overlapping, their union is the interval $[a,b]$, and the following condition is satisfied: $t_i \in [x_{i-1}, x_i] \subset (t_i - \delta(t_i), t_i + \delta(t_i))$, for $i = 1, 2, \ldots, n$.

A function $f : [a,b] \to \mathbf{R}$ is called Henstock-integrable if there exists a number I such that for every $\varepsilon > 0$ there exists a positive function $\delta : [a,b] \to \mathbf{R}$ such that every partition of the interval $[a,b]$ results in

$$\left| \sum_{i=1}^n f(t_i)\,(x_i - x_{i-1}) - I \right| < \varepsilon.$$

The number I, usually written as $\int_a^b f(t)\,dt$, is called the Henstock integral of f. For details, see Peng-Yee [19].

Improper Integrals

An integral in which the integrand is not bounded, or the interval of integration is unbounded, is said to be an improper integral. For example, the following are improper integrals:

$$\int_0^1 \frac{dx}{x} \quad \text{and} \quad \int_0^\infty \frac{dx}{1+x^4}.$$

Suppose that $f(x)$ has the singular point z in the interval (a, b), and suppose that $f(x)$ is integrable everywhere in the interval, except at the point z. The integral $\int_a^b f(x)\,dx$ is then defined to have the value

$$\lim_{\substack{\alpha\to 0\\ \beta\to 0}} \left\{ \int_a^{z-\alpha} f(x)\,dx + \int_{z+\beta}^b f(x)\,dx \right\},$$

where the limits are to be evaluated independently.

Lebesgue Integral

Let X be a space with a non-negative complete countably-additive measure μ, where $\mu(X) < \infty$. Separate X into $\{X_n\}$ so that $\cup_{n=1}^\infty X_n = X$. A simple function g is a measurable function that takes at most a countable set of values; that is $g(x) = y_n$, with $y_n \neq y_k$ for $n \neq k$, if $x \in X_n$. A simple function g is said to be summable if the series $\sum_{n=1}^\infty y_n \mu X_n$ converges absolutely; the sum of this series is the Lebesgue integral $\int_X g\,d\mu$.

A function $f : X \to \mathbf{R}$ is summable on X (denoted $f \in L_1(X, \mu)$) if there is a sequence of simple, summable functions $\{g_n\}$, uniformly convergent to f on a set of full measure, and if the limit $\lim_{n\to\infty} \int_X g_n\,d\mu$ is finite. The number I is the Lebesgue integral of the function f; this is written

$$I = \int_X f\,d\mu. \tag{13.1}$$

A simple figure can clarify how the Lebesgue integral is evaluated. Given the function $f(x)$ on (a, b), subdivide the vertical axis into $n+1$ points: $\min f \geq y_0 < y_1 < \cdots < y_n \geq \max f$. Then form the sum

$$\sum_{i=1}^n y_{i-1} \times \text{ measure}\,(x \mid y_{i-1} \leq f(x) < y_i)$$

in which the measure is the sum of the lengths of the subintervals on which the stated inequality takes place. (See Figure 13.1.) In the limit of $n \to \infty$, as the largest length $(y_i - y_{i-1})$ tends to zero, this sum becomes the Lebesgue integral of $f(x)$ from a to b.

The Lebesgue integral is a linear non-negative functional with the following properties:

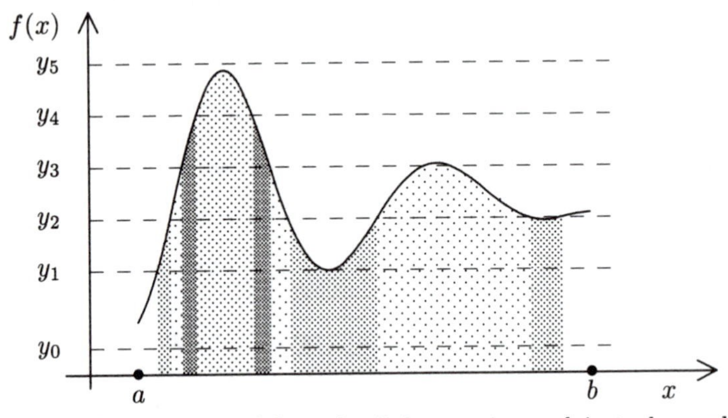

Figure 13.1 A schematic of how the Lebesgue integral is to be evaluated. Regions with similar values are shaded in the same way.

[1] If $f \in L_1(X, \mu)$ and if $\mu\{x \in X \mid f(x) \neq h(x)\} = 0$, then $h \in L_1(X, \mu)$ and $\int_X f\, d\mu = \int_X h\, d\mu$.

[2] If $f \in L_1(X, \mu)$, then $|f| \in L_1(X, \mu)$ and $\left|\int_X f\, d\mu\right| \leq \int_X |f|\, d\mu$.

[3] If $f \in L_1(X, \mu)$, $|h| \leq f$ and h is measurable, then $h \in L_1(X, \mu)$ and $\left|\int_X h\, d\mu\right| \leq \int_X f\, d\mu$.

[4] If $m \leq f \leq M$ and if f is measurable, then $f \in L_1(X, \mu)$ and $m\, \mu X \leq \int_X f\, d\mu \leq M\, \mu X$.

For functions from $\mathbf{R}^n$ to $\mathbf{R}$, if the measure used is the Lebesgue measure, then (13.1) may be written as $J = \int_{\mathbf{R}^n} f(x)\, dx$. For different measures, the functional J is called a Lebesgue–Stieltjes integral.

Every function that is Riemann integrable on a bounded interval is also Lebesgue integrable, but the converse is not true. For example, the Dirichlet function (equal to 0 for irrational arguments, but equal to 1 for rational arguments) is Lebesgue integrable but not Riemann integrable.

Alternatively, the existence of an improper Riemann integral does not imply the existence of a Lebesgue integral. For example, the integral $\int_0^\infty \sin x / x\, dx = \pi/2$ is not Lebesgue integrable because Lebesgue integrability requires that both f and $|f|$ should be integrable. In this example $\int_0^\infty |\sin x|/x\, dx = \infty$.

Riemann Integral

Let $f(x)$ be a bounded real-value function defined on the interval $I = [a, b]$. Denote a partition of I by $D = \{x_0, \ldots, x_n\}$ where $a = x_0 < x_1 < \ldots < x_n = b$ and n is finite. Let I_i denote the sub-interval $[x_i, x_{i+1}]$. Define

$$m_i = \inf_{x \in I_i} f(x), \qquad M_i = \sup_{x \in I_i} f(x).$$

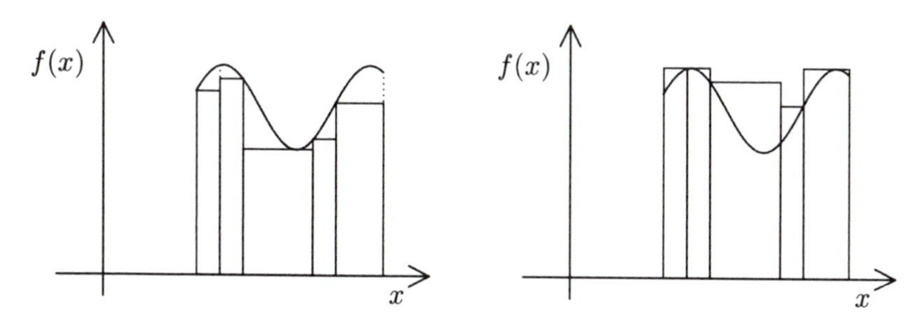

Figure 13.2 An illustration of the lower (left) and upper (right) sums of a function.

The oscillation of f on I_i is defined to be $M_i - m_i$. Now define the Darboux sums $\overline{\sigma}(D)$ and $\underline{\sigma}(D)$:

$$\overline{\sigma}(D) = \sum_{i=1}^{n} M_i (x_i - x_{i-1}), \qquad \underline{\sigma}(D) = \sum_{i=1}^{n} m_i (x_i - x_{i-1}).$$

Considering all possible partitions of D, we define

$$\text{Riemann upper integral of } f = \overline{\int_a^b} f(x)\, dx = \inf_D \overline{\sigma}(D),$$

$$\text{Riemann lower integral of } f = \underline{\int_a^b} f(x)\, dx = \sup_D \underline{\sigma}(D).$$

(See Figure 13.2.) If the Riemann upper and lower integrals of f coincide, then the common value is called the Riemann integral of f on $[a, b]$ and is denoted by $\int_a^b f(x)\, dx$. In this case, the function f is said to be *Riemann integrable*, or just *integrable.*

Darboux's theorem states that:

> **Theorem** (*Darboux*): For each $\varepsilon > 0$ there exists a positive δ such that the inequalities
>
> $$\left| \overline{\sigma}(D) - \overline{\int_a^b} f(x)\, dx \right| < \varepsilon, \qquad \left| \underline{\sigma}(D) - \underline{\int_a^b} f(x)\, dx \right| < \varepsilon,$$
>
> hold for any partition D with $\max(x_i - x_{i+1}) < \delta$, for $i = 1, 2, \ldots, n$.

From Darboux's theorem we conclude that necessary and sufficient conditions for a function $f(x)$ to be integrable on $[a, b]$ is that for each positive ε there exists a δ such that

$$\left| \sum_{j=1}^{n} f(\zeta_i)(x_j - x_{j-1}) - \int_a^b f(x)\, dx \right| < \varepsilon,$$

where $\delta(D) = \max_i(x_i - x_{i-1}) < \delta$ and ζ_i is chosen arbitrarily from I_i.

Stochastic Integrals

Let $X(t)$ be an arbitrary random process defined in some interval $a \leq t \leq b$ and let $h(t, \tau)$ be an arbitrary deterministic function defined in the same interval. Define the integral $I(\tau) = \int_a^b h(t, \tau) X(t)\, dt$. If the integral exists, then $I(\tau)$ is, itself, a random variable.

The integral can be shown to exist for each sample function $x(t)$ if

$$\int_a^b E[|h(t, \tau) X(t)|]\, dt = \int_a^b |h(t, \tau)| E[X(t)]\, dt < \infty.$$

Furthermore, when the integral exists, we can write

$$E[I(\tau)] = E\left[\int_a^b h(t, \tau) X(t)\, dt\right] = \int_a^b h(t, \tau) E[X(t)]\, dt,$$

and the interchange of the order of integration and the expectation operation is justified.

Even if the integral does not exist in the usual sense for each sample function $x(t)$ of $X(t)$, it may be possible to define the equality in some stochastic sense. (See page 186.)

Notes

[1] Other properties of an integral can be inferred from the stated properties of integrals. Let $\mathbf{I}$ denote the set of all functions integrable on the interval $I = [a, b]$. If f and g belong to $\mathbf{I}$, and α and β are arbitrary real numbers then

(A) $|f| \in \mathbf{I}$,
(B) $\alpha f + \beta g \in \mathbf{I}$,
(C) $f * g \in \mathbf{I}$,
(D) $\min\{f, g\} \in \mathbf{I}$,
(E) $\max\{f, g\} \in \mathbf{I}$,
(F) $f/g \in \mathbf{I}$ (assuming that $|g| \geq A > 0$ on I).

[2] We also have the conventions $\int_a^a f(x)\, dx = 0$ and $\int_b^a f(x)\, dx = -\int_a^b f(x)\, dx$.

[3] Botsko [3] describes a generalization of the Riemann integral that admits every derivative into the set of integrable functions. As an example, Botsko considers the function f' where

$$f(x) = \begin{cases} x^2 \sin \dfrac{1}{x^2} & \text{if } 0 < x \leq 1, \\ 1 & \text{if } x = 0. \end{cases}$$

The function f' has an unbounded derivative and is not Lebesgue integrable, but can be integrated with Botsko's integral.

[4] The Burkill integral [8] was originally introduced to determine surface areas, see Burkill [4]. In modern usage, it is used for integration of non-additive functions. The Burkill integral is less general than the subsequently introduced Kolmogorov integral; any function that is Burkill-integrable is also Kolmogorov-integrable. The name of "Burkill integral" is also given to a number of generalizations of the Perron integral [20].

[5] The Henstock integral is also known as the generalized Riemann integral. The Henstock integral and the Denjoy integral are equivalent. The restricted Denjoy integral includes the Newton integral and the Lebesgue integral.

[6] The Perron integral, the Luzin integral, the gauge integral, the Kurzweil–Henstock integral, and the special Denjoy integral are all equivalent. See Henstock [9]. In one dimension, the Perron integral is equivalent to the restricted Denjoy integral. A multi-dimensional Perron integral is described in Jurkat and Knizia [13].

[7] The Boks integral [8] is a generalization of the Lebesgue integral, first proposed by Denjoy, but studied in detail by Boks [2]. The definition starts by taking a real-valued function f defined on a segment $[a, b]$ and periodically extending it to the entire real line (with period $b - a$). The A-integral [8] is more convenient to use than the Boks integral.

[8] Other types of integrals not described in this book include:
  (A) Banach integrals, Birkhoff integrals, Bochner integrals, Denjoy integrals, Dunford integrals, Gel'fand–Pettic integrals, and harmonic integrals (see Iyanaga and Kawada [12], pages 12–15, 337–340, 627–629, 787).
  (B) Norm integrals, refinement integrals, gauge integrals, Perron integrals, absolute integrals, general Denjoy integrals, and strong variational integrals (see Henstock [9]).
  (C) Borel's integral, Daniell's integral, Denjoy integral, improper Dirichlet integrals, Harnack integrals, Hölder's integrals, Khinchin's integrals, Radon's integrals, Young's integrals, and De la Vallée–Possin's integrals (see Pesin [20]).
  (D) Curvilinear integrals are better known as line integrals, see page 164.
  (E) Fuzzy integrals (see Ichihashi *et al.* [11]).
  (F) Kolmogorov integrals (see Hazewinkel [8], page 296).

[9] It is also possible to define an integral over an algebraic structure. For example, integrals in a Grassman algebra are discussed in de Souza and Thomas [21].

[10] The Lommel integrals are specific analytical formulas for the integration of products of Bessel functions. See Iyanaga and Kawada [12], page 155.

[11] The notion of stochastic integration was first introduced by Wiener in connection with his studies of the Brownian motion process. Given a one-dimensional path $X(t)$ and any function $f(t)$, Wiener wanted to be able to define the integral $\int_0^T f(t)\,dX(t)$. This integral makes no sense as a Stieltjes sum since $X(t)$ is not a function of bounded variation.

References

[1] G. A. Bliss, *Algebraic Functions*, Dover Publications, Inc., New York, 1966.

[2] T. J. Boks, "Sur les rapports entre les méthodes de l'intégration de Riemann et de Lebesgue," *Rend. Circ. Mat. Palermo*, **45**, No. 2, 1921, pages 211–264.

[3] M. W. Botsko, "An Easy Generalization of the Riemann Integral," *Amer. Math. Monthly*, **93**, No. 9, November 1986, pages 728–732.

[4] J. C. Burkill, "Functions of Intervals," *Proc. London. Math. Soc.*, **22**, No. 2, 1924, pages 275–310.

[5] G. F. Carrier, M. Krook, and C. E. Pearson, *Functions of a Complex Variable*, McGraw–Hill Book Company, New York, 1966.

[6] P. J. Daniell, "A General Form of Integral," *Ann. of Math.*, **19**, 1918, pages 279–294.

[7] A. Denjoy, "Une extension de l'integrale de M. Lebesgue," *C. R. Acad. Sci.*, **154**, 1912, pages 859–862.

[8] M. Hazewinkel (managing ed.), *Encyclopaedia of Mathematics*, Kluwer Academic Publishers, Dordrecht, The Netherlands, 1988.

[9] R. Henstock, *The General Theory of Integration*, Oxford University Press, New York, 1991.

[10] W. V. D. Hodge, *The Theory and Application of Harmonic Integrals*, Cambridge University Press, New York, 1989.

[11] H. Ichihashi, H. Tanaka, and K. Asai, "Fuzzy Integrals Based on Pseudo-Additions and Multiplications," *J. Math. Anal. Appl.*, **130**, 1988, pages 354–364.

[12] S. Iyanaga and Y. Kawada, *Encyclopedic Dictionary of Mathematics*, MIT Press, Cambridge, MA, 1980.

[13] W. B. Jurkat and R. W. Knizia, "A Characterization of Multi-Dimensional Perron Integrals and the Fundamental Theorem," *Can. J. Math.*, **43**, No. 3, 1991, pages 526–539.

[14] A. Ya. Khinchin, "Sur une extension de l'integrale de M. Denjoy," *C. R. Acad. Sci.*, **162**, 1916, pages 287–291.

[15] S. Lang, *Introduction to Algebraic and Abelian Functions*, Addison–Wesley Publishing Co., Reading, MA, 1971.

[16] M. Lebesgue, "Leçons sur l'integration," Gauthier–Villars, Paris, Second Edition, 1928, page 105.

[17] E. J. McShane, "Integrals Devised for Special Purposes," *Bull. Amer. Math. Soc.*, No. 5, September 1963, pages 597–627.

[18] R. M. McLeod, *The Generalized Riemann Integral*, Mathematical Association of America, Providence, RI, 1980.

[19] L. Peng-Yee, "Lanzhou Lectures on Henstock Integration," World Scientific, Singapore, 1989.

[20] I. N. Pesin, "Classical and Modern Integration Theories," translated by S. Kotz, Academic Press, New York, 1970.

[21] S. M. de Souza and M. T. Thomas, "Beyond Gaussian Integrals in Grassman Algebra," *J. Math. Physics*, **31**, No. 6, June 1990, pages 1297–1299.

[22] W. Squire, *Integration for Engineers and Scientists*, American Elsevier Publishing Company, New York, 1970.

14. Caveats

Idea

There are many ways in which an integration "result" may be incorrect.

Example 1

Consider the integral

$$I(a, b) = \int_0^1 \frac{dx}{ax + b}, \tag{14.1}$$

for real x and arbitrary nonzero complex a and b. The indefinite integral has the primitive $\log(ax+b)/a$. Hence, a careless "direct" derivation would yield the result

$$I(a, b) \stackrel{?}{=} \frac{\log(a + b)}{a} - \frac{\log b}{a}. \tag{14.2}$$

The problem, of course, is that the logarithm function has a branch cut. Hence, the two logarithms in (14.2) may not be on the same Riemann sheet.

The correct way to evaluate (14.1) is to separate the region of integration into two sub-intervals, with the division point being the value where $ax + b$ may vanish. An easier way, for this integral, is to first write the integral as

$$I(a, b) = \frac{1}{a} \int_0^1 \frac{dx}{x + b/a} = \frac{1}{a} \log\left(x + \frac{b}{a}\right)\Big|_0^1.$$

No matter what the sign is of $\mathrm{Im}(b/a)$, the argument of the logarithm never crosses the cut (since x is real). Thus, the answer is

$$I(a, b) = \frac{1}{a} \left[\log\left(1 + \frac{b}{a}\right) - \log \frac{b}{a}\right].$$

Note that since $1 + b/a$ and b/a have the same imaginary part, we may combine them to obtain our final answer

$$I(a, b) = \frac{1}{a} \log \frac{a + b}{b}.$$

Example 2

Consider the simple integral

$$\int \frac{dx}{x} \stackrel{?}{=} \log x. \tag{14.3}$$

Clearly, the integrand $(1/x)$ is an odd function, yet $\log x$ is neither an even nor an odd function. Hence, there must be an error in (14.3). The error in this case is simple, the integral should be written as

$$\int \frac{dx}{x} = \log |x|.$$

Now the fact that the result is an even function is clearly indicated.

As a similar example, consider the integral

$$I = \int \frac{dx}{\sqrt{a^2 - x^2}} \stackrel{?}{=} \sin^{-1} \frac{x}{a}. \tag{14.4}$$

when $|x| \leq a$. The integrand is an even function of both x and a, so that the formula for the integral should be odd in x and even in a. However, the formula in (14.4) is odd with respect to both x and a. The correct evaluation of the integral in (14.4) can be written as $I = \sin^{-1}(x/|a|)$ or as $I = \tan^{-1} \dfrac{x}{\sqrt{a^2 - x^2}}$.

Example 3

Consider the integral $I = \int_0^{4\pi} dx/(2 + \sin x)$. Since this integrand has an elementary indefinite integral we readily find

$$\begin{aligned} \int_0^{4\pi} \frac{dx}{2 + \sin x} &= \frac{2}{\sqrt{3}} \tan^{-1} \left. \frac{2\tan\frac{x}{2} + 1}{\sqrt{3}} \right|_0^{4\pi} \\ &= \frac{2}{\sqrt{3}} \left[\tan^{-1} \frac{2\tan 2\pi + 1}{\sqrt{3}} - \tan^{-1} \frac{2\tan 0 + 1}{\sqrt{3}} \right]. \end{aligned} \tag{14.5}$$

Now we must carefully consider what branch of the inverse tangent function to take in each of these two terms. In this case, we must take

$$\begin{aligned} \tan^{-1} \frac{2\tan 2\pi + 1}{\sqrt{3}} &= \tan^{-1} \frac{1}{\sqrt{3}} = 2\pi + \frac{\pi}{6} \\ \tan^{-1} \frac{2\tan 0 + 1}{\sqrt{3}} &= \tan^{-1} \frac{1}{\sqrt{3}} = \frac{\pi}{6} \end{aligned}$$

so that $I = \dfrac{2}{\sqrt{3}} \left[\left(2\pi + \frac{\pi}{6}\right) - \left(\frac{\pi}{6}\right) \right] = \dfrac{4\pi}{\sqrt{3}}$.

Notes

[1] Very few tables of integrals use absolute value signs, such as are required in Example 2.

[2] If the principle branch of the inverse tangent function were chosen for both terms in (14.5), then the calculation would return the value zero. This should be identified immediately as being in error, since the integrand is always positive.

Observe that contour integration techniques (see page 129) can quickly yield the correct evaluation of I in Example 3.

[3] Particular care must be used when a symbolic manipulation package (see page 117) is used to compute a definite integral. Results from such packages may be incorrect for many reasons; choosing the wrong branch cut, as Example 3 illustrates, is a common error.

15. Changing Order of Integration

Applicable to Multiple integrals.

Yields

A different multiple integral.

Idea

Sometimes the order of integration in a multiple integral can be changed. This may make it easier to evaluate the integral.

Procedure

When Fubini's theorem is satisfied, the order of integration in a multiple integral may be changed. Fubini's theorem states:

> **Theorem:** (*Fubini*) If $f(x, y)$ is measurable and non-negative, and any one of the integrals
>
> $$\int_a^A \int_b^B f\, dx\, dy, \qquad \int_a^A dx \int_b^B f\, dy, \qquad \int_b^B dy \int_a^A f\, dx$$
>
> exists, then the other integrals exist, and all are equal.

For principal-value integrals, the Poincaré–Bertrand theorem may be useful (see Muskelishvili [3]):

> **Theorem:** (*Poincaré–Bertrand*) If $f(x, y)$ is analytic, then
>
> $$\begin{aligned} I(u) &= \fint_{-\infty}^{\infty} \frac{dx}{x-u} \fint_{-\infty}^{\infty} \frac{f(x,y)}{y-x}\, dy \\ &= \left(\fint_{-\infty}^{\infty} dy \fint_{-\infty}^{\infty} \frac{f(x,y)}{(x-u)(y-x)}\, dx \right) - \pi^2 f(u,u). \end{aligned}$$

This theorem is also true with weaker conditions on $f(x)$.

Example

Consider the double integral

$$I = \int_0^1 \int_0^1 \frac{y}{\left(xy^2+1\right)^2}\,dx\,dy.$$

The integrand is positive in the region of integration. In the process of evaluating the integral we will determine if it exists. Integrating first with respect to x we have

$$\begin{aligned}
I &= \int_0^1 dy \int_0^1 dx \frac{y}{\left(xy^2+1\right)^2} \\
&= \int_0^1 dy \left(\frac{-1}{y\left(xy^2+1\right)} \right)\Bigg|_{x=0}^{x=1} \\
&= \int_0^1 \frac{y}{y^2+1}\,dy \\
&= \frac{1}{2} \log\left(y^2+1\right)\Big|_{y=0}^{y=1} \\
&= \tfrac{1}{2} \log 2.
\end{aligned}$$

Since this integral exists, we know that integrating first with respect to y will give the same answer. We demonstrate this:

$$\begin{aligned}
I &= \int_0^1 dx \int_0^1 dy \frac{y}{\left(xy^2+1\right)^2} \\
&= \frac{1}{2} \int_0^1 dx \int_0^1 dz \frac{1}{(xz+1)^2} \\
&= \frac{1}{2} \int_0^1 dx \left(\frac{-1}{x(xz+1)} \right)\Bigg|_{z=0}^{z=1} \\
&= \frac{1}{2} \int_0^1 \frac{dx}{1+x} \\
&= \frac{1}{2} \log(1+x)\Bigg|_{x=0}^{x=1} \\
&= \tfrac{1}{2} \log 2.
\end{aligned}$$

Notes

[1] The order of integration can make a difference if Fubini's theorem is not satisfied. As an example, define the function $f(x,y) = (x^2 - y^2)/(x^2 + y^2)^2$, and consider integrating it over the unit square. Depending on the order of the integrations, we will obtain different answers. Integrating first with respect to y results in

$$\begin{aligned}\int_{\square} f(x,y) &= \int_0^1 dx \int_0^1 \frac{x^2 - y^2}{(x^2 + y^2)^2}\, dy = \int_0^1 dx \left(\frac{y}{x^2 + y^2} \right) \Bigg|_{y=0}^{y=1} \\ &= \int_0^1 \frac{dx}{x^2 + 1} = \frac{\pi}{4}.\end{aligned}$$

If, instead, we integrated first with respect to x we find

$$\begin{aligned}\int_{\square} f(x,y) &= \int_0^1 dy \int_0^1 \frac{x^2 - y^2}{(x^2 + y^2)^2}\, dx = \int_0^1 dy \left(-\frac{x}{x^2 + y^2} \right) \Bigg|_{x=0}^{x=1} \\ &= -\int_0^1 \frac{dy}{y^2 + 1} = -\frac{\pi}{4}.\end{aligned}$$

[2] Other examples where the requirements for Fubini's theorem are not satisfied are easy to find. For example,

$$\int_0^1 dx \int_0^1 \frac{x^2 - y^2}{(x^2 + y^2)^2}\, dy = \int_0^1 \frac{dx}{1 + x^2} = \frac{\pi}{4},$$

while

$$\int_0^1 dy \int_0^1 \frac{x^2 - y^2}{(x^2 + y^2)^2}\, dx = -\int_0^1 \frac{dy}{1 + x^2} = -\frac{\pi}{4}.$$

Another example is given by

$$\int_0^1 dy \int_0^\infty \left(e^{-xy} - 2e^{-2xy} \right) dx \neq \int_0^\infty dx \int_0^1 \left(e^{-xy} - 2e^{-2xy} \right) dy.$$

[3] The technique of interchanging integration order can be used to derive analytical formulae for integrals.

- For example, the integral $I = \iint_{\substack{0 \le x \le 1 \\ a \le y \le b}} x^y \, dx\, dy$ can be evaluated in two different ways,

$$\begin{aligned}I &= \int_a^b dy \int_0^1 x^y \, dx = \int_a^b \frac{dy}{y+1} = \log\left(\frac{1+b}{1+a} \right) \\ &= \int_0^1 dx \int_a^b x^y \, dy = \int_0^1 dx \left(\frac{x^y}{\log x} \right) \Bigg|_{y=a}^{y=b} = \int_0^1 \frac{x^b - x^a}{\log x}\, dx\end{aligned}$$

to yield the result $\displaystyle\int_0^1 \frac{x^b - x^a}{\log x}\,dx = \log\left(\frac{1+b}{1+a}\right)$.

- As a second example, the integral $\displaystyle\int\int_{\substack{0\le x,y\le 1\\ z\ge 0}}\int \frac{dx\,dy\,dz}{(1+x^2z^2)(1+y^2z^2)}$ can be used to derive the relation (see George [2], page 206)

$$\int_0^\infty \left(\frac{\tan^{-1} z}{z}\right)^2 dz = \pi \log 2.$$

[4] A more general statement of Fubini's theorem is:

Theorem: (*Fubini*) Let $X = \mathbf{R}^p$ and $Y = \mathbf{R}^p$; then the formula

$$\int\int_{X\times Y} f(\mathbf{x},\mathbf{y})\,d\mathbf{x}\,d\mathbf{y} = \int_X d\mathbf{x} \int_Y f(\mathbf{x},\mathbf{y})\,d\mathbf{y} = \int_Y d\mathbf{y} \int_X f(\mathbf{x},\mathbf{y})\,d\mathbf{x}$$

is valid in each of the following two cases:

[1] f is a measurable positive arithmetic function on $X \times Y$;

[2] f is an integrable function over $X \times Y$.

References

[1] G. Fubini, "Sugli Integrali Multipli," *Opere Scelte*, Cremonese, Vol 2, 1958, pages 243–249.

[2] C. George, "Exercises in Integration," Springer–Verlag, New York, 1984.

[3] N. I. Muskelishvili, *Singular Integral Equations*, Noordhoff, Groningen, 1953, pages 56–61.

16. Convergence of Integrals

Applicable to Single and multiple integrals.

Yields

Knowledge of whether an integral converges or diverges.

Idea

By comparing a given integral to a different integral, the convergence or divergence of the original integral may be determined.

Procedure

Most techniques that indicate convergence or divergence use some sort of integral inequality (see page 205). In this section we use the following two theorems:

> **Theorem** (*Comparison test for convergence*): Let $f(x)$ and $g(x)$ be continuous for $a < x \le b$, with $0 \le |f(x)| \le g(x)$. If $\int_a^b g(x)\,dx$ converges, then $\int_a^b f(x)\,dx$ converges and $0 \le \left|\int_a^b f(x)\,dx\right| \le \int_a^b g(x)\,dx$.

> **Theorem** (*Comparison test for divergence*): Let $f(x)$ and $g(x)$ be continuous for $a < x \le b$, with $0 \le g(x) \le f(x)$. If $\int_a^b g(x)\,dx$ diverges, then $\int_a^b f(x)\,dx$ diverges.

To use these theorems effectively, a knowledge base must be created of converging and diverging integrals. Some common integrals used for comparison include (assuming $a < b$):

(A) $\displaystyle\int_a^b \frac{dx}{(x-a)^p}$, which converges for $p < 1$ and diverges for $p \ge 1$.

(B) $\displaystyle\int_a^b \frac{dx}{(b-x)^p}$, which converges for $p < 1$ and diverges for $p \ge 1$.

(C) $\displaystyle\int_{-1}^1 \frac{dx}{x^p}$, which converges for $p < 1$ and diverges for $p \ge 1$.

(D) $\displaystyle\int_1^\infty \frac{dx}{x^p}$, which converges for $p > 1$ and diverges for $p \le 1$.

(E) $\displaystyle\int_2^\infty \frac{dx}{x(\log x)^p}$, which converges for $p > 1$ and diverges for $p \le 1$.

Example 1

Consider the integral $I = \int_{-\infty}^\infty \frac{x\sin^2 x}{x^3+1}\,dx$. For the given range of integration (i.e., for $x \in [0,\infty]$) we can easily bound the trigonometric term: $|x\sin^2 x| \le x$. The integral $J = \int_{-\infty}^\infty \frac{x}{x^3+1}\,dx$ will now be shown to converge, which then implies the convergence of I. First, we write

$$J = J_1 + J_2 = \int_0^1 \frac{x}{x^3+1}\,dx + \int_1^\infty \frac{x}{x^3+1}\,dx.$$

The integrand in J_1 is bounded above by x, and $\int_0^1 x\,dx = \frac{1}{2}$ (this shows that J_1 is convergent). The integrand in J_2 can be bounded above by x^{-2} and $\int_1^\infty x^{-2}\,dx = 1$ (this shows that J_2 is convergent). Hence, J is convergent (it is bounded above by $\frac{3}{2}$) and so is I.

Example 2

Consider the integral

$$I = \int_1^\infty \left(\int_0^a \sin(y^2 x^3)\, dy \right) dx.$$

Changing variables in the inner integral to $z = y^2 x^3$ results in

$$I = \frac{1}{2} \int_1^\infty \left(\int_0^{a^2 x^3} \frac{\sin z}{\sqrt{z}}\, dz \right) \frac{dx}{x^{3/2}}.$$

The inner integral is bounded for all a and x since $\int_0^\infty \frac{\sin z}{\sqrt{z}}\, dz$ converges. The integral $\int_1^\infty x^{-3/2}\, dx$ also converges, so we conclude that I converges.

Notes

[1] Note that we can analytically integrate J in Example 1. We find that

$$\begin{aligned} J &= \int_{-\infty}^\infty \frac{x}{x^3+1}\, dx \\ &= \left(\frac{1}{6} \log \frac{1-x+x^2}{(1+x)^2} + \frac{1}{\sqrt{3}} \tan^{-1} \frac{2x-1}{\sqrt{3}} \right) \Bigg|_{x=0}^{x=\infty} \\ &= \frac{\pi}{3\sqrt{3}} \approx 0.605. \end{aligned}$$

[2] Another useful theorem for determining whether an integral converges is Chartier's test (see Whittaker and Watson [3])

> **Theorem** (*Chartier*): If $f(x)$ decreases to zero monotonically as $x \to \infty$, and $\left| \int_a^x \phi(t)\, dt \right|$ is bounded as $x \to \infty$, then $\int_a^x f(x)\phi(x)\, dx$ converges.

(A) For example, consider $I = \int_0^\infty x^{-1} \sin x\, dx$. In this case, $f(x) = x^{-1}$ is monotonically decreasing and $\int_0^x \sin t\, dt = 1 - \cos t$, which is certainly bounded. We conclude that I converges.

(B) For a more interesting example, consider $J = \int_1^\infty x \cos(x^3 - x)\, dx$. This integral can be written as

$$J = \int_1^\infty \frac{x}{3x^2 - 1} \left(\frac{d}{dx} \sin(x^3 - x) \right) dx.$$

We recognize that $x/(3x^2 - 1)$ decreases monotonically, and that

$$\int_1^x \left(\frac{d}{dx} \sin(x^3 - x) \right) dx = \sin(x^3 - x),$$

which is bounded. We conclude that J converges.

[3] The following decomposition

$$K = \int_0^\infty \frac{\cos x \sin ax}{x}\, dx = \frac{1}{2} \int_0^\infty \frac{\sin(a+1)x}{x}\, dx + \frac{1}{2} \int_0^\infty \frac{\sin(a-1)x}{x}\, dx,$$

combined with the first example in the last note shows that K converges.

[4] Divergent integrals can sometimes be regularized to obtain a finite value, see Wong and McClure [2].

References

[1] W. Kaplan, *Advanced Calculus*, Addison–Wesley Publishing Co., Reading, MA, 1952.

[2] R. Wong and J. P. McClure, "Generalized Mellin Convolutions and their Asymptotic Expansions," *Can. J. Math*, **36**, 1984, pages 924–960.

[3] E. T. Whittaker and G. N. Watson, *A Course of Modern Analysis*, Cambridge University Press, New York, 1962.

17. Exterior Calculus

Applicable to Integration of differential forms.

Idea

The integral of a differential form is a higher-dimensional generalization of such ideas as the work of a force along a path, or the flux of a fluid across a surface.

Procedure

A real-valued function of $\mathbf{x}$, also known as a scalar field, is called a *zero-form* in exterior calculus. A *one-form* (in, say, three space) is an expression of the form $F_1\,dx + F_2\,dy + F_3\,dz$, where F_1, F_2, and F_3 are functions of x, y, and z.

A *two-form* in "standard order" is an expression of the form

$$F_1\,dy \wedge dz + F_2\,dz \wedge dx + F_3\,dx \wedge dy.$$

The inverted v's are called *carets* or *wedges.* The wedges distinguish between a two-form, such as $4y^3\,dz \wedge dx$, and objects such as $4y^3\,dz\,dx$, that occur in double integrals where the order of the differentials is not important.

In n-dimensional space, p-forms exist for $p = 0, 1, \dots, n$. The addition of two p-forms, or the multiplication of a p-form by a scalar, is another p-form.

The wedge product has the following properties:

- The wedge product is associative: $\omega \wedge (\theta \wedge \zeta) = (\omega \wedge \theta) \wedge \zeta$.
- The wedge product is anti-commutative: $\omega \wedge \theta = (-1)^{kl}\theta \wedge \omega$, for ω a k-form and θ an l-form.
- The wedge product of repeated differentials vanishes: $dx \wedge dx = 0$
- The wedge product of a p-form and a q-form is a $(p+q)$-form.

When dealing with n-dimensional space, it is useful to deal with forms in a more formal way. If we choose a linear coordinate system $\{x_1, x_2, \dots, x_n\}$ on $\mathbf{R}^n$, then each x_i is a 1-form. If the $\{x_i\}$ are linearly independent, then every 1-form ω can be written in the form $\omega = a_1x_1 + \ldots + a_nx_n$, where the $\{a_i\}$ are real functions. The value of ω on the vector ζ is equal to $\omega(\zeta) = a_1x_1(\zeta)+\ldots+a_nx_n(\zeta)$, where $x_1(\zeta), \ldots, x_n(\zeta)$, are the components of ζ in the chosen coordinate system.

A 2-form ω_2 is a function on pairs of vectors which is bilinear and skew-symmetric:

$$\omega_2(\lambda_1\zeta_1 + \lambda_2\zeta_2, \zeta_3) = \lambda_1\omega_2(\zeta_1, \zeta_3) + \lambda_2\omega_2(\zeta_2, \zeta_3),$$
$$\omega_2(\zeta_1, \zeta_2) = -\omega_2(\zeta_2, \zeta_1).$$

From this it is easy to derive that $\omega_2(\zeta, \zeta) = 0$ for every 2-form. Analogously, an exterior form of degree k, or a k-form, is a function of k vectors which is k-linear and antisymmetric.

A differentiable k-form $\omega^k|_{\mathbf{x}}$, at a point $\mathbf{x}$ of a manifold M, is an exterior k-form on the tangent space $TM_{\mathbf{x}}$ to M at the point $\mathbf{x}$. That is, it is a k-linear skew-symmetric function of k vectors $\{\zeta_1, \ldots, \zeta_k\}$ tangent to M at $\mathbf{x}$. Every differentiable k-form on the space $\mathbf{R}^n$ with a given coordinate system $\{x_1, \ldots, x_n\}$ can we written uniquely in the form

$$\omega^k = \sum_{i_1<\cdots<i_k} a_{i_1,\ldots,i_k}(\mathbf{x})\, dx_{i_1} \wedge \cdots \wedge dx_{i_k},$$

where the $\{a_{i_1,\ldots,i_k}(\mathbf{x})\}$ are smooth functions on $\mathbf{R}^n$.

A chain of dimension n on a manifold M consists of a finite collection of n-dimensional oriented cells $\sigma_1, \ldots, \sigma_r$ in M and integers $m_1, \ldots, m_r$, called multiplicities (the multiplicities can be positive, negative, or zero). A chain is denoted by $c_k = m_1\sigma_1 + \ldots + m_r\sigma_r$.

Let $\{x_1, \ldots, x_l\}$ be an oriented coordinate system on $\mathbf{R}^k$. Then every k-form on $\mathbf{R}^k$ has the form $\omega^k = \phi(\mathbf{x})\, dx_1\wedge\cdots\wedge dx_k$, where $\phi(\mathbf{x})$ is a smooth function. Let D be a bounded convex polyhedron in $\mathbf{R}^k$. We define the integral of the form ω^k on D to be

$$\int_S \omega^k = \int_D \phi(\mathbf{x})\, dx_1 \ldots dx_k,$$

where the integral on the right is understood to be the usual limit of Riemann sums. To integrate a k-form over an n-dimensional manifold, the role of the usual path of integration is replaced by a k-dimensional cell σ of M represented by a triple $\sigma = (D, f, \mathrm{Or})$ where

(A) $D \subset \mathbf{R}^k$ is a convex polyhedron;
(B) $f : D \to M$ is a differentiable map;

(C) Or represents an orientation of D on $\mathbf{R}^k$.

Then the integral of the k-form ω over the k-dimensional cell σ is the integral of the corresponding form over the polyhedron D: $\int_\sigma \omega = \int_D f^*\omega$. Here the form $f^*\omega$ is defined by

$$(f^*\omega)(\zeta_1, \ldots, \zeta_k) = \omega(f_*\zeta_1, \ldots, f_*\zeta_k),$$

where the $\{\zeta_i\}$ are tangent vectors and f_* is the differential of the map f. The integral of the form ω^k over the chain c_k is the sum of the integrals on the cells, counting multiplicities: $\int_{c_k} \omega^k = \sum_i m_i \int_{\sigma_i} \omega^k$.

Some classes of integrals can be immediately evaluated. Using the two definitions:

- A differential form ω on a manifold M is *closed* if its exterior derivative is zero: $d\omega = 0$,
- A *cycle* on a manifold M is a chain whose boundary is equal to zero,

we have the two theorems:

- The integral of a closed form ω^k over the boundary of any $(k+1)$-dimensional chain c_{k+1} is equal to zero: That is, $\int_{\partial c_{k+1}} \omega^k = 0$ if $d\omega^k = 0$;
- The integral of a differential over any cycle is equal to zero: That is, $\int_{c_{k+1}} d\omega^k = 0$ if $\partial c_{k+1} = 0$.

Example

The two-form $T = (x\,dx + y\,dz) \wedge (y\,dx - y^2\,dy)$ can be expanded into

$$T = xy\,dx \wedge dx + y^2\,dz \wedge dx - xy^2\,dx \wedge dy - y^3\,dz \wedge dy.$$

The first term vanishes because it contains a repeated differential. Writing the remaining terms in standard order results in

$$T = y^3\,dy \wedge dz + y^2\,dz \wedge dx - xy^2\,dx \wedge dy.$$

Notes

[1] As a matter of convention, the function identically equal to zero is called a p-form for every p.

[2] Stokes' formula can be stated as $\int_{\partial c} \omega = \int_c d\omega$, where c is the $(k+1)$-dimensional chain on a manifold M and ω is any k-form on M.

References

[1] L. Arnold, *Mathematical Methods of Classical Mechanics*, translated by K. Vogtmann and A. Weinstein, Second edition, Springer–Verlag, New York, 1989, Chapter 7.

[2] B. A. Dubrovin, A. T. Fomenko, and S. P. Novikov, *Modern Geometry Methods and Applications*, Springer–Verlag, New York, 1991.

[3] D. G. B. Edelen, *Applied Exterior Calculus*, John Wiley & Sons, New York, 1985.

[4] P. J. Olver, *Applications of Lie Groups to Differential Equations*, Graduate Texts in Mathematics #107, Springer–Verlag, New York, 1986.

[5] H. Whitney, *Geometric Integration Theory*, Princeton University Press, Princeton, NJ, 1957.

18. Feynman Diagrams

Applicable to Feynman diagrams are used to denote a collection of ordinary or path integrals.

Idea

Some integrals equations have a "natural" expansion scheme in terms of integrals. The diagrams are used to keep track of the terms.

Procedure

If a given differential equation is only a "small" perturbation from a linear differential equation (with a known Green's function), then we may obtain an equivalent integral equation. This integral equation may be expanded methodically into a series of integrals. Diagrams are often used to keep track of the terms.

Example

Consider the nonlinear ordinary differential equation

$$\begin{aligned} \frac{dz}{dt} &= f(t) + g(t)z^2, \\ z(0) &= 0. \end{aligned}$$

in which the nonlinear term (i.e., the $g(t)$ function) is "small." This equation may be directly integrated to obtain

$$z(t) = \int_0^t f(\tau)\,d\tau + \int_0^t g(\tau)z^2(\tau)\,d\tau. \tag{18.1}$$

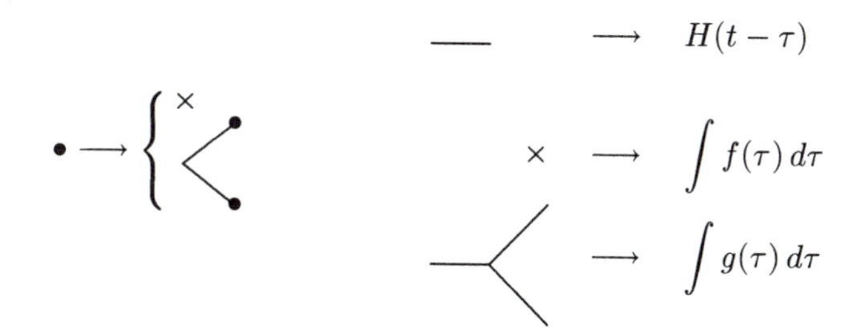

Figure 18.1 Rules for creating diagrams and rules for interpreting diagrams.

If the value of $z(t)$ from the left hand side of (18.1) is used in the right hand side, then

$$\begin{aligned} z(t) = \int_0^t f(\tau)\,d\tau + \int_0^t g(\tau)\left[\int_0^\tau f(\tau_1)\,d\tau_1\right]^2 d\tau \\ + 2\int_0^t g(\tau)\left[\int_0^\tau f(\tau_1)\,d\tau_1\right]\left[\int_0^\tau g(\tau_2)z^2(\tau_2)\,d\tau_2\right] d\tau \qquad (18.2) \\ + \int_0^t g(\tau)\left[\int_0^\tau g(\tau_2)z^2(\tau_2)\,d\tau_2\right]^2 d\tau. \end{aligned}$$

A "natural" perturbation expansion would be to keep the first two terms in the right hand side of (18.2), and assume that the last two terms are "small". If $|z(t)| \ll 1$ then this may well be the case since the last two terms involve $|z|^2$ while the first two terms involve constants.

A functional iteration technique can be used to derive (18.2) and its higher order extensions from diagrams. We need two sets of rules: One set of rules describes how the diagrams may be computed; the other set of rules describes how the diagrams are to be turned into mathematical expressions. If we use the rules in Figure 18.1, (where $H(\,)$ denotes the Heaviside function), then the first two steps in the diagrammatic solution to $z(t)$ (from (18.1)) are given by the diagrams in Figure 18.2.

Note that the third and fourth diagrams in Figure 18.2 represent the same mathematical expression since they are topologically equivalent. The purpose of the Heaviside function is to restrict the range of integration. By careful inspection, the mathematical expressions associated with the last set of diagrams will be seen to be identical to (18.2). See Zwillinger [5] for more details.

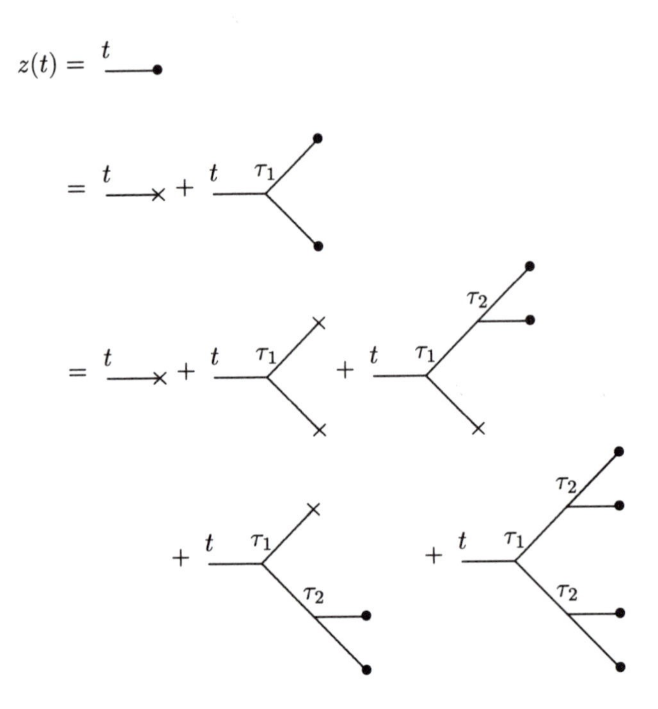

Figure 18.2 Two steps in the diagrammatic expansion of (18.1).

Notes

[1] Often an "algebra of diagrams" is created, so that diagrams can be added, subtracted and multiplied without recourse to the mathematical expression that each diagram represents. This would require amplification of the rules that were used in the example.

[2] This technique is particularly important in problems in which there is no "small" parameter. In these cases, the *formally correct* diagrammatic expansion may be algebraically approximated by exactly summing certain classes of diagrams. See Mattuck [4] for details.

References

[1] F. Battaglia and T. F. George, "A Rule for the Total Number of Topologically Distinct Feynman Diagrams," *J. Math. Physics*, **25**, No. 12, December 1984, pages 3489–3491.

[2] T. Kaneko, S. Kawabata, and Y. Shimizu, "Automatic Generation of Feynman Graphs and Amplitudes in QED," *Comput. Physics Comm.*, **43**, 1987, pages 279–295.

[3] J. Küblbeck, M. Böhm, and A. Denner, "Feyn Arts – Computer-Algebraic Generation of Feynman Graphs and Amplitudes," *Comput. Physics Comm.*, **60**, 1990, pages 165–180.

[4] R. D. Mattuck, *A Guide to Feynman Diagrams in the Many-Body Problem*, Academic Press, New York, 1976.

[5] D. Zwillinger, *Handbook of Differential Equations*, Academic Press, New York, Second Edition, 1992.

19. Finite Part of Integrals

Applicable to Divergent integrals.

Yields

A convergent expression.

Idea

Many divergent integrals diverge in a common way. By removing the diverging piece of an integral, a finite term is left.

Procedure

Given the integral

$$I = \int_a^b (x-a)^{\lambda-n} f(x)\, dx, \tag{19.1}$$

we assume that $f(a) \neq 0$ and $-1 < \lambda \leq 0$. Clearly, this integral diverges for all integral value of n.

However, when $f \in C^n[s,r]$ and $\lambda \neq 0$, we define the finite part of I to be the value of

$$\lim_{t \to a^+} \left[\int_t^b (x-a)^{\lambda-n} f(x)\, dx - g(t)(t-a)^{\lambda-n+1} \right], \tag{19.2}$$

where $g(t)$ is any function in $C^n[s,r]$ such that the above limit exists. Denote the finite part of the integral I by $⨍_a^b (x-a)^{\lambda-n} f(x)\, dx$. The limit in (19.2) can be explicitly evaluated to find (for $\lambda \neq 0$):

$$\begin{aligned} ⨍_a^b (x-a)^{\lambda-n} f(x)\, dx &= \sum_{k=0}^{n-1} \frac{f^{(k)}(a)(b-a)^{\lambda-n+k+1}}{(\lambda-n+k+1)k!} \\ &\quad + \frac{1}{(n-1)!} \int_a^b (x-a)^{\lambda-n} \int_a^x (x-y)^{n-1} f^{(n)}(y)\, dy\, dx. \end{aligned} \tag{19.3}$$

For $\lambda = 0$ we find:

$$\begin{aligned} ⨍_a^b (x-a)^{-n} f(x)\, dx &= \sum_{k=0}^{n-2} \frac{f^{(k)}(a)(b-a)^{-n+k+1}}{(-n+k+1)k!} + \frac{\log(b-a) f^{(n-1)}(a)}{(n-1)!} \\ &\quad + \frac{1}{(n-1)!} \int_a^b (x-a)^{-n} \int_a^x (x-y)^{n-1} f^{(n)}(y)\, dy\, dx. \end{aligned} \tag{19.4}$$

Some special cases are easy to express in terms of a Cauchy principal value type limit. For example, if $w(x)$ is a non-negative weight function integrable on the interval (a,b), c is in the interval (a,b), and $f(x)$ is differentiable in a neighborhood of c with its derivative satisfying a Lipschitz condition, then (see Paget [8])

$$\fint \frac{w(x)f(x)}{(x-c)^2}\,dx = \lim_{\varepsilon\to 0}\left[\left(\int_a^{c-\varepsilon}+\int_{c+\varepsilon}^b\right)\frac{w(x)f(x)}{(x-c)^2}\,dx - 2\frac{w(c)f(c)}{\varepsilon}\right]. \tag{19.5}$$

Example

Consider the finite-part integral $J = \fint_0^x \frac{f(y)}{y^2}\,dy$. This can be written as

$$J = \fint_0^x \frac{f(y)-f(0)-yf'(0)}{y^2}\,dy + f(0)\fint_0^x \frac{dy}{y^2} + f'(0)\fint_0^x \frac{dy}{y}.$$

For finite-part integrals, we have the usual relations

$$\fint_0^x \frac{dy}{y^2} = -\frac{1}{x} \qquad \text{and} \qquad \fint_0^x \frac{dy}{y} = \log x. \tag{19.6}$$

Combining these results, and assuming that $f(y)$ possesses a continuous second derivative on the interval $[0,x]$, we have

$$J = \int_0^x \frac{f(y)-f(0)-yf'(0)}{y^2}\,dy - \frac{f(0)}{x} + f'(0)\log x. \tag{19.7}$$

Notes

[1] This is also called *Hadamard's finite part of an integral.*

[2] The finite parts of integrals have different properties from usual integrals. Some of these properties are (from Davis and Rabinowitz [2]):

(A) $\fint$ is a consistent extension of the concept of regular integrals.

(B) $\fint$ is additive with respect to the union of integration intervals and is invariant with respect to translation.

(C) $\fint$ is a continuous linear functional of the integrand.

(D) $\fint_0^\infty x^{-\alpha}\,dx = 0$ if $\alpha > 1$.

(E) $\fint_0^1 x^{-1}\,dx = 0$ (this is consistent with (19.6)).

(F) For $\alpha > 1$, we have $\fint_0^1 x^{-\alpha}\,dx = 1/(1-\alpha) < 0$. Hence, if $f(x) > 0$, it may be that $\fint f(x)\,dx < 0$.

[3] Kutt [7] has derived quadrature rules for the evaluation of finite-part integrals of the form $\fint_a^b \frac{x-s}{|x-s|^{k+1}} f(x)\,dx$, when $s \in (a,b)$ and $k = 1, 2, \ldots$.

[4] It is possible to write some finite-part integrals as derivatives of principal-value integrals. For example:

$$\int_a^b \frac{w(x)f(x)}{(x-c)^2}\,dx = \frac{d}{dc}\int_a^b \frac{w(x)f(x)}{x-c}\,dx.$$

This is the basis for the numerical technique in Paget [8].

[5] Davis and Rabinowitz [2] give the example

$$\int_0^1 \frac{y^{-2}\,dy}{\sqrt{(y-2)^2+1}} = \frac{2\sqrt{5}}{25}\left[\log\left(10\sqrt{10}-30\right)-1\right]-\frac{\sqrt{2}}{5}.$$

[6] Other classes of finite-part integrals also exist. For example, there are theories developed for each of the integrals

$$\int_a^b \frac{f(x)\log^m(x-a)}{(x-a)^{n-\lambda}}\,dx \quad \text{and} \quad \int_a^b \frac{f(x)}{\sin^{n-\lambda}(x-a)}\,dx$$

where m is a positive integer.

[7] Finite part integrals arise naturally in fluid mechanics, solid mechanics, and electromagnetic theory. See, for example, Ioakimidis [4].

References

[1] B. Bialecki, "A Sinc Quadrature Rule for Hadamard Finite-Part Integrals," *Numer. Math.*, **57**, 1990, pages 263–269.

[2] P. J. Davis and P. Rabinowitz, *Methods of Numerical Integration*, Second Edition, Academic Press, Orlando, Florida, 1984, pages 11–15 and 188–190.

[3] N. I. Ioakimidis, "On the Gaussian Quadrature Rule for Finite-Part Integrals with a First-Order Singularity," *Comm. Appl. Numer. Meths.*, **2**, 1986, pages 123–132.

[4] N. I. Ioakimidis, "Application of Finite-Part Integrals to the Singular Integral Equations of Crack Problems in Plane and Three-Dimensional Elasticity," *Acta. Mech.*, **45**, 1982, pages 31–47.

[5] N. Ioakimidis, "On Kutt's Gaussian Quadrature Rule for Finite-Part Integrals," *Appl. Num. Math.*, **5**, No. 3, 1989, pages 209–213.

[6] N. I. Ioakimidis and M. S. Pitta, "Remarks on the Gaussian Quadrature Rule for Finite-Part Integrals with a Second-Order Singularity," *Computer Methods in Appl. Mechanics and Eng.*, **69**, 1988, pages 325–343.

[7] H. R. Kutt, "The Numerical Evaluation of Principal Value Integrals by Finite-Part Integration," *Numer. Math.*, **24**, 1975, pages 205–210.

[8] D. F. Paget, "The Numerical Evaluation of Hadamard Finite-Part Integrals," *Numer. Math.*, **36**, 1981, pages 447–453.

[9] G. Tsamasphyros, and G. Dimou, "Gauss Quadrature Rules for Finite Part Integrals," *Internat. J. Numer. Methods Engrg.*, **30**, No. 1, 1990, pages 13–26.

20. Fractional Integration

Applicable to Fractional integrals.

Procedure

The Riemann–Liouville fractional derivative of order v is defined by

$$\frac{d^{-v}f}{d(x-c)^{-v}} = {}_cD_x^v f(x) = \frac{1}{\Gamma(v)}\int_c^x (x-t)^{v-1} f(t)\,dt. \tag{20.1}$$

(This is sometimes represented by ${}_cI_x^v$.) The fractional derivative has the following properties:

[1] The operation of order zero leaves a function unchanged: ${}_cD_x^0 f(x) = f(x)$.

[2] The law of exponents for integration of arbitrary order holds: ${}_cD_x^v\,{}_cD_x^u f(x) = {}_cD_x^{u+v} f(x)$.

[3] The fractional operator is linear: ${}_cD_x^v[af(x)+bg(x)] = a\,{}_cD_x^v f(x) + b\,{}_cD_x^v g(x)$ (assuming, for $v<0$, that f and g are analytic).

[4] The operation ${}_cD_x^v f(x)$ yields $f^{(v)}(x)$, the v-th derivative of f, when v is a positive integer. If v is a negative integer, say $v=-n$, then ${}_cD_x^{-n} f(x)$ is the same as the ordinary n-fold integration of $f(x)$. (The integration constants are chosen so that ${}_cD_x^{-n} f(x)$ vanishes, along with its first $n-1$ derivatives, at $x=c$.)

[5] If $f(z)$ is an analytic function of the complex variable z, the function ${}_cD_z^v f(z)$ is an analytic function of v and z.

Many common functions can be written as fractional integrals of other functions. For example, we find

$$J_p\left(\sqrt{u}\right) = \frac{u^{-p/2}}{2^p\sqrt{\pi}}\,{}_0D_u^{-(p+1/2)}\left(\frac{\cos\sqrt{u}}{\sqrt{u}}\right)$$

$${}_2F_1(a,b;g;x) = \frac{x^{1-g}\Gamma(g)}{\Gamma(b)}\,{}_0D_x^{-(g-b)}\left(x^{b-1}(1-x)^{-a}\right).$$

It is also possible to represent some ordinary integrals as fractional integrals in non-obvious ways. For example, from Oldham and Spanier [3] we have (page 182)

$$\int_0^{x^{1/p}} f\left(x - xz^p\right)dz = \Gamma\left(\frac{p+1}{p}\right) x^{-1/p}\frac{d^{-1/p}f}{dx^{-1/p}}$$

or, when $x=1$

$$\int_0^1 f\left(1-z^p\right)dz = \Gamma\left(\frac{p+1}{p}\right)\left.\frac{d^{-1/p}f}{dx^{-1/p}}\right|_{x=1}.$$

For example, we find

$$\begin{aligned}\int_0^1 \sin\left(\sqrt{1-z^2}\right) dz &= \Gamma\left(\frac{3}{2}\right) \frac{d^{-1/2}}{dx^{-1/2}} \sin\left(\sqrt{x}\right)\Big|_{x=1} \\ &= \frac{\sqrt{\pi}}{2}\left(\sqrt{\pi x} J_1\left(\sqrt{x}\right)\right)\big|_{x=1} \\ &= \frac{\pi}{2} J_1(1) \\ &\approx .69122984\ldots.\end{aligned}$$

Notes

[1] A semi-integral is a half integral, that is $v = -\frac{1}{2}$ in (20.1). Table 20 contains a short table of semi-integrals.

[2] Another definition of a fractional derivative is Weyl's integral

$$_xK_\infty^v = \frac{1}{\Gamma(v)} \int_x^\infty (t-x)^{v-1} f(t)\, dt$$

for $v > 0$. To compute fractional integrals, let m be the smallest positive integer such that $v < m$ and define $r = m - v$. Then

$$_xK_\infty^{-v} = \frac{d^m}{dx^m}\left(\frac{1}{\Gamma(r)} \int_x^\infty (t-x)^{r-1} f(t)\, dt\right).$$

Note that we also have

$$_cD_x^{-v} = \frac{d^m}{dx^m}\left(\frac{1}{\Gamma(r)} \int_c^x (x-t)^{r-1} f(t)\, dt\right)$$

when m is the smallest positive integer such that $v < m$ and $r = m - v$.

[3] Osler [4] has established the fractional integral generalization of Leibniz's rule:

$$\frac{d^q[fg]}{dx^q} = \int_{-\infty}^{\infty} \binom{q}{\lambda+\gamma} \frac{d^{q-\gamma-\lambda} f}{dx^{q-\gamma-\lambda}} \frac{d^{\gamma+\lambda} g}{dx^{\gamma+\lambda}}\, d\lambda$$

where γ is arbitrary.

References

[1] A. Erdélyi, "Axially Symmetric Potentials and Fractional Integration," *J. Soc. Indust. Appl. Math.*, **13**, No. 1, March 1965, pages 216–228.

[2] F. G. Lether, D. M. Cline and O. Evans, "An Error Analysis for the Calculation of Semiintegrals and Semiderivatives by the RL Algorithm," *Appl. Math. and Comp.*, **17**, 1985, pages 45–67.

[3] K. B. Oldham and J. Spanier, *The Fractional Calculus*, Academic Press, New York, 1974.

[4] T. J. Osler, "Leibniz Rule for Fractional Derivatives Generalized and an Application to Infinite Series," *SIAM J. Appl. Math.*, **18**, No. 3, May 1970, pages 658–674.

[5] B. Ross, *Fractional Calculus and Its Applications*, Proceedings of the International Conference at the University of New Haven, June 1974, Springer–Verlag, New York, Lecture Notes in Mathematics #457, 1975.

Table 20. A short table of semi-integrals.

f	$\dfrac{d^{-1/2}f}{dx^{-1/2}}$
0	0
C, a constant	$2C\sqrt{\dfrac{x}{\pi}}$
$x^{-1/2}$	$\sqrt{\pi}$
x	$\dfrac{4x^{2/3}}{3\sqrt{\pi}}$
$x^n,\ n = 0, 1, 2, \ldots$	$\dfrac{(n!)^2(4x)^{n+1/2}}{(2n+1)!\sqrt{\pi}}$
$x^p,\ p > -1$	$\dfrac{\Gamma(p+1)}{\Gamma(p+\frac{3}{2})}x^{p+1/2}$
$\sqrt{1+x}$	$\sqrt{\dfrac{x}{\pi}} + \dfrac{(1+x)\tan^{-1}\left(\sqrt{x}\right)}{\sqrt{\pi}}$
$\dfrac{1}{\sqrt{1+x}}$	$\dfrac{2}{\sqrt{\pi}}\arctan\left(\sqrt{x}\right)$
$\dfrac{1}{1+x}$	$\dfrac{2\sinh^{-1}\left(\sqrt{x}\right)}{\sqrt{\pi(1+x)}}$
e^x	$e^x\operatorname{erf}\left(\sqrt{x}\right)$
$e^x\operatorname{erf}\left(\sqrt{x}\right)$	$e^x - 1$
$\sin\left(\sqrt{x}\right)$	$\sqrt{\pi x}J_1\left(\sqrt{x}\right)$
$\cos\left(\sqrt{x}\right)$	$\sqrt{\pi x}H_{-1}\left(\sqrt{x}\right)$
$\sinh\left(\sqrt{x}\right)$	$\sqrt{\pi x}I_1\left(\sqrt{x}\right)$
$\cosh\left(\sqrt{x}\right)$	$\sqrt{\pi x}L_{-1}\left(\sqrt{x}\right)$
$\dfrac{\sin\left(\sqrt{x}\right)}{\sqrt{x}}$	$\sqrt{\pi}H_0\left(\sqrt{x}\right)$
$\dfrac{\cos\left(\sqrt{x}\right)}{\sqrt{x}}$	$\sqrt{\pi}J_0\left(\sqrt{x}\right)$
$\log x$	$2\sqrt{\dfrac{x}{\pi}}\left[\log(4x) - 2\right]$
$\dfrac{\log x}{\sqrt{x}}$	$\sqrt{\pi}\log\left(\dfrac{x}{4}\right)$

21. Liouville Theory

Applicable to Indefinite one-dimensional integrals.

Yields

Knowledge of whether the integral can be integrated in closed form in terms of "elementary functions."

Idea

The elementary functions are defined to be

[1] rational functions,
[2] algebraic functions (explicit and implicit),
[3] exponential and logarithmic functions,
[4] functions generated by a finite combination of the preceding classes.

The derivative of an elementary function is an elementary function. The question that Liouville addressed is: "When is the integral of an elementary function an elementary function?"

Procedure

The theory underlying the evaluation of integrals is complex. We need the following definitions:

- Let K be a field of functions. The function θ is an *elementary generator* over K if
  - θ is algebraic over K, i.e., θ satisfies a polynomial equation with coefficients in K;
  - θ is an exponential over K, i.e., there is a ζ in K such that $\theta' = \zeta'\theta$, which is an algebraic way of saying that $\theta = \exp\zeta$;
  - θ is a logarithm over K, i.e., there is a ζ in K such that $\theta' = \zeta'/\zeta$, which is an algebraic way of saying that $\theta = \log\zeta$.
- Let K be a field of functions. An over-field $K(\theta_1, \ldots, \theta_n)$ of K is called a *field of elementary functions over* K if every θ_i is an elementary generator over K. A function is *elementary over* K if it belongs to a field of elementary functions over K.

Note that, with $K = \mathbf{C}$ (the complex numbers), trigonometric functions (and their inverses) as well as rational functions are elementary. For example $\cos z = \frac{1}{2}\left(e^{iz} + e^{-iz}\right)$. That is, the cosine function is made up of complex multiplications, complex exponentiations, and rational operations.

The fundamental theorem in this area is

> **Theorem:** (*Liouville's principle*) Let f be a function from a function field K. If f has an integral elementary over K, it has an integral of the following form:
>
> $$\int f = v_0 + \sum_{i=1}^{n} c_i \log v_i,$$

where v_0 belongs to K, the $\{v_i\}$ belong to $\widehat{K}$, an extension of K by a finite number of constants algebraic over K, and the $\{c_i\}$ belong to $\widehat{K}$ and are constant.

For example:

[1] If an integral of an algebraic function y is elementary, then the integral must be of the form

$$\int y\,dx = R_0(x) + \sum_k c_k \log R_k(x)$$

where the $\{R_k\}$ are all rational functions and the number of terms in the sum is finite but undetermined.

[2] If y is given by the solution of an Nth-degree polynomial equation, and the integral of y is purely algebraic, then the integral must be of the form

$$\int y\,dx = \sum_{k=1}^{N-1} R_k(x) y^k$$

where the $\{R_k\}$ are all rational functions.

[3] If the integral $I = \int f e^g\,dx$ is elementary, and f and g are elementary, then I is of the form $I = Re^g$, where R is a rational function of x, f, and g.

Risch proved an extension to the above theorem that incorporates logarithms and exponentials:

> **Theorem:** (*Risch*) Let $K = C(x, \theta_1, \theta_2, \ldots, \theta_n)$ be a field of functions, where C is a field of constants and each θ_i is a logarithm or an exponential of an element of $C(x, \theta_1, \theta_2, \ldots, \theta_{i-1})$ and is transcendental over $C(x, \theta_1, \theta_2, \ldots, \theta_{i-1})$. Moreover, the field of constants of K must be C. There is an algorithm which, given an element f of K, either gives an elementary function over K that is the integral of f or proves that f has no elementary integral over K.

Davenport proved an extension to the above theorem that incorporates some algebraic functions:

> **Theorem:** (*Davenport*) Let $K = C(x, y, \theta_1, \theta_2, \ldots, \theta_n)$ be a field of functions, where C is a field of constants, y is algebraic over $C(x)$, and each θ_i is a logarithm or an exponential of an element of $C(x, y, \theta_1, \theta_2, \ldots, \theta_{i-1})$, and is transcendental over $C(x, y, \theta_1, \theta_2, \ldots, \theta_{i-1})$. Moreover, the field of constants of K must be C. There is an algorithm that, given an element f of K, either gives an elementary function over K that is the integral of f or proves that f has no elementary integral over K.

There is, as yet, no theorem that will allow algebraic extensions that depend on logarithms or exponential functions. Note that the theorems above are for integrating elementary functions when the integrals also have to be elementary. There are two ways of expanding the above results:

- Allow a larger or different class of functions to appear in the integrand (in this case the class of functions allowed in the evaluated integral may be the same or different).
- Allow a larger or different class of functions to appear in the evaluated integral, such as those created from elementary functions and logarithmic integrals, error functions, or dilogarithms.

See Baddoura [1], Cherry [3] and [4], and Davenport, Siret, and Tournier [6] for details.

Example 1

Using the above statements, we can resolve the question "Is the integral $I = \int e^{x^2}/x \, dx$ elementary?" If it is, then by statement number 3 (above), the integral must be of the form $I = Re^{x^2}$ where R is a rational function of x, x^{-1}, and x^2. That is, R is a rational function of x. Equating the two expressions for I, and differentiating, results in

$$\frac{e^{x^2}}{x} = R'e^{x^2} + 2xRe^{x^2} \quad \text{or} \quad x(R' + 2xR) = 1. \tag{21.1}$$

Since R is a rational function of x, it can be written in the form $R = N(x)/D(x)$, where $N(x)$ is a polynomial of degree n and $D(x)$ is a polynomial of degree d. Evaluating the differential equation for R in (21.1), using N and D, results in

$$xDN' - xD'N + 2x^2ND = D^2. \tag{21.2}$$

Each term in this equation is a polynomial, the degrees of the terms are $d + n$, $d + n$, $d + n + 2$, and $2d$. The first two terms in (21.2) cannot equal each other unless $N = D$ (which implies that R is a constant, which does not work). Clearly, the third term cannot have same degree as the first term. The fourth term cannot balance both the first and third terms, so no values of n and d will work. We conclude that I does not have an elementary integral.

Example 2

As another application of statement number 3, we investigate the integral $I = \int e^{-x^2}\,dx$. If this integral is elementary, then the integral must be of the form $I = Re^{-x^2}$ where R is a rational function of x. Equating the two expressions for I and differentiating results in

$$e^{-x^2} = R'e^{-x^2} + Re^{-x^2}(-2x).$$

This can be written as $1 = R' - 2Rx$. If we write $R(x) = P(x)/Q(x)$, where $P(x)$ and $Q(x)$ are relatively prime polynomials, then we obtain the relation

$$Q(Q - P' + 2xP) = -PQ'. \tag{21.3}$$

$Q(x)$ must have at least one root, β. Hence, $Q(x)$ can be written in the form $Q(x) = (x-\beta)^r T(x)$, where r is a positive integer and $T(\beta) \neq 0$. Since Q and P are relatively prime, $P(\beta) \neq 0$. Now we have a contradiction: β is a root of the left-hand side of (21.3), with a multiplicity of at least r, but β is a root of the right-hand side of (21.3), with a multiplicity of at most $r - 1$. Hence, the assumption that $\int e^{-x^2}\,dx$ has an elementary integral must be incorrect.

Example 3

Consider the integral $I = \int xe^{-x^2}\,dx$. Assuming the integral has the form $I = Re^{-x^2}$ results in the following differential equation: $R' - 2xR = x$. This has the solution $R = -\frac{1}{2}$, so $\int xe^{-x^2}\,dx$ is integrable.

Notes

[1] The symbolic computer language MAPLE will optionally, in the course of trying to evaluate an indefinite integral, print out information similar to the steps in the above examples. This makes it easy to monitor the computation and verify the result.

[2] The integration procedures described in this section are actually used in computer languages that can perform indefinite integration. See the section beginning on page 117.

[3] Even if an indefinite integral is not elementary, it sometimes possible to evaluate a definite integral with the same integrand. For example, even though $\int e^{-x^2}\,dx$ is not elementary (see Example 2), the definite integral $\int_{-\infty}^{\infty} e^{-x^2}\,dx$ is easily shown to be equal to $\sqrt{\pi}$ (see for example, page 115).

[4] Sometimes, in practice, it seems as if most integrals of interest are not elementary. It is certainly true that many simple-looking integrals are not elementary, such as $\displaystyle\int \frac{dx}{\log x}$, $\displaystyle\int \frac{\sin x}{x}\,dx$, $\int \sin x^2\,dx$, $\int \sqrt{\sin x}\,dx$, and $\int \sqrt{1 - x^3}\,dx$.

References

[1] J. Baddoura, "Integration in Finite terms and Simplification with Dilogarithms: A Progress Report," in E. Kaltoflen and S. M. Watt (eds.), *Computers and Mathematics*, Springer–Verlag, New York, 1990, pages 166–181.

[2] M. Bronstein, "Symbolic Integration: Towards Practical Algorithms," in *Computer Algebra and Differential Equations*, Academic Press, New York, 1990, pages 59–85.

[3] G. W. Cherry, "An Analysis of the Rational Exponential Integral," *SIAM J. Comput*, **18**, No. 5, October 1989, pages 893–905.

[4] G. W. Cherry, "Integration in Finite Terms with Special Functions: The Error Function," *J. Symbolic Comp.*, **1**, 1985, pages 283–302.

[5] J. H. Davenport, "The Risch Differential Equation Problem," *SIAM J. Comput*, **15**, No. 4, November 1986, pages 903–918.

[6] J. H. Davenport, Y. Siret, and E. Tournier, *Computer Algebra: Systems and Algorithms for Algebraic Computation*, Academic Press, New York, 1988, pages 165–186.

[7] K. O. Geddes and L. Y. Stefanus, "On the Risch–Norman Integration Method and Its Implementation in Maple," *Proceedings of ISSAC '89*, ACM, New York, 1989, pages 218–227.

[8] P. H. Knowles, "Integration of Liouvillian Functions with Special Functions," *SYMSAC '86*, ACM, New York.

[9] J. Moses, "Symbolic Integration, the Stormy Decade," *Comm. ACM*, **14**, 1971, pages 548–560.

[10] R. D. Richtmyer, "Integration in Finite Terms: A Method for Determining Regular Fields for the Risch Algorithm," *Lett. Math. Phys.* **10**, No. 2–3, 1985, pages 135–141.

[11] R. H. Risch, "The Problem of Integration in Finite Terms," *Trans. Amer. Math. Soc.*, **139**, 1969, pages 167–183.

[12] R. H. Risch, "The Solution of the Problem of Integration in Finite Terms," *Bull. Amer. Math. Soc.*, **76**, 1970, pages 605–608.

[13] J. F. Ritt, "On the Integrals of Elementary Functions," *Trans. Amer. Math. Soc.*, **25**, 1923, pages 211–222.

[14] J. F. Ritt, *Integration in Finite Terms: Liouville's Theory of Elementary Methods*, Columbia University Press, New York, 1948.

[15] W. Squire, *Integration for Engineers and Scientists*, American Elsevier Publishing Company, New York, 1970, pages 41–49.

22. Mean Value Theorems

Applicable to Single and multiple integrals.

Yields

Information about a function integrated over a region.

Idea

The integral of a function can sometimes be written as some value of that function times the integral of a simplier function.

Procedure

There are several theorems that yield information about the mean value of a quantity; we indicate only a few of them.

First Mean Value Formula

Assuming that $a \leq b$, we have

$$(b-a) f_{\max} \geq \int_a^b f(x)\, dx \geq (b-a) f_{\min}.$$

If $f(x)$ is continuous then we have

$$\int_a^b f(x)\, dx = (b-a) f(\widehat{x}), \tag{22.1}$$

where $a \leq \widehat{x} \leq b$.

This theorem is sometimes written as (assuming, again, that $f(x)$ is continuous)

$$\int_a^b f(x) g(x)\, dx = f(\widehat{x}) \int_a^b g(x)\, dx.$$

Second Mean Value Formula

If f is a decreasing positive function on $[a, b]$, and g an integrable function on this interval, then

$$\left| \int_a^b f g \right| \leq f(a) \sup_{a \leq x \leq b} \left| \int_a^x g \right|.$$

Mean Value Theorem for Double Integrals

If $f(x, y)$ is continuous on a compact region R, with area A, there exists a point (ζ, η) in the interior or R such that

$$\iint\limits_R f(x, y)\, dA = f(\zeta, \eta) A. \tag{22.2}$$

Example

Consider the integral $I = \int_0^\pi x \sin x \, dx$. The exact evaluation is $I = \sin x - x \cos x\big|_0^\pi = \pi$. From (22.1), we can write $I = \pi \hat{x} \sin \widehat{x}$, for at least one value of $\widehat{x}$ in the range $[0, \pi]$. In this case, we find that $\widehat{x} \approx 1.1141$ is one such value.

Notes

[1] There are two theorems that are also called the second mean value theorems (see Gradshteyn and Ryzhik [2], page 211):

(A) If $f(x)$ is monotonic and non-negative in the interval (a, b) (with $a < b$), and if $g(x)$ is integrable over that interval, then there exists at least one point ζ in the interval such that

$$\int_a^b f(x) g(x) \, dx = f(a) \int_a^\zeta g(x) \, dx.$$

(B) If, in addition to the requirements in the last statement, $f(x)$ is non-decreasing, then there exists at least one point ζ in the interval such that

$$\int_a^b f(x) g(x) \, dx = f(b) \int_\zeta^b g(x) \, dx.$$

[2] There is also a mean value theorem from complex analysis (see Iyanaga and Kawada [3], page 624):

Let u be a harmonic function (i.e., $\nabla^2 u = 0$), let D be the domain of definition of u, and let S be the boundary of D. Then the mean value of u on the surface or the interior of any ball in D is equal to the value of u at the center of the ball. That is

$$u(P) = \frac{1}{\tau_n r^n} \int_{B(P,r)} u \, d\tau = \frac{1}{\sigma_n r^{n-1}} \int_{S(P,r)} u \, d\sigma.$$

where τ_n and σ_n are the volume and surface area of a unit ball in $\mathbf{R}^n$, $B(P, r)$ is the open ball with center at P and radius r, $S(P, r)$ is the spherical surface with center at P and radius r, $d\tau$ is the volume element, and $d\sigma$ is an element of surface area.

References

[1] W. Kaplan, *Advanced Calculus*, Addison–Wesley Publishing Co., Reading, MA, 1952.

[2] I. S. Gradshteyn and I. M. Ryzhik, *Tables of Integrals, Series, and Products*, Academic Press, New York, 1980.

[3] S. Iyanaga and Y. Kawada, *Encyclopedic Dictionary of Mathematics*, MIT Press, Cambridge, MA, 1980.

23. Path Integrals

Idea

Path integrals are integrals of functionals. Feynman path integrals are specific path integrals for the propagator of the Schrödinger equation; they have the form $G = \int e^{iS/\hbar}$, where S is the action.

Procedure 1

We illustrate the general procedure by considering a specific case.

Consider a collection of particles that perform Brownian motion in one dimension and do not interact with one another. Define $c(x,t)\,dx$ to be the number (or concentration) of particles in an interval of size dx about the point x at time t. It is well known that the concentration satisfies the partial differential equation

$$\frac{\partial c}{\partial t} = D\frac{\partial^2 c}{\partial x^2}, \tag{23.1}$$

where D is the diffusion coefficient.

If one particle starts out at x_0 at time t_0, then the concentration has the initial condition $c(x_0, t_0) = \delta(t - t_0)$. (The function $\delta(\,)$ denotes Dirac's delta function.) The solution of (23.1) with this initial condition is (see Zwillinger [11]):

$$c(x,t) = \frac{1}{\sqrt{4\pi D(t-t_0)}}\exp\left\{-\frac{(x-x_0)^2}{4D(t-t_0)}\right\}. \tag{23.2}$$

This gives the likelihood of finding the particle at the point x at any time $t \geq t_0$. The likelihood of finding the particle near x, with an uncertainty of dx, is just $c(x,t)\,dx$. (In mathematics books, $c(x,t)$ would be called the Green's function of (23.1); in physics books it would be called the propagator of a Brownian particle.)

Now divide the interval (t_0, t) into $N+1$ equal intervals of length ε: $t_1 < t_2 < \ldots < t_N$ (so that $t_{i+1} - t_i = \varepsilon$). We would like to know the probability that the particle that started at (x_0, t_0) is near the point x_1 (with uncertainty dx_1) at time t_1, near the point x_2 (with uncertainty dx_2) at time $t_2, \ldots$, near the point x_N (with uncertainty dx_N) at time t_N, and near the point $x := x_{N+1}$ (with uncertainty dx) at time $t := t_{N+1}$. See Figure 23. Because the Brownian particle is memoryless, this probability is equal to the product of (23.2) taken over successive subintervals:

$$\text{Prob} = \left(\frac{1}{\sqrt{4\pi D(t-t_0)}}\right)^{N+1}\exp\left\{-\frac{1}{4D}\sum_{j=0}^{N}\frac{(x_{j+1}-x_j)^2}{\varepsilon}\right\}\prod_{j=1}^{N+1} dx_j. \tag{23.3}$$

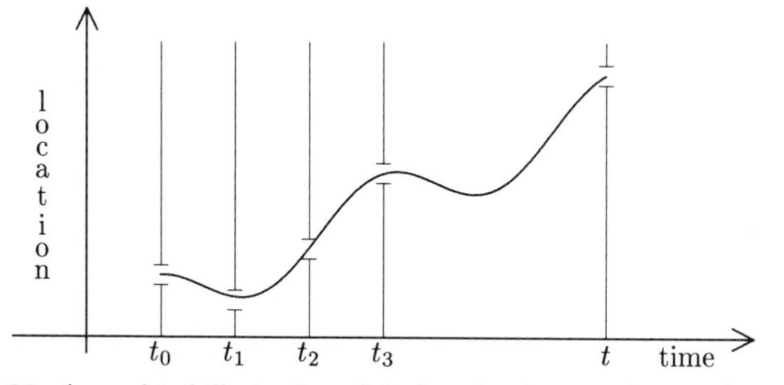

Figure 23. A graphical illustration of the locational constraints on the particle.

If the limit of infinite discretization is taken (i.e., $\varepsilon \to 0$ and $N \to \infty$, while $(N+1)\varepsilon = t - t_0$), then (23.3) can be interpreted as the probability that the particle follows a particular path $x(\tau)$ from (x_0, t_0) to (x, t), where $x(\tau_j) = x_j$. In this limit, the exponential in (23.3) can be written in the compact form:

$$\exp\left\{-\frac{1}{4D}\int_{t_0}^{t}\left(\frac{dx}{d\tau}\right)^2 d\tau\right\}. \tag{23.4}$$

Using (23.4) in (23.3) we obtain the path integral

$$\text{Prob} = \int_{x_0,t_0}^{x,t} \exp\left\{-\frac{1}{4D}\int_{t_0}^{t}\left(\frac{dx}{d\tau}\right)^2 d\tau\right\} d[x(\tau)]. \tag{23.5}$$

The value of this path integral is, however, given in (23.2):

$$\text{Prob} = \frac{1}{\sqrt{4\pi D(t-t_0)}} \exp\left\{-\frac{(x-x_0)^2}{4D(t-t_0)}\right\}. \tag{23.6}$$

Note that the particular path taken, $x(\tau)$, does not enter into the path integral in (23.5). This is because the integral is considering all possible paths from the initial point to the final point.

It is worth emphasizing that (23.5) is only a notational way to indicate the limit of a discrete number of integrations. The terms in (23.5) can be individually interpreted as follows:

(A) The symbol $d[x(\tau)]$ denotes $\prod_{j=1}^{N} dx_j$ and restricts the $x(\tau)$ appearing in the integral to be constrained by $x(t_0) = x_0$; $x_j < x(t_j) < x_j + dx_j$ (for $j = 1, 2, \ldots, N$); and $x(t_{N+1}) = x(t) = x$. These limits for the $x(\tau)$ function are indicated by the limits on the first integral sign.

(B) Each dx_j is to be integrated from $-\infty$ to ∞, and an implicit normalization factor is present

$$\int d[x(\tau)] \Longrightarrow \left(\frac{1}{\sqrt{4\pi D\varepsilon}}\right)^{N+1} \int_{-\infty}^{\infty} dx_1 \int_{-\infty}^{\infty} dx_1 \cdots \int_{-\infty}^{\infty} dx_N,$$

where $\varepsilon = (t_{j+1} - t_j)/N$.

(C) The exponent in (23.5) is to be interpreted as

$$\int_{t_0}^{t} \left(\frac{dx}{d\tau}\right)^2 d\tau \Longrightarrow \sum_{j=0}^{N} \frac{(x_{j+1}-x_j)^2}{\varepsilon}.$$

(D) After all the integrations have been performed, the limit $N \to \infty$ (which corresponds to $\varepsilon \to 0$) must be taken.

Procedure 2

Consider a classical system that starts in state x_a at time t_a and ends in state x_b at time t_b. The evolution of the system can be described by a set of variables that are functions of time; these variables describe a "path" from state x_a to state x_b. The classical system will almost certainly have a Lagrangian L that depends on the path of the system $L = L[\text{path}]$. The action S is defined as the time integral of the Lagrangian

$$S[\text{path}] = \int_{t_a}^{t_b} L[\text{path}]\, dt. \tag{23.7}$$

In quantum mechanics, the wave function ψ evolves according to $\psi(b) = \int P(b \mid a)\psi(a)\, dx_a$ where $P(b \mid a)$ is the propagator from x_a to x_b. The propagator can be obtained by solving the Schrödinger equation $i\hbar P_t = HP$, where H is the Hamiltonian and $\hbar$ denotes Planck's constant divided by 2π. The propagator can also be obtained by the Feynman path integral

$$P(b \mid a) = \int_{x_a,t_a}^{x_b,t_b} \exp\left\{\frac{i}{\hbar} S[\text{path}]\right\} d[\text{path}]. \tag{23.8}$$

This second method of obtaining the propagator is more useful in some circumstances, such as in dissipative systems that have a Lagrangian but not a Hamiltonian. See Wiegel [9] for more details. The path differential measure takes the form $d[\text{path}] \Longrightarrow A_N \prod_{j=1}^{N-1} d\mathbf{x}_j$, where A_N is an appropriate normalization constant.

Example 1

We indicate two ways in which the integral

$$I = \int_{x=0,t=0}^{0,t} \exp\left\{-\frac{1}{4D}\int_0^t \left(\frac{dx}{d\tau}\right)^2 d\tau\right\} d[x(\tau)] \tag{23.9}$$

may be evaluated. This integral is the one in (23.5), with $x_0 = x = t_0 = 0$, but here we presume that we do not know its value. The integral in (23.9) can be written, by the four rules given above, as

$$\begin{aligned} I = \lim_{N\to\infty} \left(\frac{1}{\sqrt{4\pi D\varepsilon}}\right)^{N+1} \int_{-\infty}^{\infty} dx_0 \int_{-\infty}^{\infty} dx_1 \cdots \int_{-\infty}^{\infty} dx_N \\ \times \exp\left\{-\frac{1}{4D\varepsilon}\sum_{j=0}^{N}(x_{j+1}-x_j)^2\right\}, \end{aligned} \tag{23.10}$$

where ε is related to N by $\varepsilon = t/(N+1)$.

Example 1.A

The quadratic terms appearing in the exponent in (23.10) can be written as (note that $x_0 = 0$ and $x_{N+1} = x = 0$)

$$\sum_{j=0}^{N} (x_{j+1} - x_j)^2 = \sum_{k,l=1}^{N} x_k M_{k,l}^{(N)} x_l, \tag{23.11}$$

where $M^{(N)}$ is the tri-diagonal matrix

$$M^{(N)} = \begin{pmatrix} 2 & -1 & & & & 0 \\ -1 & 2 & -1 & & & \\ & -1 & 2 & -1 & & \\ & & \ddots & \ddots & \ddots & \\ & & & -1 & 2 & -1 \\ 0 & & & & -1 & 2 \end{pmatrix}.$$

It is straightforward to show that $\det M^{(N)} = N + 1$. Since $M^{(N)}$ is Hermitian, its eigenvalues are real. Denoting its eigenvalues by $\{\lambda_j\}$, the sum in (23.11) may be written as $\sum_{j=1}^{N} \lambda_j y_j^2$, where the $\{y_j\}$ are linearly related to the $\{x_j\}$. Again, since $M^{(N)}$ is Hermitian, the Jacobian of this transformation is unity. Hence, (23.10) may be written as

$$\begin{aligned} I &= \lim_{N\to\infty} \left(\frac{1}{\sqrt{4\pi D\varepsilon}}\right)^{N+1} \int_{-\infty}^{\infty} dy_0 \int_{-\infty}^{\infty} dy_1 \cdots \int_{-\infty}^{\infty} dy_N \\ &\qquad\qquad \times \exp\left\{-\frac{1}{4D\varepsilon} \sum_{j=0}^{N} \lambda_j y_j^2\right\} \\ &= \lim_{N\to\infty} \left(\frac{1}{\sqrt{4\pi D\varepsilon}}\right)^{N+1} \prod_{j=1}^{N} \sqrt{\frac{4\pi D\varepsilon}{\lambda_j}} \\ &= \lim_{N\to\infty} \left(\frac{1}{\sqrt{4\pi D\varepsilon \det M^{(N)}}}\right) = \lim_{N\to\infty} \left(\frac{1}{\sqrt{4\pi Dt}}\right) = \frac{1}{\sqrt{4\pi Dt}}. \end{aligned} \tag{23.12}$$

This is precisely the value in (23.6), when $x = x_0 = t_0 = 0$.

Example 1.B

The integrations in (23.10) can also be performed one at a time. The formula

$$\int_{-\infty}^{\infty} e^{-a(x-x_\mathrm{L})^2-b(x-x_\mathrm{U})^2}\,dx = \sqrt{\frac{\pi}{a+b}}\exp\left\{-\frac{ab}{a+b}(x_\mathrm{L}-x_\mathrm{U})^2\right\}$$

can be used to evaluate the integrals sequentially. First, choose $x_\mathrm{L} = x_0$, $x = x_1$, and $x_\mathrm{U} = x_2$. Then the x_1-integral appearing in (23.10) can be evaluated to obtain

$$\frac{1}{\sqrt{4\pi D\varepsilon}}\int_{-\infty}^{\infty}\exp\left\{-\frac{(x_1-x_0)^2+(x_1-x_2)^2}{4D\varepsilon}\right\}dx_1$$
$$= \frac{1}{\sqrt{2}}\exp\left\{-\frac{(x_0-x_2)^2}{8D\varepsilon}\right\}$$

This expression has the same form as a term in (23.10), but with ε replaced by 2ε throughout. Now the integration with respect to x_3 is carried out, using $x_\mathrm{L} = x_2$ and $x_\mathrm{U} = x_4$. Then the integrations with respect to x_5, x_7, ..., are carried out. The variables left at this point are $\{x_0, x_2, x_4, x_6, \ldots\}$.

Now the integrals with respect to x_2, x_6, x_{10}, ..., are carried out, and again the same form is obtained, but now 2ε has been replaced by 4ε. This operation is recursively applied. (Without loss of generality, we can assume that $N+1$ is a power of two.) The end result is

$$I = \lim_{N\to\infty}\frac{1}{\sqrt{4\pi D(N+1)\varepsilon}} = \frac{1}{\sqrt{4\pi Dt}}.$$

Notes

[1] The main task of evaluating a path integral is to evaluate the multiple Riemann integrals and then take the $N \to \infty$ limit. This is only possible in a few cases, such as when

- the integrals involved are Gaussians or
- a recursive formula is available to carry out the successive integrations.

The propagators for a free particle and for the harmonic oscillator of constant frequency fall in the first category. We quote here the standard propagators:

- For a free particle of mass m

$$P(b \mid a) = \left(\frac{m}{2\pi i\hbar T}\right)^{3/2}\exp\left\{\frac{im}{2\hbar T}(x_b - x_a)^2\right\},$$

where $T = t_b - t_a$.

- For a free particle of mass m and frequency ω

$$P(b \mid a) = \left(\frac{m\omega}{2\pi i\hbar\sin\omega T}\right)^{3/2}\exp\left\{\frac{im\omega}{2\hbar\sin\omega T}\left([x_b^2 + x_a^2]\cos\omega T - 2x_b\cdot x_a\right)\right\}.$$

[2] When a Feynman path integral is constructed, the action is rewritten, using the principle of canonical quantization, with the substitutions

$$q \to \widehat{q} = q \qquad p \to \widehat{p} = -i\hbar\frac{\partial}{\partial q},$$

where q is a generalized coordinate, and p is a conjugate momentum.

[3] When $L = L(\mathbf{x}, \dot{\mathbf{x}}, t)$, the action functional in (23.7) can be discretized using the midpoint rule to obtain

$$S_N[\mathbf{x}] = \varepsilon \sum_{j=1}^{N} L\left(\frac{\mathbf{x}_j + \mathbf{x}_{j-1}}{2}, \frac{\mathbf{x}_j - \mathbf{x}_{j-1}}{2\varepsilon}, j\varepsilon\right).$$

For general L, the midpoint rule is required for the propagator as given by (23.8) to satisfy Schrödinger's equation. However, in some cases, other formulas for the derivatives may be used, see Khandekar and Lawande [4].

[4] When the classical Lagrangian has no explicit time dependence, the propagator can be written in the form

$$P(b \mid a) = \sum_{n} \exp\left\{-\frac{iE_n(t_b - t_a)}{\hbar}\right\} \phi_n^*(x_a)\phi_n(x_b),$$

where the $\{E_k\}$ and the $\{\phi_k\}$ are the complete set of energy eigenvalues and eigenfunctions: $H\phi_k = E_k\phi_k$. This expansion is known as the Feynman–Kac expansion theorem.

[5] Note the oscillatory nature of the integral in (23.8) for small $\hbar$. From the stationary phase technique (see page 226) it is clear that the major contributor to the propagator comes from paths for which $\delta S = 0$ and $\delta^2 S > 0$. The equation $\delta S = 0$ is Hamilton's principle; its solution is the classical path of the system. This equation is also equivalent to the usual Euler–Lagrange equations. If the integrand is expanded about the classical path, and terms out to second order are kept, then the resultant propagator is equivalent to the WKB approximation of the Euler–Lagrange equations (see Zwillinger [11]).

[6] Example 1.B has the rudiments of a renormalization group calculation, see Zwillinger [11].

[7] Other terms used for path integrals are *Feynman path integrals* and *functional integrals*.

References

[1] S. Albeverio and R. Høegh–Krohn, *Mathematical Theory of Feynman Path Integrals*, Springer–Verlag, New York, 1976.

[2] R. P. Feynman and A. R. Hibbs, *Quantum Mechanics and Path Integrals*, McGraw–Hill Book Company, New York, 1965.

[3] T. Kaneko, S. Kawabata, and Y. Shimizu, "Automatic Generation of Feynman Graphs and Amplitudes in QED," *Comput. Physics Comm.*, **43**, 1987, pages 279–295.

[4] D. C. Khandekar and S. V. Lawande, "Feynman Path Integrals: Some Exact Results and Applications," *Physics Reports*, **137**, No. 2–3, 1986, pages 115–229.

[5] P. K. MacKeown, "Evaluation of Feynman Path Integrals by Monte Carlo Methods," *Am. J. Phys.*, **53**, No. 9, September 1985, pages 880–885.

[6] L. S. Schulman, *Techniques and Applications of Path Integration*, John Wiley & Sons, New York, 1981.

[7] R. G. Stuart and A. Góngora-T., "Algebraic Reduction of One-Loop Feynman Diagrams to Scalar Integrals. II," *Comput. Physics Comm.*, **56**, 1990, pages 337–350.
[8] G. J. van Oldenborgh, "FF – A Package to Evaluate One-Loop Feynman Diagrams," *Comput. Physics Comm.*, **66**, 1991, pages 1–15.
[9] F. W. Wiegel, *Introduction to Path-Integral Methods in Physics and Polymer Science*, World Scientific, Singapore, 1986.
[10] P. Zhang, "Simpson's Rule of Discretized Feynman Path Integration," *J. Scientific Comput.*, **6**, No. 1, March 1991, pages 57–60.
[11] D. Zwillinger, *Handbook of Differential Equations*, Academic Press, New York, Second Edition, 1992.

24. Principal Value Integrals

Applicable to A formally divergent definite integral.

Yields

A convergent integral.

Idea

Many integrals are improper because of the limiting process involved. By restricting the way limits are taken in a definite integral, a convergent expression may sometimes be obtained.

Procedure

Suppose that $f(x)$ has the singular point z in the interval (a, b), and suppose that $f(x)$ is integrable everywhere in the interval, except at the point z. By definition (see page 53), the integral $\int_a^b f(x)\,dx$ has the value

$$\lim_{\substack{\alpha\to 0\\ \beta\to 0}} \left\{ \int_a^{z-\alpha} f(x)\,dx + \int_{z+\beta}^{b} f(x)\,dx \right\}, \tag{24.1}$$

where the limits are to be evaluated independently. If the value in (24.1) does not exist, but the value

$$\lim_{\gamma\to 0} \left\{ \int_a^{z-\gamma} f(x)\,dx + \int_{z+\gamma}^{b} f(x)\,dx \right\} \tag{24.2}$$

does exist, then this value is the *principal value*, or the *Cauchy principal value*, of the improper integral I. This is denoted $I = \fint_a^b f(x)\,dx$. Using the relation $\fint_a^c = \fint_a^b + \fint_b^c$, we can define the principal value of an integral with multiple singularities.

Example

Consider the integral $I = \int_0^3 dx/(x-1)$. Because of the singularity at $x = 1$, this is an improper integral. Hence, I does not exist in the usual sense. However, I does have a principal value which is easy to determine:

$$\begin{aligned}
⨍_0^3 \frac{dx}{x-1} &= \lim_{\gamma\to 0}\left\{\int_0^{1-\gamma}\frac{dx}{x-1}+\int_{1+\gamma}^3\frac{dx}{x-1}\right\} \\
&= \lim_{\gamma\to 0}\left\{\log|x-1|\Big|_0^{1-\gamma}+\log|x-1|\Big|_{1+\gamma}^{3}\right\} \\
&= \lim_{\gamma\to 0}\left\{\Big(\log|\gamma|-\log|1|\Big)+\Big(\log|2|-\log|\gamma|\Big)\right\} \\
&= \lim_{\gamma\to 0}\{\log 2\} = \log 2.
\end{aligned}$$

Notes

[1] Suppose that $f(x)$ is continuous on the interval $[a,b]$, and vanishes only at the point z in this interval. If the first and second derivatives of f exist in a region containing the point z, and $f'(z) \neq 0$, then the integral $\displaystyle\int_a^b \frac{dx}{f(x)}\,dx$ is improper, but exists in a principal value sense. For the example presented above, $f(x) = x-1$ and $f'(x) = 1$.

[2] The variables in a principal value integral may be changed under some mild restrictions. We have (see Davis and Rabinowitz [4], page 22):

> **Theorem**: If $z(\alpha) = a$ and $z(\beta) = b$, and if, on the interval $[\alpha, \beta]$, $z(\zeta)$ is monotonic, $z'(\zeta)$ does not vanish, and $z(\zeta)$ has a continuous second derivative, then
>
> $$⨍_a^b \frac{f(t)}{t-x}\,dt = ⨍_\alpha^\beta \frac{f(z(\zeta))z'(\zeta)}{z(\zeta)-m(\xi)}\,d\zeta,$$
>
> where $x = m(\xi)$.

[3] The integral $\int_{-\infty}^{\infty} f(x)\,dx$ may be written in two forms:

(A) $\int_{-\infty}^{\infty} f(x)\,dx = \lim_{R_1\to\infty}\int_{-R_1}^{0} f(x)\,dx + \lim_{R_2\to\infty}\int_0^{R_2} f(x)\,dx$

(B) $\int_{-\infty}^{\infty} f(x)\,dx = \lim_{R\to\infty}\left(\int_{-R}^{0} f(x)\,dx + \int_0^{R} f(x)\,dx\right)$.

The integral $\int_{-\infty}^{\infty} f(x)\,dx$ is *convergent* when the first set of limits converge. If the second formulation converges, but not the first, then the integral is *convergent in the sense of Cauchy* and it is denoted $⨍_{-\infty}^{\infty} f(x)\,dx$. For example, $\int_{-\infty}^{\infty} x\,dx$ converges in the sense of Cauchy, but the integral is not convergent.

[4] Davies *et al.* [3] defines a higher-order principal value to be given by:

$$I(u) = ⨍_{-\infty}^{\infty} \frac{f(x)}{(x-u)^n}\,dx = \frac{1}{(n-1)!}⨍_{-\infty}^{\infty} \frac{1}{(x-u)}\,\frac{d^{(n-1)}f(x)}{dx^{(n-1)}}\,dx.$$

[5] Many methods for the numerical evaluation of principal value integrals may be found in the references. For example, in the work by Hunter and Smith [7], the location of the poles do not need to be known before the numerical algorithm is run.

(A) Mastroianni [10] considers numerical methods for integrals of the form
$$\int\!\!\int \frac{f(x,y)}{(x-s)(y-t)} w_1(x) w_2(y)\, dx\, dy.$$

(B) Rabinowitz [12] considers numerical methods for integrals of the form
$$\int_{-1}^{1} \frac{\omega(x) f(x)}{x-\lambda}\, dx \text{ with } \omega(x) = (1-x^2)^{\mu-1/2} \text{ for } 0 \le \mu \le 2.$$

References

[1] G. Criscuolo and G. Mastroianni, "On the Uniform Convergence of Gaussian Quadrature Rules for Cauchy Principal Value Integrals," *Numer. Math.*, **54**, 1989, pages 445–461.

[2] G. Criscuolo and G. Mastroianni, "On the Convergence of Product Formulas for the Numerical Evaluation of Derivatives of Cauchy Principal Value Integrals," *SIAM J. Math. Anal.*, **25**, No. 3, June 1988, pages 713–727.

[3] K. T. R. Davies, R. W. Davies, and G. D. White, "Dispersion Relations for Causal Green's Functions: Derivations Using the Poincaré–Bertrand Theorem and its Generalizations," *J. Math. Physics*, **31**, No. 6, June 1990, pages 1356–1373.

[4] P. J. Davis and P. Rabinowitz, *Methods of Numerical Integration*, Second Edition, Academic Press, Orlando, Florida, 1984, pages 11–15.

[5] G. A. Gazonas, "The Numerical Evaluation of Cauchy Principal Value Integrals via the Fast Fourier Transform," *Int. J. Comp. Math.*, **18**, 1986, pages 277–288.

[6] A. Gerasoulis, "Piecewise-Polynomial Quadratures for Cauchy Singular Integrals," *SIAM J. Numer. Anal.*, **23**, No. 4, 1986, pages 891–902.

[7] D. B. Hunter and H. V. Smith, "The Evaluation of Cauchy Principal Value Integrals Involving Unknown Poles," *BIT*, **29**, No. 3, 1989, pages 512–517.

[8] H. R. Kutt, "The Numerical Evaluation of Principal Value Integrals by Finite-Part Integration," *Numer. Math.*, **24**, 1975, pages 205–210.

[9] J. N. Lyness, "The Euler–Maclaurin Expansion for the Cauchy Principal Value Integral," *Numer. Math.*, **46**, No. 4, 1985, pages 611–622.

[10] G. Mastroianni, "On the Convergence of Product Formulas for the Evaluation of Certain Two-Dimensional Cauchy Principal Value Integrals," *Math. of Comp.*, **52**, No. 185, January 1989, pages 95–101.

[11] P. Rabinowitz, "Numerical Evaluation of Cauchy Principal Value Integrals with Singular Integrands," *Math. of Comp.*, **55**, No. 191, July 1990, pages 265–276.

[12] P. Rabinowitz, "A Stable Gauss–Kronrod Algorithm for Cauchy Principal-Value Integrals," *Comput. Math. Appl. Part B*, **12**, No. 5–6, 1986, pages 1249–1254.

[13] P. Theocaris, N. I. Ioakimidis, and J. G. Kazantzakis, "On the Numerical Evaluation of Two-Dimensional Principal Value Integrals," *Int. J. Num. Methods Eng.*, **15**, 1980, pages 629–634.

25. Transforms: To a Finite Interval

Applicable to Integrals that have an infinite limit of integration.

Yields

An exact reformulation.

Idea

Many transformations map an integral on an infinite domain to an integral on a finite domain. Such a transformation may be needed before a numerical approximation scheme is used.

Procedure

Given the integral

$$I = \int_0^\infty f(x)\,dx \tag{25.1}$$

a numerical approximation technique of the form

$$I \approx \sum_{j=1}^{N} w_j f(x_j) \tag{25.2}$$

may not be appropriate because of a fundamental indeterminacy in (25.1). If S is any positive number, then

$$I = \int_0^\infty f(x)\,dx = S \int_0^\infty f(Sx)\,dx.$$

Hence, the region where most of the "mass" of the integrand lies may have been (inadvertently) scaled to be outside of the range of the $\{x_i\}$. Without some knowledge of the shape of $f(x)$, it may be difficult to determine appropriate values for N and the $\{x_i\}$ in (25.2).

One approach is to transform I to a finite domain and then use a numerical integration scheme on the finite domain. It is generally easier to understand the form of an integrand on a finite domain, than it is on an infinite domain.

Under the transformation $t = t(x)$, we have $I = \int_0^1 f[x(t)]\,|dx/dt|\,dt$ (see page 109). Table 25 contains several useful transformations of (25.1).

Table 25. Several transformations of the integral $\int_0^\infty f(x)\,dx$, with an infinite integration range, to an integral with a finite integration range.

$t(x)$	$x(t)$	$\dfrac{dx}{dt}$	Finite interval integral
e^{-x}	$-\log t$	$-\dfrac{1}{t}$	$\displaystyle\int_0^1 \frac{f(-\log t)}{t}\,dt$
$\dfrac{x}{1+x}$	$\dfrac{t}{1-t}$	$\dfrac{1}{(1-t)^2}$	$\displaystyle\int_0^1 f\left(\frac{t}{1-t}\right)\frac{dt}{(1-t)^2}$
$\tanh x$	$\dfrac{1}{2}\log\dfrac{1+t}{1-t}$	$\dfrac{1}{1-t^2}$	$\displaystyle\int_0^1 f\left(\frac{1}{2}\log\frac{1+t}{1-t}\right)\frac{dt}{1-t^2}$

Notes

[1] Another useful transformation, for any positive value of S, is:

$$\int_0^\infty f(x)\,dx = S\int_0^1 \left[f(St) + \frac{1}{t^2}f\left(\frac{S}{t}\right)\right]\,dt.$$

[2] The transformation $\displaystyle\int_a^b f(t)\,dt = \int_{1/b}^{1/a} \frac{1}{t^2}f\left(\frac{1}{t}\right)\,dt$, valid for $ab > 0$, is useful in the two cases:

- $b \to \infty$ with $a > 0$;
- $a \to -\infty$ with $b < 0$.

[3] If an integral has a known power-law singularity at an endpoint, then a transformation can be made to remove it. For example, suppose that $f(x)$ diverges as $(x-a)^\alpha$ near $x = a$ (where $-1 < \alpha < 0$). Then we can use

$$\int_a^b f(t)\,dt = \frac{1}{1-\alpha}\int_0^{(b-a)^{1-\alpha}} t^{\alpha/(1-\alpha)} f\left(t^{1/(1-\alpha)} + a\right)\,dt,$$

where $b > a$. If, instead, $f(x)$ diverges as $(b-x)^\beta$ near $x = b$ (where $-1 < \beta < 0$), and $b > a$, then we can use

$$\int_a^b f(t)\,dt = \frac{1}{1-\beta}\int_0^{(b-a)^{1-\beta}} t^{\beta/(1-\beta)} f\left(b - t^{1/(1-\beta)}\right)\,dt.$$

[4] The change of variables in this section can be performed analytically, or a numerical quadrature routine can perform the change.

Reference

[1] W. Squire, *Integration for Engineers and Scientists*, American Elsevier Publishing Company, New York, 1970, pages 174–176.

26. Transforms: Multidimensional Integrals

Applicable to Definite multidimensional integrals.

Yields

A reformulation into a one-dimensional integral.

Idea

Some classes of multidimensional integrals can be written in terms of a single integral.

Procedure

The following listing contains many multidimensional integrals each involving a general function. These multidimensional integrals can be written in terms of a single integral.

Two-Dimensional Integrals

$$\int_0^{\pi} d\omega \int_0^{\infty} f'(p\cosh x + q\cos\omega\sinh x)\sinh x\,dx = -\frac{\pi[\operatorname{sign} p]}{\sqrt{p^2-q^2}} f\left([\operatorname{sign} p]\sqrt{p^2-q^2}\right)$$

if $p^2 > q^2$ and $\lim_{x\to\infty} f(x) = 0$ (see Gradshteyn and Ryzhik [2], 4.620.1, page 618).

$$\int_0^{2\pi} d\omega \int_0^{\infty} f'\left(p\cosh x + (q\cos\omega + r\sin\omega)\sinh x\right)\sinh x\,dx = -\frac{2\pi[\operatorname{sign} p]}{\sqrt{p^2-q^2-r^2}} f\left([\operatorname{sign} p]\sqrt{p^2-q^2-r^2}\right)$$

if $p^2 > q^2+r^2$ and $\lim_{x\to\infty} f(x) = 0$ (see Gradshteyn and Ryzhik [2], 4.620.2, page 618).

$$\int_0^\pi \int_0^\pi f'\left(\frac{p - q\cos x}{\sin x \sin y} + r \cot y\right) \frac{dx\,dy}{\sin x \sin^2 y}$$
$$= -\frac{2\pi[\operatorname{sign} p]}{\sqrt{p^2 - q^2 - r^2}} f\left([\operatorname{sign} p]\sqrt{p^2 - q^2 - r^2}\right)$$

if $p^2 > q^2 + r^2$ and $\lim_{x\to\infty} f(x) = 0$ (see Gradshteyn and Ryzhik [2], 4.620.3, page 618).

$$\int_{-\infty}^{\infty} dx \int_{-\infty}^{\infty} f'\left(p\cosh x \cosh y + q \sinh x \cosh y + r \sinh y\right) \cosh y\,dy$$
$$= -\frac{2\pi[\operatorname{sign} p]}{\sqrt{p^2 - q^2 - r^2}} f\left([\operatorname{sign} p]\sqrt{p^2 - q^2 - r^2}\right)$$

if $p^2 > q^2 + r^2$ and $\lim_{x\to\infty} f(x) = 0$ (see Gradshteyn and Ryzhik [2], 4.620.4, page 618).

$$\int_0^\infty dx \int_0^\pi f\left(p\cosh x + q \cos\omega \sinh x\right) \sinh^2 x \sin\omega\,d\omega$$
$$= 2\int_0^\infty f\left([\operatorname{sign} p]\sqrt{p^2 - q^2}\cosh x\right) \sinh^2 x\,dx$$

if $\lim_{x\to\infty} f(x) = 0$ (see Gradshteyn and Ryzhik [2], 4.620.5, page 619).

$$\int_0^\infty \int_0^\infty f\left(a^2x^2 + b^2y^2\right) dx\,dy = \frac{\pi}{4ab}\int_0^\infty x f(x)\,dx$$

if $\lim_{x\to\infty} f(x) = 0$ (see Gradshteyn and Ryzhik [2], 4.623, page 619).

$$\int\limits_{\substack{x\ge 0,\quad y\ge 0\\ \left(\frac{x}{a}\right)^p + \left(\frac{y}{b}\right)^q \le 1}}\!\!\int x^{\alpha-1} y^{\beta-1} f\left[\left(\frac{x}{a}\right)^p + \left(\frac{y}{b}\right)^q\right] dx\,dy$$
$$= \frac{a^\alpha b^\beta}{pq}\Gamma\begin{bmatrix}\alpha/p, & \beta/q\\ \alpha/p + \beta/q\end{bmatrix}\int_0^1 f(t) t^{\alpha/p + \beta/q - 1}\,dt$$

if $a, b, \alpha, \beta, p, q > 0$ (see Prudnikov, Brychov, and Marichev [3], 3.1.2.1, page 565).

$$\int\limits_{x^2+y^2\le 1}\!\!\int f(\alpha x + \beta y + \gamma)\,dx\,dy = 2\int_{-1}^1 \sqrt{1-t^2}\, f\left(t\sqrt{\alpha^2+\beta^2} + \gamma\right) dt$$

if $a, b > 0$ and α, β, and γ are real (see Prudnikov, Brychov, and Marichev [3], 3.1.2.2, page 566).

$$\int\limits_{|x|+|y|\leq 1}\int f(x+y)\,dx\,dy = \int_{-1}^{1} f(t)\,dt$$

(see Prudnikov, Brychov, and Marichev [3], 3.1.2.3, page 566).

$$\int_0^1\int_0^1 f(xy)(1-x)^{\alpha-1}y^{\alpha}(1-y)^{\beta-1}\,dx\,dy = B(\alpha,\beta)\int_0^1 f(t)(1-t)^{\alpha+\beta-1}\,dt$$

if $\operatorname{Re}\alpha > 0$ and $\operatorname{Re}\beta > 0$ (see Prudnikov, Brychov, and Marichev [3], 3.1.2.4, page 566).

$$\int_0^\infty\int_0^\infty f(ax+by)e^{-px-qy}\,dx\,dy$$
$$= \frac{1}{pq(aq-bp)}\left[aq\int_0^\infty f(t)e^{-pt/a}\,dt - bp\int_0^\infty f(t)e^{-qt/b}\,dt\right]$$

if $a, b, p, q > 0$ (see Prudnikov, Brychov, and Marichev [3], 3.1.3.1, page 567).

$$\int_0^\infty\int_0^\infty f(|x-y|)e^{-px-qy}\,dx\,dy$$
$$= \frac{1}{pq(p+q)}\left[p\int_0^\infty f(t)e^{-qt}\,dt + q\int_0^\infty f(t)e^{-pt}\,dt\right]$$

if $p, q > 0$ (see Prudnikov, Brychov, and Marichev [3], 3.1.3.2, page 567).

$$\int_0^\infty\int_0^\infty f(xy)e^{-px-qy}\,dx\,dy = 2\int_0^\infty K_0\left(2\sqrt{pqt}\right)f(t)\,dt$$

if $p, q > 0$ (see Prudnikov, Brychov, and Marichev [3], 3.1.3.3, page 567).

$$\int_0^\infty\int_0^\infty \frac{e^{-px-qy}}{\sqrt{x+y}}f\left(\frac{xy}{x+1}\right)dx\,dy = \frac{\sqrt{\pi}\,(\sqrt{p}+\sqrt{q})}{\sqrt{pq}}\int_0^\infty e^{-(\sqrt{p}+\sqrt{q})^2 t}f(t)\,dt$$

if $p, q > 0$ (see Prudnikov, Brychov, and Marichev [3], 3.1.3.4, page 567).

Three-Dimensional Integrals

$$\int_0^\infty dx\int_0^{2\pi} d\omega\int_0^\pi f\left(p\cosh x + (q\cos\omega + r\sin\omega)\sin\theta\sinh x\right)\sinh^2 x\sin\theta\,d\theta$$
$$= 4\int_0^\infty f\left([\operatorname{sign} p]\sqrt{p^2-q^2-r^2}\cosh x\right)\sinh^2 x\,dx$$

if $p^2 > q^2+r^2$ and $\lim_{x\to\infty} f(x) = 0$ (see Gradshteyn and Ryzhik [2], 4.620.6, page 619).

$$\int_0^\infty dx \int_0^{2\pi} d\omega \int_0^\pi f\left(p\cosh x + [(q\cos\omega + r\sin\omega)\sin\theta + s\cosh\theta]\sinh x\right) \\ \times \sinh^2 x \sin\theta\, d\theta \\ = 4\pi \int_0^\infty f\left([\operatorname{sign} p]\sqrt{p^2 - q^2 - r^2 - s^2}\cosh x\right)\sinh^2 x\, dx$$

if $p^2 > q^2 + r^2 + s^2$ and $\lim_{x\to\infty} f(x) = 0$ (see Gradshteyn and Ryzhik [2], 4.620.7, page 619 or Prudnikov, Brychov, and Marichev [3], 3.2.3.4, page 584).

$$\int\int_{\substack{x\geq 0,\ y\geq 0,\ z\geq 0 \\ \left(\frac{x}{a}\right)^p + \left(\frac{y}{b}\right)^q + \left(\frac{z}{c}\right)^r}} \int x^{\alpha-1}y^{\beta-1}z^{\gamma-1} f\left[\left(\frac{x}{a}\right)^p + \left(\frac{y}{b}\right)^q + \left(\frac{z}{c}\right)^r\right] dx\, dy\, dz \\ = \frac{a^\alpha b^\beta c^\gamma}{pqr}\Gamma\left[\begin{matrix}\alpha/p,\ \beta/q,\ \gamma/r \\ \alpha/p + \beta/q + \gamma/r\end{matrix}\right]\int_0^1 f(t)t^{\alpha/p+\beta/q+\gamma/r-1}\, dt$$

if $a, b, c, p, q, r, \alpha, \beta, \gamma > 0$ and the integral on the right-hand side converges absolutely (see Prudnikov, Brychov, and Marichev [3], 3.2.2.1, page 583).

$$\int_0^a \int_0^{a-x} \int_0^{a-x-y} x^{\alpha-1}y^{\beta-1}z^{\gamma-1} f(x+y+z)\, dx\, dy\, dz \\ = \Gamma\left[\begin{matrix}\alpha,\ \beta,\ \gamma \\ \alpha+\beta+\gamma\end{matrix}\right]\int_0^a f(t)t^{\alpha+\beta+\gamma-1}\, dt$$

if $a, \alpha, \beta, \gamma > 0$ and the integral on the right-hand side converges absolutely (see Prudnikov, Brychov, and Marichev [3], 3.2.2.3, page 583).

Multi-Dimensional Integrals

$$\underbrace{\int\int\cdots\int}_{\sum t_i \leq 1} f\left(\sum t_i\right)\prod\left(t_i^{a_i-1}\, dt_i\right) = \frac{\prod\Gamma(a_i)}{\Gamma\left(\sum a_i\right)}\int_0^1 f(x)x^{\sum a_i - 1}\, dx$$

This is known as the Dirichlet reduction (see Squire [6], pages 82–83).

$$\underbrace{\int\int\cdots\int}_{\substack{t_i\geq 0\\ t_1+t_2+\cdots+t_n\leq 1}} f\left(\sum_{i=1}^{n-s} t_i\right)\prod \phi_i(t_i)\,dt_i$$
$$=\int_0^1 \left(f(u)\left[\phi_1(u)\star\phi_2(u)\star\cdots\star\phi_{n-s}(u)\right]\right)\star\phi_{n-s+1}\star\cdots\star\phi_n(u)\,du$$

where the stars represent convolutions (see Sivazlian [5]).

$$\int_a^x g(x_1)\,dx_1\int_a^{x_1} g(x_2)\,dx_2\cdots\int_a^{x_{n-1}} g(x_n)\,dx_n\int_a^{x_n} f(t)\,dt$$
$$=\frac{1}{n!}\int_a^x f(t)\left(\int_t^x g(y)\,dy\right)^n dt$$

(see Prudnikov, Brychov, and Marichev [3], 3.3.1.1, page 585).

$$\iint_R\cdots\int f\left[\left(\frac{x_1}{q_1}\right)^{\alpha_1}+\left(\frac{x_2}{q_2}\right)^{\alpha_2}+\ldots+\left(\frac{x_n}{q_n}\right)^{\alpha_n}\right]x_1^{p_1-1}x_2^{p_2-1}\ldots x_n^{p_n-1}\,d\mathbf{x}$$
$$=\frac{q_1^{p_1}q_2^{p_2}\ldots q_n^{p_n}}{\alpha_1\alpha_2\ldots\alpha_n}\frac{\Gamma\left(\frac{p_1}{a_1}\right)\Gamma\left(\frac{p_2}{a_2}\right)\ldots\Gamma\left(\frac{p_n}{a_n}\right)}{\Gamma\left(\frac{p_1}{\alpha_1}+\frac{p_2}{\alpha_2}+\ldots+\frac{p_n}{\alpha_n}\right)}\int_1^\infty f(z)z^{\left(\frac{p_1}{\alpha_1}+\frac{p_2}{\alpha_2}+\ldots+\frac{p_n}{\alpha_n}-1\right)}\,dz$$

where R is the region $R:\begin{cases} x_1\geq 0 \quad x_2\geq 0 \quad \ldots \quad x_n\geq 0\\ \left(\frac{x_1}{q_1}\right)^{\alpha_1}+\left(\frac{x_2}{q_2}\right)^{\alpha_2}+\ldots+\left(\frac{x_n}{q_n}\right)^{\alpha_n}\geq 1\end{cases}$

when the integral on the right converges absolutely (see Gradshteyn and Ryzhik [2], 4.635.1, page 620).

$$\iint_S\cdots\int f\left[\left(\frac{x_1}{q_1}\right)^{\alpha_1}+\left(\frac{x_2}{q_2}\right)^{\alpha_2}+\ldots+\left(\frac{x_n}{q_n}\right)^{\alpha_n}\right]x_1^{p_1-1}x_2^{p_2-1}\ldots x_n^{p_n-1}\,d\mathbf{x}$$
$$=\frac{q_1^{p_1}q_2^{p_2}\ldots q_n^{p_n}}{\alpha_1\alpha_2\ldots\alpha_n}\frac{\Gamma\left(\frac{p_1}{a_1}\right)\Gamma\left(\frac{p_2}{a_2}\right)\ldots\Gamma\left(\frac{p_n}{a_n}\right)}{\Gamma\left(\frac{p_1}{\alpha_1}+\frac{p_2}{\alpha_2}+\ldots+\frac{p_n}{\alpha_n}\right)}\int_0^1 f(z)z^{\left(\frac{p_1}{\alpha_1}+\frac{p_2}{\alpha_2}+\ldots+\frac{p_n}{\alpha_n}-1\right)}\,dz$$

where S is the region $S:\begin{cases} x_1\geq 0 \quad x_2\geq 0 \quad \ldots \quad x_n\geq 0\\ \left(\frac{x_1}{q_1}\right)^{\alpha_1}+\left(\frac{x_2}{q_2}\right)^{\alpha_2}+\ldots+\left(\frac{x_n}{q_n}\right)^{\alpha_n}\leq 1\end{cases}$

when the integral on the right converges absolutely, and the numbers q_i, α_i, and p_i are positive. (see Gradshteyn and Ryzhik [2], 4.635.2, page 621).

$$\iint\limits_R \cdots \int f(x_1 + x_2 + \ldots + x_n) \frac{x_1^{p_1-1} x_2^{p_2-1} \ldots x_n^{p_n-1}\, d\mathbf{x}}{(q_1 x_1 + q_2 x_2 + \ldots q_n x_n + r)^{p_1+p_2+\ldots+p_n}}$$
$$= \frac{\Gamma(p_1)\Gamma(p_2)\ldots\Gamma(p_n)}{\Gamma(p_1 + p_2 + \ldots + p_n)} \int_0^1 f(x) \frac{x^{p_1+p_2+\ldots+p_n-1}}{(q_1 x + r)^{p_1}(q_2 x + r)^{p_2} \ldots (q_n x + r)^{p_n}}\, dx$$

where R is the region $R: \begin{cases} x_1 \geq 0, \quad x_2 \geq 0, \quad \ldots, \quad x_n \geq 0 \\ x_1 + x_2 + \ldots + x_n \leq 1 \end{cases}$, $f(x)$ is continuous on $(0, 1)$ and $q_i \geq 0$ and $r > 0$ (see Gradshteyn and Ryzhik [2], 4.637, page 622).

$$\iint\limits_{x_1^2+x_2^2+\ldots+x_n^2 \leq R^2} \cdots \int f\left(\sqrt{x_1^2 + x_2^2 + \ldots + x_n^2}\right) d\mathbf{x} = \frac{2\pi^{n/2}}{\Gamma\left(\frac{n}{2}\right)} \int_0^R x^{n-1} f(x)\, dx$$

if $f(x)$ is continuous on the interval $(0, R)$ (see Gradshteyn and Ryzhik [2], 4.642, page 623).

$$\int_0^1 \int_0^1 \cdots \int_0^1 f(x_1 x_2 \ldots x_n)(1 - x_1)^{p_1-1}(1 - x_2)^{p_2-1} \ldots (1 - x_n)^{p_n-1}$$
$$\times\, x_2^{p_1} x_3^{p_1+p_2} \ldots x_n^{p_1+p_2+\ldots+p_{n-1}}\, d\mathbf{x}$$
$$= \frac{\Gamma(p_1)\Gamma(p_2)\ldots\Gamma(p_n)}{\Gamma(p_1 + p_2 + \ldots + p_n)} \int_0^1 f(x)(1 - x)^{p_1+p_2+\ldots+p_n-1}\, dx$$

when the integral on the right converges absolutely (see Gradshteyn and Ryzhik [2], 4.643, page 623).

$$\iint\limits_R \cdots \int \frac{f(p_1 x_1 + p_2 x_2 \cdots + p_n x_n)}{\sqrt{1 - x_1^2 - x_2^2 - \ldots - x_{n-1}}}\, d\mathbf{x}$$
$$= \frac{\pi^{(n-1)/2}}{\Gamma\left(\frac{n-1}{2}\right)} \int_0^\pi f\left(\sqrt{p_1^2 + p_2^2 + \ldots + p_n^2} \cos x\right) \sin^{n-2} x\, dx$$

if R is the region $R = x_1^2 + x_2^2 + \ldots + x_{n-1}^2 \leq 1$, $n \geq 3$, and $f(x)$ is continuous on the interval $[-\beta, \beta]$, with $\beta = \sqrt{p_1^2 + p_2^2 + \ldots + p_n^2}$ (see Gradshteyn and Ryzhik [2], 4.644, page 624).

Notes

[1] See also the section on how to change variables on page 109.

[2] Schwartz [4] considers a numerical technique for integrals of the form $\left(\prod_{i=1}^{d} \int g_i(x_i)\,dx_i\right) F\left(\sum_{i=1}^{d} f_i(x_i)\right)$. His technique is based on the use of integral transforms.

References

[1] R. Y. Denis and R. A. Gustafson, "An $SU(N)N$ Q-Beta Integral Transformation and Multiple Hypergeometric Series Identities," *SIAM J. Math. Anal.*, **23**, No. 2, March 1992.

[2] I. S. Gradshteyn and I. M. Ryzhik, *Tables of Integrals, Series, and Products*, Academic Press, New York, 1980.

[3] A. P. Prudnikov, Yu. A. Brychov, and O. I. Marichev, *Integrals and Series*, Volume 1, translated by N. M. Queen, Gordon and Breach, New York, 1990. C. Schwartz, "Numerical Integration in Many Dimensions. I," *J. Math. Physics*, **26**, No. 5, May 1985, pages 951–954.

[4] B. D. Sivazlian, "A Class of Multiple Integrals," *SIAM J. Math. Anal.*, **2**, No. 1, February 1971, pages 72–75.

[5] W. Squire, *Integration for Engineers and Scientists*, American Elsevier Publishing Company, New York, 1970.

27. Transforms: Miscellaneous

Idea

This section has a collection of transformations of integrals.

Some Specific Forms for One-Dimensional Integrals

When an integrand has a specific form, there are some common transformations that might be useful. For example, if the integrand:

(A) is a rational function of $\sin x$ and $\cos x$, introduce $u = \tan(x/2)$ so that

$$\sin x = \frac{2u}{1+u^2}, \qquad \cos x = \frac{1-u^2}{1+u^2}, \qquad dx = \frac{2\,du}{1+u^2}.$$

For example, the integral $J = \int \frac{\sin^2 x}{\cos^4 x + \cos x \sin^3 x}\, dx$ becomes

$$\begin{aligned}
J &= \int \frac{8u^2(u^2+1)}{(u^2-1)(u^2-2u-1)(u^4+2u^3+2u^2-2u+1)}\, du \\
&= \int \left[\frac{4u^3+6u^2+4u-2}{u^4+2u^3+2u^2-2u+1} + \frac{2u-2}{3(u^2-2u-1)} - \frac{1}{u+1} - \frac{1}{u-1}\right] du \\
&= \tfrac{1}{3}\log\left(u^4+2u^3+2u^2-2u+1\right) + \tfrac{1}{3}\log\left(3(u^2-2u-1)\right) \\
&\qquad - \log(u+1) - \log(u-1) \\
&= \frac{1}{3}\log\left(1 - \frac{8u^3}{(u^2-1)^3}\right) \\
&= \tfrac{1}{3}\log\left(1+\tan^3 x\right).
\end{aligned}$$

(B) is a rational function of $\sinh x$ and $\cosh x$, introduce $u = \tanh(x/2)$ so that

$$\sinh x = \frac{2u}{1-u^2}, \qquad \cosh x = \frac{1+u^2}{1-u^2}, \qquad dx = \frac{2\,du}{1-u^2}.$$

(C) is a rational function of x and $\sqrt{1-x^2}$, introduce $x = \cos v$ to get to case (A).

(D) is a rational function of x and $\sqrt{x^2-1}$, introduce $x = \cosh v$ to get to case (B).

(E) is a rational function of x and $\sqrt{x^2+1}$, introduce $u = x + \sqrt{x^2+1}$ so that

$$x = \frac{1}{2}\left(u - \frac{1}{u}\right), \quad \sqrt{x^2+1} = \frac{1}{2}\left(u + \frac{1}{u}\right), \quad dx = \frac{1}{2}\left(u + \frac{1}{u^2}\right) du.$$

(F) is a rational function of x and $\sqrt{ax^2+bx+c}$, then the substitution

$$v = \frac{2ax+b}{\sqrt{|4ac-b^2|}} \iff x = \frac{v\sqrt{|4ac-b^2|}-b}{2a}$$

will reduce the integral to one of the above cases.

(G) is a rational function of x and $u = \sqrt{\frac{ax+b}{cx+d}}$, then use the substitution

$$u = \sqrt{\frac{ax+b}{cx+d}} \iff x = -\frac{b-u^2d}{a-u^2c}.$$

Integrals of the form $\int x^m(a+bx^n)^p\, dx$, where m, n and p are rational numbers, can be expressed in terms of elementary functions in each of the following cases (see Gradshteyn and Ryzhik [3] 2.202, page 71): p is an integer; $(m+1)/n$ is an integer; and $p+(m+1)/n$ is an integer.

Three useful transformations of definite integrals from Gradshteyn and Ryzhik [3] (3.032.1–3.032.3) are

$$\int_0^{\pi/2} f(\sin x)\,dx = \int_0^{\pi/2} f(\cos x)\,dx$$
$$\int_0^{2\pi} f(p\cos x + q\sin x)\,dx = 2\int_0^{\pi} f\left(\sqrt{p^2+q^2}\cos x\right)\,dx$$
$$\int_0^{\pi/2} f(\sin 2x)\cos x\,dx = \int_0^{\pi/2} f\left(\cos^2 x\right)\cos x\,dx.$$

The Slobin transformation is (see Squire [6], page 95)

$$\int_{-\infty}^{\infty} f\left(t-\frac{1}{t}\right)\,dt = \int_{-\infty}^{\infty} f(t)\,dt.$$

The Wolstenholme transformation is (see Squire [6], page 95)

$$\int_1^a f\left(t^2+\frac{a^2}{t^2}\right)\frac{dt}{t} = \int_1^a f\left(t+\frac{a^2}{t}\right)\frac{dt}{t}.$$

Two unnamed transformations for definite integrals (from Squire [6] page 94) are:

$$\int_0^{\pi} xf\left(\sin x, \cos^2 x\right)\,dx = \frac{\pi}{2}\int_0^{\pi} f\left(\sin x, \cos^2 x\right)\,dx$$
$$\int_0^{\infty} f\left(x^n + x^{-n}\right)\frac{\tan^{-1} x}{x}\,dx = \frac{\pi}{2}\int_0^{1} f\left(x^n + x^{-n}\right)\frac{dx}{x}.$$

Dealing with Singularities

Often, integrable singularities can cause a numerical routine to have trouble approximating an integral. While there are no universal principles to be applied, some simple techniques can be useful.

Consider the integral $I = \int_0^1 \frac{f(x)}{\sqrt{x}}\,dx$. Assuming that $f(0)$ is finite, I has an integrable singularity. But a computer routine evaluating this integrand near $x=0$ may have trouble.

(A) If the $f(x)$ term was not present, then we would note the indefinite integral $\int dx/\sqrt{x} = 2\sqrt{x}$. Since the form $\sqrt{x}$ appears, it might be reasonable to transform the integral to be on this scale. Changing variables via $u = \sqrt{x}$ changes I to $I = 2\int_0^1 f(u^2)\,du$. This integral no longer has a singularity at the origin.

(B) Another technique is to subtract out a singularity. For example, if $f(0)$ has the finite value A, then I can be written as

$$I = \int_0^1 \frac{f(x) - A + A}{\sqrt{x}}\,dx = \int_0^1 \frac{f(x) - A}{\sqrt{x}}\,dx + 2A.$$

This new integrand has more smoothness than the original integrand, thus should be easier to evaluate numerically.

(C) Another technique is to break the integral up into several sub-integrals. For the above integral we might try

$$I = J_1 + J_2 = \int_0^{\varepsilon} \frac{f(x)}{\sqrt{x}}\,dx + \int_{\varepsilon}^1 \frac{f(x)}{\sqrt{x}}\,dx,$$

where ε is large enough so that a numerical routine does not have trouble with J_1, say $\varepsilon = 10^{-4}$. Then the integral J_2 has a narrow region of integration; expansion methods, such as Taylor series, may yield a good numerical approximation.

Of course, commercial integration software can automatically detect power singularities, and apply an appropriate (numerical) transformation to avoid computational difficulties.

Products of Special Functions

Piquette [5] considers integrals of the form $I = \int f(x) \prod_{i=1}^m R_{\mu_i}^{(i)}(x)\,dx$, where $R_{\mu_i}^{(i)}(x)$ is of the i-th type of a special function of order μ, obeying the set of recurrence relations

$$\begin{aligned} R_{\mu+1}^{(i)} &= a_\mu(x) R_\mu^{(i)} + b_\mu(x) R_{\mu-1}^{(i)}, \\ \frac{dR_\mu^{(i)}}{dx} &= c_\mu(x) R_\mu^{(i)} + d_\mu(x) R_{\mu-1}^{(i)}. \end{aligned} \tag{27.1}$$

Here $\{a_\mu, b_\mu, c_\mu, d_\mu\}$ are known functions. Most of the special functions of physics, such as the Bessel functions, Legendre functions, Hermite functions, and Laguerre functions have recursion relations in the form of (27.1).

The indefinite integral I is assumed to have the form

$$I = \sum_{p_1=0}^{1} \sum_{p_2=0}^{1} \cdots \sum_{p_m=0}^{1} A_{p_1 p_2 \dots p_m}(x) \prod_{i=1}^{m} R_{\mu_i + p_i}^{(i)}(x).$$

Then the values of the $\{A_{\mathbf{p}}\}$ are determined, sometimes with the help of a differential equation. Piquette [5] obtains, for example, the following integral involving Legendre functions

$$\begin{aligned}\int x\left[P_{1/3}(x)\right]^3 dx &= \left(\tfrac{125}{12}x^4 - \tfrac{14}{3}x^2 - \tfrac{5}{12}\right)\left[P_{1/3}(x)\right]^3 \\ &\quad + (-4+20x^2)P_{1/3}(x)\left[P_{4/3}(x)\right]^3 \\ &\quad + (9x-25x^3)\left[P_{1/3}(x)\right]^2 P_{4/3}(x) - \tfrac{16}{3}x\left[P_{4/3}(x)\right]^2.\end{aligned}$$

A Jacobian of Unity

Sometime a judicious change of variables in a multiple integral may make it easier to evaluate. Finding the correct change of variables may sometimes be done by solving a partial differential equation.

For example, changing variables in the usual way in two dimensions results in (see page 109)

$$I = \iint f(x,y)\,dx\,dy = \iint F(u,v)\left|\frac{\partial(x,y)}{\partial(u,v)}\right| du\,dv, \tag{27.2}$$

where $F(u,v) := f\left(x(u,v), y(u,v)\right)$. If u is chosen so that $F(u,v) = F(u)$, and v is chosen so that

$$\left|\frac{\partial(x,y)}{\partial(u,v)}\right|^{-1} = \frac{\partial u}{\partial x}\frac{\partial v}{\partial y} - \frac{\partial u}{\partial y}\frac{\partial v}{\partial x} = 1, \tag{27.3}$$

then (27.2) becomes $I = \iint F(u)\,du\,dv$, which may be easier to integrate. Note that equation (27.3) for v is a partial differential equation.

Example

Suppose we have the integral $I = \int_1^2 \int_3^4 \log\left(\frac{x^2-y^2}{4}\right) dx\,dy$. Choosing $u = x^2 - y^2$ simplifies (27.3) to $2x\frac{\partial v}{\partial y} + 2y\frac{\partial v}{\partial x} = 1$ which has as a solution $v = \frac{1}{2}\log(x+y)$.

Using the u and v variables defined above, I becomes

$$I = \left\{\int_5^8 \int_{v_1(u)}^{v_3(u)} + \int_8^{12} \int_{v_2(u)}^{v_3(u)} + \int_{12}^{15} \int_{v_2(u)}^{v_4(u)}\right\} \log\frac{u}{4}\,dv\,du,$$

where

$$\begin{aligned} v_1(u) &= \tfrac{1}{2}\log\left(3+\sqrt{9-u}\right) \\ v_2(u) &= \tfrac{1}{2}\log\left(1+\sqrt{1+u}\right) \\ v_3(u) &= \tfrac{1}{2}\log\left(2+\sqrt{4+u}\right) \\ v_4(u) &= \tfrac{1}{2}\log\left(4+\sqrt{16-u}\right). \end{aligned}$$

It is now straightforward to evaluate this integral.

See Morris [4] or Squire [6] pages 12–13 for details on this method.

Notes

[1] If $R(\sin x, \cos x)$ satisfies the relation $R(\sin x, \cos x) = -R(-\sin x, \cos x)$ then it is normally convenient to make the substitution $z = \cos x$. This results in the formula

$$\int R(\sin x, \cos x)\, dx = -\int R\left(\sqrt{1-z^2}, z\right) \frac{dz}{\sqrt{1-z^2}}.$$

[2] If $R(\sin x, \cos x)$ satisfies the relation $R(\sin x, \cos x) = -R(\sin x, -\cos x)$ then it is normally convenient to make the substitution $z = \sin x$. This results in the formula

$$\int R(\sin x, \cos x)\, dx = \int R\left(z, \sqrt{1-z^2}\right) \frac{dz}{\sqrt{1-z^2}}.$$

[3] If $R(\sin x, \cos x)$ satisfies the relation $R(\sin x, \cos x) = R(-\sin x, -\cos x)$ then it is normally convenient to make the substitution $z = \tan x$. This results in the formula

$$\int R(\sin x, \cos x)\, dx = \int R\left(\frac{z}{\sqrt{1+z^2}}, \frac{1}{\sqrt{1+z^2}}\right) \frac{dz}{1+z^2}.$$

For example, we find (compare to the first example in this section)

$$\begin{aligned}
J &= \int \frac{\sin^2 x}{\cos^4 x + \cos x \sin^3 x}\, dx \\
&= \int \frac{z^2}{(z+1)(z^2-z+1)}\, dz \\
&= \int \left[\frac{2z-1}{3(z^2-z+1)} + \frac{1}{3(z+1)}\right] dz \\
&= \tfrac{1}{3}\log\left(z^2-z+1\right) + \tfrac{1}{3}\log(z+1) \\
&= \tfrac{1}{3}\log\left(1+z^3\right) \\
&= \tfrac{1}{3}\log\left(1+\tan^3 x\right).
\end{aligned}$$

References

[1] M. Abramowitz and I. A. Stegun, *Handbook of Mathematical Functions*, National Bureau of Standards, Washington, DC, 1964

[2] M. S. Ashbaugh, "On Integrals of Combination of Solutions of Second-Order Differential Equations," *J. Phys. A: Math. Gen.*, **19**, 1986, pages 3701–3703.

[3] I. S. Gradshteyn and I. M. Ryzhik, *Tables of Integrals, Series, and Products*, Academic Press, New York, 1980.

[4] W. L. Morris, "A New Method for the Evaluation of $\iint_A f(x,y)\, dy\, dx$," *Amer. Math. Monthly*, **43**, 1936, pages 358–362.

[5] J. C. Piquette, "Applications of a Technique for Evaluating Indefinite Integrals Containing Products of the Special Functions of Physics," *SIAM J. Math. Anal.*, **20**, No. 5, September 1989, pages 1260–1269.

[6] W. Squire, *Integration for Engineers and Scientists*, American Elsevier Publishing Company, New York, 1970.

III

Exact Analytical Methods

28. Change of Variable

Applicable to Definite integrals.

Yields

A reformulation.

Idea

A re-parameterization of an integral may make it easier to evaluate.

Procedure

Given the integral

$$I = \int_{x=a}^{x=b} f(x)\,dx, \tag{28.1}$$

if there exists a transformation of the form $x = u(y)$, which can be uniquely inverted on the interval $a \leq x \leq b$ to give $y = v(x)$, then

$$I = \int_{y=v(a)}^{y=v(b)} f(u(y))\,dx = \int_{y=v(a)}^{y=v(b)} f(u(y))u'(y)\,dy. \tag{28.2}$$

It is sometimes easier to evaluate (28.2) than to evaluate (28.1) directly.

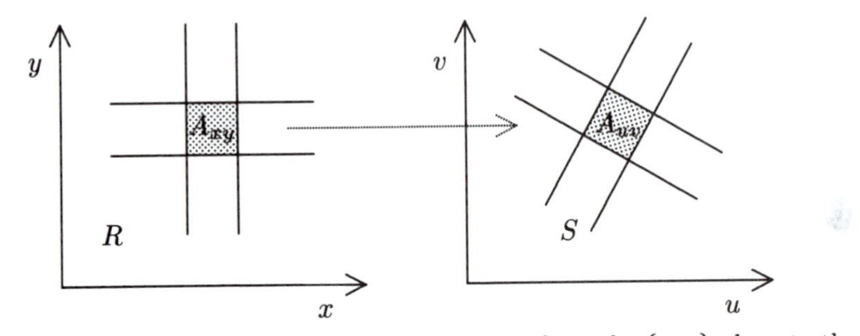

Figure 28.1 Pictorial representation of a mapping from the $\{x, y\}$ plane to the $\{u, v\}$ plane.

Similar results hold in multiple dimensions. If there exists a one-to-one transformation from the region R in the xy-plane to the region S in the uv-plane, defined by the continuously differentiable functions $x = x(u, v)$, $y = y(u, v)$; and if the Jacobian

$$J = J(u, v) = \frac{\partial(x, y)}{\partial(u, v)} = \begin{vmatrix} x_u & y_u \\ x_v & y_v \end{vmatrix} \tag{28.3}$$

is nonzero in the region of interest, then

$$\iint_R f(x, y)\, dA_{x,y} = \iint_S f(x(u, v), y(u, v))|J(u, v)|\, dA_{u,v} \tag{28.4}$$

(see Figure 28.1.) A listing of some common two-dimensional transformations is in Table 28.1.

In an orthogonal coordinate system, let $\{\mathbf{a}_i\}$ denote the unit vectors in each of the three coordinate directions, and let $\{u_i\}$ denote distance along each of these axes. The coordinate system may be designated by the *metric coefficients* $\{g_{11}, g_{22}, g_{33}\}$, defined by

$$g_{ii} = \left(\frac{\partial x_1}{\partial u_i}\right)^2 + \left(\frac{\partial x_2}{\partial u_i}\right)^2 + \left(\frac{\partial x_3}{\partial u_i}\right)^2, \tag{28.5}$$

where $\{x_1, x_2, x_3\}$ represent rectangular coordinates. Using the metric coefficients defined in (28.5), we define $g = g_{11}g_{22}g_{33}$; the magnitude of the Jacobian of the transformation is then given by $|J| = \sqrt{g}$. Moon and Spencer [4] list the metric coefficients for 43 different orthogonal coordinate systems. (These consist of 11 general systems, 21 cylindrical systems, and 11 rotational systems.)

Operations with orthogonal coordinate systems are sometimes written in terms of the $\{h_i\}$ functions, instead of the $\{g_{ii}\}$ terms. Here, $h_i = \sqrt{g_{ii}}$, so that $|J| = \sqrt{g} = h_1h_2h_3$. For example, cylindrical polar coordinates are defined by $x_1 = r\cos\phi$, $x_2 = r\sin\phi$, and $x_3 = z$. Therefore, using $\{u_1 = r$, $u_2 = \theta$, $u_3 = z\}$, we find $h_1 = 1$, $h_2 = r$, and $h_3 = 1$. From this we compute the magnitude of the Jacobian to be $|J| = r$. A listing of some common three-dimensional transformations, and their Jacobians, is in Table 28.2.

Example 1

Given the integral

$$I = \int_0^{\pi/2} \frac{\sin x \cos x}{1+\sin^2 x}\, dx, \tag{28.6}$$

we choose to change variables by $x = u(y) = \sin^{-1} y$. This function can be uniquely inverted (on the range $0 \le x \le \frac{\pi}{2}$) via $y = v(x) = \sin x$. Thus,

$$dx = u'(y)\, dy = \frac{dy}{\sqrt{1-y^2}} = \frac{dy}{\sqrt{1-\sin^2 x}} = \frac{dy}{\cos x}.$$

This could also have been obtained directly by $dy = v'(x)\, dx = \cos dx$. Our change of variable then permits (28.6) to be written as

$$I = \int_{y=0}^{y=1} \frac{y \cos x}{1+y^2} \frac{dy}{\cos x} = \int_{y=0}^{y=1} \frac{y}{1+y^2} dy = \frac{1}{2} \log(1+y^2)\bigg|_{y=0}^{y=1} = \tfrac{1}{2} \log 2. \tag{28.7}$$

Example 2

Consider the integral $I = \int_{-1}^{1} x^2\, dx$. If we choose to change variables by $y = v(x) = x^2$, then we note that $v(x)$ cannot be uniquely inverted on the given interval of integration. Hence, we must break up the integral I into pieces, with $v(x)$ uniquely invertible on each piece. If we write I as

$$I = I_1 + I_2 = \int_{-1}^{0} x^2\, dx + \int_0^1 x^2\, dx,$$

then we can use the change of variables $\{y = x^2,\ x = -\sqrt{y}\}$ on I_1 and the change of variables $\{y = x^2,\ x = \sqrt{y}\}$ on I_2 to obtain

$$I_1 = \int_{-1}^{0} x^2\, dx = \int_1^0 y \frac{dy}{-2\sqrt{y}} = \frac{1}{2} \int_0^1 \sqrt{y}\, dy = \frac{1}{2} \left(\frac{2}{3} y^{3/2} \right)\bigg|_0^1 = \frac{1}{3}$$

and

$$I_2 = \int_0^1 x^2\, dx = \int_0^1 y \frac{dy}{2\sqrt{y}} = \frac{1}{2} \int_0^1 \sqrt{y}\, dy = \frac{1}{3}.$$

We conclude that $I = I_1 + I_2 = \frac{1}{3} + \frac{1}{3} = \frac{2}{3}$.

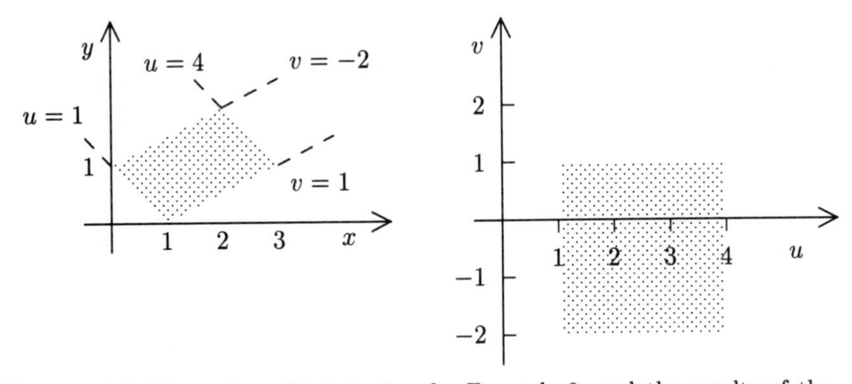

Figure 28.2 The region of integration for Example 2, and the results of the coordinate transformation $u = x + y$, $v = x - 2y$.

Example 3

Suppose we wish to evaluate the integral $I = \iint\limits_R (x+y)^2 \, dx \, dy$ where R is the parallelogram shown in Figure 28.2. The sides of R are straight lines with equations of the form

$$x + y = c_1, \qquad x - 2y = c_2,$$

with the values $c_1 = 1$, $c_1 = 4$, $c_2 = -2$, and $c_2 = 1$. If we introduce the new coordinates

$$u = x + y, \qquad v = x - 2y$$

then the region R corresponds to the rectangle $1 \le u \le 4$, $-2 \le v \le 1$. This mapping is clearly one-to-one, and the Jacobian is given by

$$J = \frac{\partial(x,y)}{\partial(u,v)} = \frac{1}{\dfrac{\partial(u,v)}{\partial(x,y)}} = \frac{1}{\begin{vmatrix} 1 & 1 \\ 1 & -2 \end{vmatrix}} = -\frac{1}{3}.$$

Therefore, we find

$$I = \int_1^4 \int_{-2}^1 (x+y)^2 |J| \, dv \, du = \int_1^4 \int_{-2}^1 \frac{u^2}{3} \, dv \, du = 21.$$

Table 28.1 Some common two-dimensional transformations.

$$\int_0^\alpha dx \int_0^{\beta x/\alpha} f(x,y)\, dy = \int_0^\beta dy \int_{\alpha y/\beta}^\alpha f(x,y)\, dx.$$

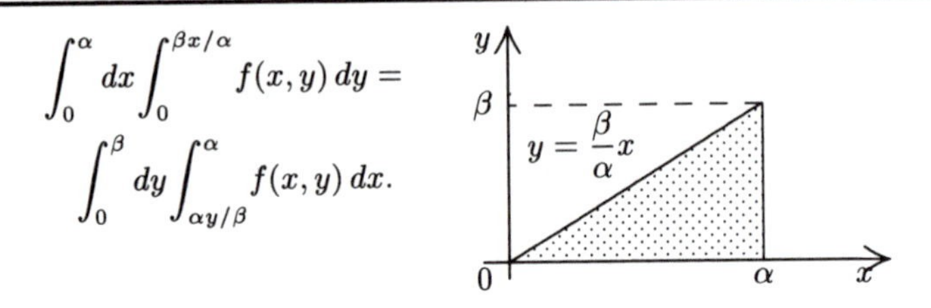

Assuming that $\alpha = \delta/(\beta+\gamma)$, $\alpha > 0$, $\beta > 0$, $\gamma > 0$ then we have

$$\int_0^{\alpha} dx \int_{\beta x}^{\delta-\gamma x} f(x,y)\, dy = \int_0^{\alpha\beta} dy \int_0^{y/\beta} f(x,y)\, dx + \int_{\alpha\beta}^{\delta} dy \int_0^{(\delta-y)/\gamma} f(x,y)\, dx.$$

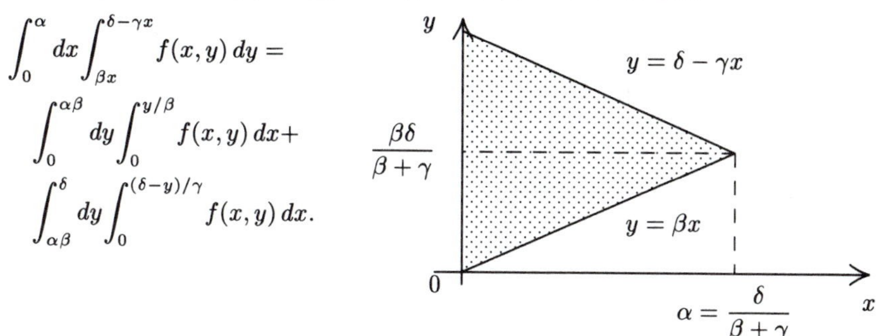

Note this second representation has the integral separated into two pieces.

$$\int_0^{R} dx \int_0^{\sqrt{R^2-x^2}} f(x,y)\, dy = \int_0^{R} dy \int_0^{\sqrt{R^2-y^2}} f(x,y)\, dx = \int_0^{\pi/2} d\psi \int_0^{R} f(r\cos\psi, r\sin\psi) r\, dr.$$

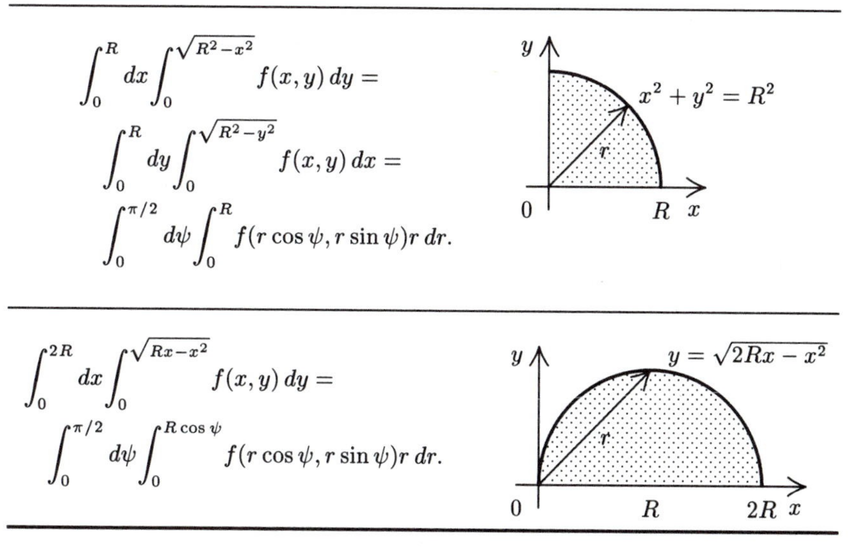

$$\int_0^{2R} dx \int_0^{\sqrt{Rx-x^2}} f(x,y)\, dy = \int_0^{\pi/2} d\psi \int_0^{R\cos\psi} f(r\cos\psi, r\sin\psi) r\, dr.$$

Table 28.2 Some common three-dimensional transformations (see Moon and Spencer [4]).

(A) **Bispherical Coordinates**

$$x_1 = a u_3 \frac{\sqrt{1-u_2^2}}{u_1-u_2}, \qquad x_2 = a\frac{\sqrt{(1-u_2^2)(1-u_3^2)}}{u_1-u_2}, \qquad x_3 = \frac{\sqrt{u_1^2-1}}{u_1-u_2}$$

$$h_1 = \frac{a}{(u_1-u_2)\sqrt{u_1^2-1}}, \; h_2 = \frac{a}{(u_1-u_2)\sqrt{1-u_2^2}}, \; h_3 = \frac{a}{u_1-u_2}\sqrt{\frac{1-u_2^2}{1-u_3^2}}.$$

(B) **Cylindrical Polar Coordinates** (here $J = r$)

$$x_1 = r\cos\phi, \qquad x_2 = r\sin\phi, \qquad x_3 = z$$
$$h_1 = 1, \qquad h_2 = r, \qquad h_3 = 1.$$

(C) **Elliptic Cylinder Coordinates** (here $J = \dfrac{u_1^2 - c^2 u_2^2}{1 - u_2^2}$)

$$x_1 = u_1 u_2, \qquad x_2 = \sqrt{(u_1^2 - c^2)(1 - u_2^2)}, \qquad x_3 = u_3$$

$$h_1 = \sqrt{\frac{u_1^2 - c^2 u_2^2}{u_1^2 - c^2}}, \qquad h_2 = \sqrt{\frac{u_1^2 - c^2 u_2^2}{1 - u_2^2}}, \qquad h_3 = 1.$$

(D) **Ellipsoidal Coordinates**

$$x_1 = \sqrt{\frac{\left(u_1^2 - a^2\right)\left(u_2^2 - a^2\right)\left(u_3^2 - a^2\right)}{a^2(a^2 - b^2)}}$$

$$x_2 = \sqrt{\frac{\left(u_1^2 - b^2\right)\left(u_2^2 - b^2\right)\left(u_3^2 - b^2\right)}{b^2(b^2 - a^2)}}, \qquad x_3 = \frac{u_1 u_2 u_3}{ab}$$

$$h_1 = \sqrt{\frac{\left(u_1^2 - u_2^2\right)\left(u_1^2 - u_3^2\right)}{\left(u_1^2 - a^2\right)\left(u_1^2 - b^2\right)}}$$

$$h_2 = \sqrt{\frac{\left(u_2^2 - u_1^2\right)\left(u_2^2 - u_3^2\right)}{\left(u_2^2 - a^2\right)\left(u_2^2 - b^2\right)}}, \qquad h_3 = \sqrt{\frac{\left(u_3^2 - u_1^2\right)\left(u_3^2 - u_2^2\right)}{\left(u_3^2 - a^2\right)\left(u_3^2 - b^2\right)}}.$$

(E) **Parabolic Cylinder Coordinates** (here $J = u_1^2 + u_2^2$)

$$x_1 = \tfrac{1}{2}\left(u_1^2 - u_2^2\right), \qquad x_2 = u_1 u_2, \qquad x_3 = u_3$$

$$h_1 = h_2 = \sqrt{u_1^2 + u_2^2}, \qquad h_3 = 1.$$

(F) **Paraboloidal Coordinates**

$$x_1 = \sqrt{\frac{\left(u_1^2 - a^2\right)\left(u_2^2 - a^2\right)\left(u_3^2 - a^2\right)}{a^2 - b^2}}$$

$$x_2 = \sqrt{\frac{\left(u_1^2 - b^2\right)\left(u_2^2 - b^2\right)\left(u_3^2 - b^2\right)}{b^2 - a^2}}, \qquad x_3 = \frac{1}{2}\left(u_1^2 + u_2^2 + u_3^2 - a^2 - b^2\right)$$

$$h_1 = \sqrt{\frac{\left(u_1^2 - u_2^2\right)\left(u_1^2 - u_3^2\right)}{\left(u_1^2 - a^2\right)\left(u_1^2 - b^2\right)}}$$

$$h_2 = u_2\sqrt{\frac{\left(u_3^2 - u_1^2\right)\left(u_3^2 - u_2^2\right)}{\left(u_2^2 - a^2\right)\left(u_2^2 - b^2\right)}}, \qquad h_3 = u_3\sqrt{\frac{\left(u_3^2 - u_1^2\right)\left(u_3^2 - u_2^2\right)}{\left(u_3^2 - a^2\right)\left(u_3^2 - b^2\right)}}.$$

(G) **Rotational Parabolic Coordinates** (here $J = u_1 u_2 \sqrt{u_1^2 + u_2^2}$)

$$x_1 = u_1 u_2 u_3, \qquad x_2 = u_1 u_2 \sqrt{1 - u_3^3}, \qquad x_3 = \tfrac{1}{2}(u_1^2 - u_2^2)$$

$$h_1 = h_2 = \sqrt{u_1^2 + u_2^2}, \qquad h_3 = \frac{u_1 u_2}{\sqrt{1 - u_3^2}}.$$

(H) **Rotational Prolate Spheroidal Coordinates**

$$x_1 = \sqrt{(u_1^2 - a^2)(1 - u_2^2)}, \quad x_2 = \sqrt{(u_1^2 - a^2)(1 - u_2^2)(1 - u_3^2)}, \quad x_3 = u_1 u_2$$

$$h_1 = \sqrt{\frac{u_1^2 - a^2 u_2^2}{u_1^2 - a^2}}, \quad h_2 = \sqrt{\frac{u_1^2 - a^2 u_2^2}{1 - u_2^2}}, \quad h_3 = \sqrt{\frac{\left(u_1^2 - a^2\right)\left(1 - u_2^2\right)}{1 - u_3^2}}.$$

(I) **Rotational Oblate Spheroidal Coordinates**

$x_1 = u_3\sqrt{(u_1^2+a^2)(1-u_2^2)}$, $x_2 = \sqrt{(u_1^2+a^2)(1-u_2^2)(1-u_3^2)}$, $x_3 = u_1 u_2$

$$h_1 = \sqrt{\frac{u_1^2+a^2u_2^2}{u_1^2+a^2}}, \quad h_2 = \sqrt{\frac{u_1^2+a^2u_2^2}{1-u_2^2}}, \quad h_3 = \sqrt{\frac{\left(u_1^2+a^2\right)\left(1-u_2^2\right)}{1-u_3^2}}.$$

Notes

[1] For spherical coordinates in n dimensions, we have the mapping $\{x_i \mid i = 1,\ldots,n\} \to \{r, \theta_i \mid i = 1,\ldots,n-1\}$ via

$$\begin{aligned}
x_1 &= r\sin\theta_1 \sin\theta_2 \cdots \sin\theta_{n-2}\sin\theta_{n-1} \\
x_2 &= r\sin\theta_1 \sin\theta_2 \cdots \sin\theta_{n-2}\cos\theta_{n-1} \\
x_3 &= r\sin\theta_1 \sin\theta_2 \cdots \cos\theta_{n-2} \\
&\vdots \\
x_{n-1} &= r\sin\theta_1\cos\theta_2 \\
x_n &= r\cos\theta_1 .
\end{aligned}$$

Hence, the Jacobian is given by $J(r,\boldsymbol{\theta}) = (-1)^{n(n-1)/2} r^{n-1} \sin^{n-2}\theta_1 \sin^{n-3}\theta_2 \cdots \sin^2\theta_{n-1}\sin\theta_{n-2}$.

[2] The following two changes of variables are frequently useful:

$$\begin{aligned}
\int_\alpha^\beta f(x)\,dx &= (\beta-\alpha)\int_0^1 f([\beta-\alpha]t+\alpha)\,dt \\
&= (\beta-\alpha)\int_0^\infty f\left(\frac{\alpha+\beta t}{1+t}\right)\frac{dt}{(1+t)^2}.
\end{aligned}$$

[3] The easiest derivation of the value of $I = \int_{-\infty}^{\infty} e^{-x^2}\,dx$ is obtained from a clever change of variables. Multiplying I by itself results in

$$\begin{aligned}
I^2 &= \left(\int_{-\infty}^{\infty} e^{-x^2}\,dx\right)\left(\int_{-\infty}^{\infty} e^{-y^2}\,dy\right) \\
&= \int_{-\infty}^{\infty}\int_{-\infty}^{\infty} e^{-(x^2+y^2)}\,dx\,dy \\
&= \int_0^\infty \int_0^{2\pi} e^{-r^2}\,r\,d\theta\,dr \\
&= \left(\int_0^\infty e^{-r^2}\,r\,dr\right)\left(\int_0^{2\pi} d\theta\right) \\
&= \left(-\tfrac{1}{2}e^{-r^2}\Big|_{r=0}^{r=\infty}\right)\left(\theta\Big|_{\theta=0}^{\theta=2\pi}\right) = \pi .
\end{aligned}$$

where we have changed from rectangular coordinates to polar coordinates via $\{r^2 = x^2+y^2,\ \tan\theta = x/y\}$, which introduced the Jacobian $J = r$. From this we conclude that $I = \sqrt{\pi}$.

[4] As Flanders [1] shows, there is another application of the transformation in the last note to the Fresnel integrals. Define

$$F(t) = \int_0^\infty e^{-tx^2} \cos x^2 \, dx, \qquad G(t) = \int_0^\infty e^{-tx^2} \sin x^2 \, dx$$

so that the Fresnel integrals are given by $F_0 = F(0) = \int_0^\infty \cos x^2 \, dx$ and $G_0 = G(0) = \int_0^\infty \sin x^2 \, dx$. We now compute

$$\begin{aligned} F^2(t) - G^2(t) &= \left(\int_0^\infty e^{-tx^2} \cos x^2 \, dx\right)\left(\int_0^\infty e^{-ty^2} \cos y^2 \, dy\right) \\ &\qquad - \left(\int_0^\infty e^{-tx^2} \sin x^2 \, dx\right)\left(\int_0^\infty e^{-ty^2} \sin y^2 \, dy\right) \\ &= \int\!\!\int_{x\geq 0, y\geq 0} e^{-t(x^2+y^2)} \left[\cos x^2 \cos y^2 - \sin x^2 \sin y^2\right] dx\, dy \\ &= \int_0^\infty \int_0^{\pi/2} e^{-tr^2} (\cos r^2) r \, dr\, d\theta \\ &= \frac{\pi}{4} \int_0^\infty r^{-tz} \cos z \, dz = \frac{\pi}{4} \frac{t}{1+t^2}, \end{aligned} \tag{28.8}$$

where the substitutions $z = r^2 = x^2 + y^2$ were used. Similarly, it is possible to derive that

$$2F(t)G(t) = \int_0^\infty \int_0^\infty e^{-t(x^2+y^2)} \sin(x^2+y^2) \, dx\, dy = \frac{\pi}{4} \frac{1}{1+t^2}. \tag{28.9}$$

Taking the limit $t \to 0$ in both (28.8) and (28.9) results in the simultaneous equations $\left\{F_0^2 - G_0^2 = 0, \quad 2F_0G_0 = \pi/4\right\}$. It is easy to show that $G_0 > 0$, so we conclude that $F_0 = G_0 = \sqrt{\pi/8}$.

[5] A precise statement of when a change of variables can be performed is contained in the following theorem:

> **Theorem**: Let V be an open set of $\mathbf{R}^p$ and u a diffeomorphism of V onto $u(V)$ (i.e., u is a bijection of V onto $u(V)$ such that u and u^{-1} are continuously differentiable). Then the formula
>
> $$\int_{u(V)} f(\mathbf{x}) \, d\mathbf{x} = \int_V f(u(\mathbf{x})) \, |J(\mathbf{x})| \, d\mathbf{x}$$
>
> is valid in each of the following two cases (see George [2], page 22):
>
> (A) f is a measurable positive arithmetic function on $u(V)$;
>
> (B) f is an integrable function over V.

References

[1] H. Flanders, "On the Fresnel Integrals," *Amer. Math. Monthly*, **89**, April 1982, pages 264–266.

[2] C. George, *Exercises in Integration*, Springer–Verlag, New York, 1984.

[3] R. G. Helsel and T. Radó, "The Transformation of Double Integrals," *Trans. Amer. Math. Soc.*, **54**, 1943, pages 83–102.

[4] P. Moon and D. E. Spencer, *Field Theory Handbook*, Springer–Verlag, New York, 1961.

29. Computer Aided Solution

Applicable to Indefinite or definite integrals.

Yields

An exact evaluation of an indefinite or definite integral.

Idea

Some commercial computer algebra packages include a symbolic integrator.

Procedure

Find an computer system that runs any of the following computer languages: Axiom [7], Derive [2], Macsyma [4], MAPLE [1], Mathematica [8], or REDUCE [5]. Then learn how to use the language, and find the routine that symbolically evaluates integrals.

To compare some of these different languages, a test suite of indefinite and definite integrals was created, and the integration routine from each language was run on the test suite. The indefinite integrals tested were

I1: $\int \sin x \, dx$ I2: $\int \sqrt{\tan x} \, dx$

I3: $\displaystyle\int \frac{x}{x^3-1} \, dx$ I4: $\displaystyle\int \frac{x}{\sin^2 x} \, dx$

I5: $\displaystyle\int \frac{\log x}{\sqrt{x+1}} \, dx$ I6: $\displaystyle\int \frac{x}{\sqrt{1+x}+\sqrt{1-x}} \, dx$

I7: $\int e^{-ax^2} \, dx$ I8: $\displaystyle\int \frac{x}{\log^3 x} \, dx$

I9: $\displaystyle\int \frac{\sin x}{x^2} \, dx$ I10: $\displaystyle\int \frac{dx}{2+\cos x}$.

Note that integrals I7, I8, and I9 are not elementary integrals; their evaluation depends on the error function, the exponential integral, and the cosine integral. Integral I10 is discussed on page 192.

The definite integrals tested were

D1: $\displaystyle\int_0^{4\pi} \frac{dx}{2+\cos x}$ D2: $\displaystyle\int_{-\infty}^{\infty} \frac{\sin x}{x} \, dx$

D3: $\displaystyle\int_0^\infty \frac{e^{-x}}{\sqrt{x}}\,dx$ D4: $\displaystyle\int_0^\infty \frac{x^2 e^{-x}}{1-e^{-2x}}\,dx$

D5: $\int_0^\infty e^{-x^2}\log^2 x\,dx$ D6: $\int_1^\infty e^{-x}x^3\log^2 x\,dx$

D7: $\displaystyle\int_0^\infty \frac{x^2}{1+x^3}\,dx$ D8: $\displaystyle\int_{-1}^1 \frac{dx}{x^2}$

D9: $\int_1^\infty e^{-x}x^{1/3}\,dx$.

Observe that none of these definite integrals, except integrals D7 and D8, have elementary indefinite integrals. Additionally, note also that integral D7 is divergent. Note that integral D8, as written, is improper and has no well defined value. If integral D8 were written as a principle value integral, $J = \fint_{-1}^1 x^{-2}\,dx$, then $J = 2$. A routine that evaluates definite integrals by first determining the indefinite integral would return this value.

Example 1

The following is the result of running the test suite on a 386 computer running Derive. In Derive the integration function has the name `INT`. The Derive results were obtained by creating an input file, running that file, and then storing the output. Using this procedure, each input line is followed by the corresponding output line. The author changed some of the output appearing below to make it fit on the page better.

Derive is intended to be used interactively in a graphical mode, and the graphical display of the results below looks much better than the corresponding printed output. For example, in the graphics mode, the square-root function displays the square-root sign, not the function name `SQRT`.

```
i1=INT(SIN(x),x)
i1=-COS(x)

i2=INT(SQRT(TAN(x)),x)
i2=SQRT(2)*ATAN(SQRT(2)*SQRT(TAN(x))+1)/2+SQRT(2)*ATAN(SQRT(2)*
SQRT(TAN(x))-1)/2-SQRT(2)*LN((COS(x)*(SQRT(2)*SQRT(TAN(x))+1)+SIN(x))/
COS(x))/4+SQRT(2)*LN((SIN(x)-COS(x)*(SQRT(2)*SQRT(TAN(x))-1))/COS(x))/4

i3=INT(x/(x^3-1),x)
i3=SQRT(3)*ATAN(SQRT(3)*(2*x+1)/3)/3-LN(x^2+x+1)/6+LN(x-1)/3

i4=INT(x/SIN(x)^2,x)
i4=LN(SIN(x))-x*COT(x)

i5=INT(LOG(x)/SQRT(x+1),x)
i5=-4*LN((SQRT(x+1)-1)/SQRT(x))+2*SQRT(x+1)*LN(x)-4*SQRT(x+1)

i6=INT(x/(SQRT(1+x)-SQRT(1-x)),x)
i6=((x+1)^(3/2)-(1-x)^(3/2))/3

i7=INT(#e^(-a*x^2),x)
i7=SQRT(pi)*ERF(SQRT(a)*x)/(2*SQRT(a))

i8=INT(x/LOG(x)^3,x)
i8=-x^2/LN(x)-x^2/(2*LN(x)^2)+2*INT(x/LN(x),x)
```

```
i9=INT(SIN(x)/x^2,x)
i9=INT(COS(x)/x,x)-SIN(x)/x

i10=INT(1/(2+COS(x)),x)
i10=-2*SQRT(3)*ATAN(SIN(x)/(COS(x)+1))/3+2*SQRT(3)*ATAN(SQRT(3)*SIN(x)/
(3*(COS(x)+1)))/3+SQRT(3)*x/3

d1=INT(1/(2+COS(x)),x,0,4*pi)
d1=4*SQRT(3)*pi/3

d2=INT(SIN(x)/x,x,-inf,inf)
d2=INT(SIN(x)/x,x,-inf,inf)

d3=INT(#e^(-x)/SQRT(x),x,0,inf)
d3=INT(#e^(-x)/SQRT(x),x,0,inf)

d4=INT(x^2*#e^(-x)/(1-#e^(-2*x)),x,0,inf)
d4=INT(x^2*#e^x/(#e^(2*x)-1),x,0,inf)

d5=INT(#e^(-x^2)*LOG(x)^2,x,0,inf)
d5=INT(#e^(-x^2)*LN(x)^2,x,0,inf)

d6=INT(x^3*#e^(-x^2)*LOG(x)^2,x,1,inf)
d6=INT(x^3*#e^(-x^2)*LN(x)^2,x,1,inf)

d7=INT(x^2/(1+x^3),x,0,inf)
d7=inf

d8=INT(1/x^2,x,-1,1)
d8=-2

d9=INT(#e^(-x)*x^(1/3),x,1,inf)
d9=INT(#e^(-x)*x^(1/3),x,1,inf)
```

In the above the following special notation has been used: `inf` for ∞ and `#e` is used to specify the base of natural logarithms $e \approx 2.718$.

From the above we observe that Derive could not evaluate the following integrals from the test suite: D2, D3, D4, D5, D6, and D9. Derive did not recognize the cosine integral in I9, but left the answer in terms of this special function. Finally, Derive made an error on integral D8 by not identifying the singularity at $x = 0$.

Example 2

The following is the result of running the test suite on a SPARCstation running Macsyma. In Macsyma the integration function has the name `integrate`. Note that `(C1)`, `(C2)`, ... are input lines ("command" lines) and that `(D1)`, `(D2)`, ... are output lines ("display" lines). The symbol `%` refers to the last expression. The author changed some of the output appearing below to make it fit on the page better.

```
(C1) I1: integrate( sin(x), x);
(D1)                                - COS(X)
```

```
(C2) I2: integrate( sqrt(tan(x)), x);
        LOG(TAN(X) + SQRT(2) SQRT(TAN(X)) + 1)
(D2) 2 (- --------------------------------------
                     4 SQRT(2)

   LOG(TAN(X) - SQRT(2) SQRT(TAN(X)) + 1)
 + --------------------------------------
                4 SQRT(2)

       2 SQRT(TAN(X)) + SQRT(2)          2 SQRT(TAN(X)) - SQRT(2)
   ATAN(------------------------)    ATAN(------------------------)
               SQRT(2)                           SQRT(2)
 + ------------------------------ + ------------------------------)
            2 SQRT(2)                         2 SQRT(2)

(C3) I3: integrate( x/(x^3-1),x);
                                          2 X + 1
                 2                   ATAN(-------)
            LOG(X  + X + 1)               SQRT(3)      LOG(X - 1)
(D3)      - --------------- + ------------- + ----------
                   6                SQRT(3)           3

(C4) I4: integrate( x/sin(x)^2, x);
                                             X
(D4)                  LOG(SIN(X)) - ------
                                          TAN(X)

(C6) I5: integrate( log(x)/sqrt(x+1), x);
                                          LOG(SQRT(X + 1) + 1)
(D6) 2 (SQRT(X + 1) LOG(X) - 2 (- --------------------
                                                   2

                                        LOG(SQRT(X + 1) - 1)
                                      + -------------------- + SQRT(X + 1)))
                                                 2

(C7) I7: integrate( exp(-a*x^2), x);
                          SQRT(%PI) ERF(SQRT(A) X)
(D7)                      ------------------------
                                 2 SQRT(A)

(C8) I8: integrate( x/(log(x))^3, x);
                                    /
                                    [    X
(D8)                                I ------- dX
                                    ]    3
                                    / LOG (X)

(C9) I6: integrate( x/(sqrt(1+x)+sqrt(1-x)),x);
                         /
                         [              X
(D9)                     I ------------------------- dX
                         ] SQRT(X + 1) + SQRT(1 - X)
                         /

(C10) block([algebraic:true],ratsimp(first(%)));
                          SQRT(1 - X) - SQRT(X + 1)
(D10)                   - -------------------------
                                      2
```

```
(C11) integrate(%,x);
                                      3/2            3/2
                           2 (X + 1)        2 (1 - X)
                         - ------------ - ------------
                                3              3
(D11)                  - ----------------------------
                                        2

(C12) I10: integrate( 1/(2+cos(x)),x);
                                          SIN(X)
                         2 ATAN(--------------------)
                                SQRT(3) (COS(X) + 1)
(D12)                    ----------------------------
                                   SQRT(3)

(C13) I9: integrate( sin(x)/x^2, x);
                                /
                                [ SIN(X)
(D13)                           I ------ dX
                                ]   2
                                /  X

(C14) D1: integrate( 1/(2+cos(x)),x,0,4*%pi);
                                4 SQRT(3) %PI
(D14)                           -------------
                                      3

(C15) D2: integrate( sin(x)/x,x,-inf,inf);
(D15)                                  %PI

(C16) D7: integrate( x^2/(1+x^3),x,0,inf);
Error: Integral is divergent

(C17) D3: integrate( exp(-x)/sqrt(x),x,0,inf);
(D17)                             SQRT(%PI)

(C18) D5: integrate( exp(-x^2)*log(x)^2,x,0,inf);
                     2                                          5/2
      SQRT(%PI) LOG (2)   SQRT(%PI) %GAMMA LOG(2)   %PI
(D18) ----------------- + ----------------------- + ------
              2                      2               16

                                                             2
                                                       %GAMMA  SQRT(%PI)
                                                     + -----------------
                                                               8

(C19) D6: integrate( exp(-x)*log(x)^2*x^3,x,1,inf);
                              INF
                              /
                              [     3   - X     2
(D19)                         I    X  %E     LOG (X) dX
                              ]
                              /
                               1

(C20) D9: integrate( exp(-x)*x^(1/3),x,1,inf);
                                          4
(D20)                               GAMMA(-, 1)
                                          3
```

```
(C21) D4: integrate( exp(-x)*x^2/(1-exp(-2*x)),x,0,inf);
                       2        X
                      X  LOG(%E  + 1)          X            X
(D21) (limit      - --------------- - LI (%E ) + X LI (%E )
       X -> INF           2                3            2

                                                 2            X
                    X                X         X  LOG(1 - %E )    7 ZETA(3)
         + LI (- %E ) - X LI (- %E ) + ---------------) + ---------
              3             2                    2                4

(C22) D8: integrate( 1/x^2,x,-1,1);
Error: Integral is divergent
```

In the above the following special notation has been used: ERF for the error function, %GAMMA for Euler's constant, GAMMA() for the incomplete gamma function, LI for the logarithmic integral of different orders, %E for the base of natural logarithms e, and %PI for π.

From the above we observe that Macsyma could not evaluate the following integrals from the test suite: I8, I9, and D6. The result for I10 is correct, but only for a limited range. Macsyma could evaluate integral I6 only after the integrand was simplified. Macsyma left the result for integral D4 in the form of a limit. Finally, Macsyma successfully identified the singularity at $x = 0$ in integral D8.

Example 3

In MAPLE the integration function has the name int. The MAPLE results were obtained by creating an input file, running that file, and then storing the output. In this case all of the input is followed by all of the output. The author changed some of the output appearing below to make it fit on the page better. MAPLE is intended to be be used interactively, and the graphical display of the results below looks much better than the corresponding printed output.

```
<< INPUT >>
   I1 := int( sin(x), x)
   I2 := int( sqrt(tan(x)), x)
   I3 := int( x/(x^3-1), x)
   I4 := int( x/sin(x)^2, x)
   I5 := int( log(x)/sqrt(x+1), x)
   I6 := int( x/(sqrt(1+x)+sqrt(1-x)), x)
   I7 := int( exp(-a*x^2), x)
   I8 := int( x/(log(x))^3, x)
   I9 := int( sin(x)/x^2, x)
   I10:= int( 1/(2+cos(x)), x)
   D1 := int( 1/(2+cos(x)), x=0..4*Pi)
   D2 := int( sin(x)/x, x=-infinity..infinity)
   D3 := int( exp(-x)/sqrt(x), x=0..infinity)
   D4 := int( exp(-x)*x^2/(1-exp(-2*x)), x=0..infinity)
   D5 := int( exp(-x^2)*log(x)^2, x=0..infinity)
   D6 := int( exp(-x)*log(x)^2*x^3, x=1..infinity)
   D7 := int( x^2/(1+x^3), x=0..infinity)
   D8 := int( 1/x^2, x=-1..1)
   D8B:= int( 1/x^2, x=-1..1, continuous)
   D9 := int( exp(-x)*x^(1/3), x=1..infinity)
```

<<OUTPUT>>

$$I1 := -\cos(x)$$

$$I2 := 1/2 \left(\arctan\left(\frac{2^{1/2} \tan(x)^{1/2}}{1 - \tan(x)}\right) - \ln\left(\frac{\tan(x) + 2^{1/2} \tan(x)^{1/2} + 1}{(1 + \tan(x)^2)^{1/2}}\right) \right) 2^{1/2}$$

$$I3 := 1/3 \ln(x-1) - 1/6 \ln(x^2 + x + 1) + 1/3\ 3^{1/2} \arctan(1/3\ (2\ x + 1)\ 3^{1/2})$$

$$I4 := -x \cot(x) + \ln(\sin(x))$$

$$I5 := 2\ (x+1)^{1/2} \ln(x) - 4\ (x+1)^{1/2} - 2 \ln((x+1)^{1/2} - 1) + 2 \ln((x+1)^{1/2} + 1)$$

$$I6 := 1/3\ (x + 1)^{3/2} - 1/3\ (x - 1)^{3/2}\ I$$

$$I7 := 1/2 \frac{Pi^{1/2}\ \mathrm{erf}(a^{1/2}\ x)}{a^{1/2}}$$

$$I8 := -1/2 \frac{x^2}{\ln(x)^2} - \frac{x^2}{\ln(x)} + 2\ \mathrm{Ei}(2 \ln(x))$$

$$I9 := -\frac{\sin(x)}{x} + \mathrm{Ci}(x)$$

$$I10 := 2/3\ 3^{1/2} \arctan(1/3 \tan(1/2\ x)\ 3^{1/2})$$

$$D1 := \int_0^{4\ Pi} \frac{1}{2 + \cos(x)}\, dx$$

$$D2 := \int_{-\mathrm{infinity}}^{\mathrm{infinity}} \frac{\sin(x)}{x}\, dx$$

$$D3 := Pi^{1/2}$$

$$D4 := 7/4\ \mathrm{Zeta}(3)$$

```
                 5/2          1/2     2          1/2                1/2      2
D5 := 1/16 Pi     + 1/8 Pi     gamma  + 1/2 Pi    gamma ln(2) + 1/2 Pi    ln(2)

                         D6 := 2 MeijerG(4, 4, 1)

                             D7 := infinity

                                     1
                                     /
                                    |    1
                            D8 :=   |   ---- dx
                                    |    2
                                   /    x
                                  -1

                               D8B := -2

                          D9 := GAMMA(4/3, 1)
```

In the above the following special notation has been used: `Ci` for the cosine integral, `Ei` for the exponential integral, `erf` for the error function, `gamma` for Euler's constant, `GAMMA()` for the incomplete gamma function, `infinity` for ∞, `MeijerG` for Meijer's G function, `Pi` for π, and `Zeta` for the zeta function $\zeta(x)$.

From the above we observe that MAPLE could not evaluate the following integrals from the test suite: D1 and D2. The result for I10 is correct, but only for a limited range. For integral D8, MAPLE recognized that the integrand was discontinuous and so did not evaluate it. When told that the integrand was continuous (which is incorrect), MAPLE then obtained the answer that would be obtained by just following symbolic rules.

Example 4

In Mathematica the integration function has the name `Integrate`. The following Mathematica results were obtained by creating an input file, running that file, and then storing the output. This output was then manipulated so that each input line is followed by the corresponding output line. Mathematica for windows, the version run for the test suite, is intended to be used interactively in a graphical mode, and the graphical display of the results below looks much better than the corresponding printed output.

```
I1: Integrate[ Sin[x], x]
I1:-Cos[x]

I2: Integrate[ Sqrt[Tan[x]], x]
I2:Integrate[Sqrt[Tan[x]], x]

I3: Integrate[ x/(x^3-1),x]
          1 + 2 x
   ArcTan[-------]                                2
          Sqrt[3]     Log[-1 + x]   Log[1 + x + x ]
I3:--------------- + ----------- - ---------------
       Sqrt[3]             3              6
```

```
I4: Integrate[ x/Sin[x]^2, x]
I4:-(x Cot[x]) + Log[Sin[x]]

I5: Integrate[ Log[x]/Sqrt[x+1], x]
I5:-4 Sqrt[1 + x] + 4 ArcTanh[Sqrt[1 + x]] +

  2 Sqrt[1 + x] Log[x]

I6: Integrate[ x/(Sqrt[1+x]+Sqrt[1-x]),x]
                1   x    1   x
I6:Sqrt[1 - x] (- - -) + (- + -) Sqrt[1 + x]

I7: Integrate[ Exp[-a*x^2], x]
                3   3    3   3
   Sqrt[Pi] Erf[Sqrt[a] x]
I7:-----------------------
         2 Sqrt[a]

I8: Integrate[ x/(Log[x])^3, x]
                                   2          2
                                  x          x
I8:2 ExpIntegralEi[2 Log[x]] - --------- - ------
                                      2    Log[x]
                               2 Log[x]

I9: Integrate[ Sin[x]/x^2, x]
                        Sin[x]
I9:CosIntegral[x] - ------
                          x

I10: Integrate[ 1/(2+Cos[x]),x]
                  x
              Tan[-]
                  2
    2 ArcTan[-------]
              Sqrt[3]
I10:-----------------
          Sqrt[3]

D1: Integrate[ 1/(2+Cos[x]), {x,0,4 Pi}]
    4 Pi
D1:-------
   Sqrt[3]

D2: Integrate[ Sin[x]/x, {x,-Infinity,Infinity}]
D2:Pi

D3: Integrate[ Exp[-x]/Sqrt[x], {x,0,Infinity}]
D3:Sqrt[Pi]

D4: Integrate[ Exp[-x]*x^2/(1-Exp[-2*x]), {x,0,Infinity}]
   7 Zeta[3]
D4:---------
       4

D5: Integrate[ Exp[-x^2]*Log[x]^2, {x,0,Infinity}]
                                   2    2
D5:(Sqrt[Pi] (2 EulerGamma  + Pi  + 8 EulerGamma Log[2] +

              2
      8 Log[2] )) / 16
```

```
D6: Integrate[ Exp[-x]*Log[x]^2*x^3, {x,1,Infinity}]
D6:Indeterminate

D7: Integrate[ x^2/(1+x^3), {x,0,Infinity}]
D7:Infinity

D8: Integrate[ 1/x^2, {x,-1,1}]
D8:-2

D9: Integrate[ Exp[-x]*x^(1/3), {x,1,Infinity}]
           4           4
D9:Gamma[-] - Gamma[-, 0, 1]
           3           3
```

In the above the following special notation has been used: `CosIntegral` for the cosine integral, `Erf` for the error function erf, `ExpIntegralEi` for the exponential integral $E_1(x)$, `EulerGamma` for Euler's constant γ, `Gamma()` for the gamma function $\Gamma(a) = \int_0^\infty t^{a-1}e^{-t}\,dt$ and for the generalized incomplete gamma function $\Gamma(a,b,c) = \int_b^c t^{a-1}e^{-t}\,dt$, `Pi` for π, and `Zeta` for the zeta function $\zeta(x)$.

From the above we observe that Mathematica could not evaluate the following integrals from the test suite: I2 and D6. Additionally, the result for I10 is correct, but only for a limited range. Perhaps because of this error, the result for integral D1 is incorrect. Finally, Mathematica made an error on integral D8 by not identifying the singularity at $x = 0$.

Example 6

Since REDUCE cannot perform definite integration, that part of the test suite was not run in REDUCE. The following is the result of running the indefinite integrals in the test suite on a SPARCstation 2. In REDUCE the integration function has the name `int`. In this listing, each input line is followed by the corresponding output line.

```
i1 := int( sin(x), x);

I1 := - COS(X)

i2 := int( sqrt(tan(x)), x);

I2 := INT(SQRT(TAN(X)),X)

i3 := int( x/(x^3-1),x);

I3 :=

 1                    2*X + 1        1         2                 1
---*SQRT(3)*ATAN(---------) - ---*LOG(X  + X + 1) + ---*LOG(X - 1)
 3                    SQRT(3)        6                           3

i4 := int( x/sin(x)^2, x);

              -1                        X  2                  X
I4 :=  - SIN(X)   *COS(X)*X - LOG(TAN(---)   + 1) + LOG(TAN(---))
                                        2                     2
```

```
i5 := int( log(x)/sqrt(x+1), x);

I5 := 2*(SQRT(X + 1)*LOG(X) - 2*SQRT(X + 1) - LOG(SQRT(X + 1) - 1)

          + LOG(SQRT(X + 1) + 1))

i6 := int( exp(-a*x^2), x);

              1
I6 := INT(-------,X)
                2
             A*X
            E

i7 := int( x/(log(x))^3, x);

               X                   -1  2     1          -2  2
I7 := 2*INT(--------,X) - LOG(X)   *X   - ---*LOG(X)   *X
             LOG(X)                        2

i8 := int( x/(sqrt(1+x)+sqrt(1-x)),x);

          1                       1
I8 :=  - ---*SQRT( - X + 1)*X + ---*SQRT( - X + 1)
          3                       3

          1                    1
       + ---*SQRT(X + 1)*X + ---*SQRT(X + 1)
          3                    3

i9 := int( 1/(2+cos(x)),x);

                                    X
                               TAN(---)
       2                            2
I9 := ---*SQRT(3)*ATAN(----------)
       3                  SQRT(3)

i10:= int( sin(x)/x^2, x);

              SIN(X)
I10 := INT(--------,X)
                 2
                X
```

In the above the following special notation has been used: E for the base of natural logarithms e. From the above we observe that REDUCE could not evaluate the following indefinite integrals from the test suite: I2, I7, and I9.

Notes

[1] The version numbers of the computer languages used above were Derive 2.08, MACSYMA 417.100, MAPLE 5.0, Mathematica 2.1, REDUCE 3.4.

[2] All of the above computer packages can perform operations on integrals that the system cannot simplify. That is, if a system cannot simplify the integrals $I = \int^x f(x)\,dx$ or $J = \int_a^b g(x)\,dx$, it will still be able to differentiate I and numerically evaluate J.

[3] In DERIVE, presently, the only definite integrals that can be computed are those for which an indefinite integral can be computed.

[4] Sometimes a symbolic algebra system can determine a specific integral, yet it does not do so without coaching because the result is a mess. For example, the integral $\int dx/(x^3+x+1)$ can clearly be integrated (the roots of a cubic can be found analytically, then partial fraction decomposition can be used). Yet Macsyma will not, unaided, carry out these steps. See Golden [3].

[5] Packages that can handle a wider variety of integrals are constantly being created. The theory underlying the algorithms is described in the section on Liouville theory (see page 77).

[6] One must be wary when using computer algebra packages, as errors can sometimes arise. For example, the integral $I = \int x^k\,dx$ is problematic since there are two different forms for the answer depending on whether or not k is equal to -1. Some systems will return $x^{k+1}/(k+1)$ without any further information; others will ask the user whether or not k is equal to -1; still others will look to explicitly declared domains to determine if k could be equal to -1.

If Derive cannot determine if k could be -1, then it returns the expression $\left(x^{k+1}-1\right)/(k+1)$ for the evaluation of I (see Stoutemyer [6]). This expression tends to $\log x$ as $k \to -1$. Hence, this expression gives the correct result for any specific value of k, provided that a limit is used instead of simple substitution.

[7] Most of the commercial computer algebra packages also include a numeric integrator.

[8] Thanks are extended to Jeffrey Golden for running the test suite in Macsyma, Tony Hearn for running the test suite in REDUCE, and Richard Pavelle for running the test suite in MAPLE.

References

[1] B. W. Char, K. O. Geddes, G. H. Gonnet, B. L. Leong, M. B. Monagan, and S. M. Watt, MAPLE V Library Reference Manual, Springer–Verlag, New York, 1991.

[2] DERIVE, Soft Warehouse, Inc., 3660 Waialae Avenue, Suite 304, Honolulu, HI 96816.

[3] J. P. Golden, "Messy Indefinite Integrals," *MACSYMA Newsletter*, **7**, No. 3, Symbolics, Inc., Burlington, MA, July 1990, pages 9–17.

[4] *VAX UNIX MACSYMA Reference Manual*, Symbolics Inc., Cambridge, MA, 1985.

[5] G. Rayna, *REDUCE: Software for Algebraic Computation*, Springer–Verlag, New York, 1987.

[6] David R. Stoutemyer, "Crimes and Misdemeanors in the Computer Algebra Trade," *Notices of the AMS*, **38**, No. 7, September 1991, pages 778–785.
[7] R. Sutor (ed.), *Axiom User Guide*, Numerical Algorithms Group, Inc., 1400 Opus Place, Suite 200, Downers Grove, IL, 1991.
[8] S. Wolfram, *Mathematica: A System of Doing Mathematics by Computer*, Second Edition, Addison–Wesley Publishing Co., Reading, MA, 1991.

30. Contour Integration

Applicable to Contour integrals, or integrals reducible to contour integrals.

Yields

An exact evaluation in terms of residues.

Idea

A definite integral can sometimes be written as a contour integral. Cauchy's residue theorem may then be used to evaluate the contour integral in terms of residues.

Procedure

Often, a definite integral can be manipulated into a contour integral. A contour integral is an integral for which the integration path is a closed contour in the complex plane. Contour integrals can often be evaluated by using Cauchy's residue theorem. One statement of the theorem is

> **Theorem:** Let $f(z)$ be analytic in a simply connected domain D except for finitely many points $\{\alpha_j\}$ at which f may have isolated singularities. Let C be a simple closed contour, traversed in the positive sense (see Notes, below), that lies in D and does not pass through any point α_j. Then
>
> $$\int_C f(z)\,dz = 2\pi i \sum \text{residue of } f \text{ at } \alpha_j \tag{30.1}$$
>
> where the sum is extended over all the points $\{\alpha_j\}$ that are inside the contour C (see Figure 30.1).

If $f(z)$, when expanded in a Laurent series about the point α, has the form

$$f(z) = \frac{a_{-m}}{(z-\alpha)^m} + \frac{a_{-m+1}}{(z-\alpha)^{m-1}} + \ldots + \frac{a_{-1}}{(z-\alpha)} + g(\alpha) \tag{30.2}$$

where $g(z)$ is analytic at α, then the residue of f at the point α is the term a_{-1}. A straightforward way to determine this term, when f has the expansion indicated in (30.2), is by

$$a_{-1} = \frac{1}{(n-1)!}\frac{d^{n-1}}{dz^{n-1}}\left[(z-\alpha)^n f(z)\right]\bigg|_{z=\alpha} \tag{30.3}$$

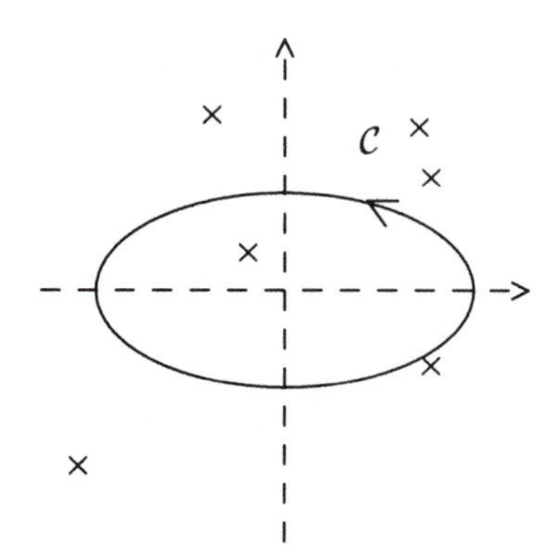

Figure 30.1 A contour C that contains a singularity. The singularities are indicated by crosses.

where $n \geq m$. For computational ease, though, it is best to take $n = m$ if the value of m is known.

If the contour C passes through a simple pole, then the contour can sometimes be deformed around the pole and then the following theorem can be used:

> **Theorem:** Let $f(z)$ have a simple pole at $z = \alpha$ with residue a_{-1} and let $A = A(\varepsilon, \phi)$ be an arc of the circle $|z - \alpha| = \varepsilon$ that subtends an angle ϕ at the center of the circle. Then $\lim_{\varepsilon \to 0+} \int_A f(z)\, dz = i\phi a_{-1}$.

If we had integrated over the entire circle (i.e., $\phi = 2\pi$) then the residue theorem states that the answer would be $2\pi i a_{-1}$, without any limiting process involved. This theorem indicates that integrating over a fraction of the circle yields a corresponding fraction of $2\pi i a_{-1}$, provided the pole is simple (i.e., $m = 1$ in (30.2)) and the limit $\varepsilon \to 0$ is taken.

There are many examples in this section, and each one attempts to demonstrate a different method or technique that can be used when evaluating contour integrals.

Example 1

Consider the integral

$$I = \int_0^{2\pi} \frac{d\theta}{2 + \cos\theta}. \tag{30.4}$$

If we make the change of variable $z = e^{i\theta}$ then $dz = ie^{i\theta} d\theta$ or $d\theta = -iz^{-1}\, dz$. Hence, using $\cos\theta = \frac{1}{2}\left(e^{i\theta} + e^{-i\theta}\right) = \frac{1}{2}\left(z + z^{-1}\right)$, we can write (30.4) as

$$\begin{aligned} I &= \int \frac{-iz^{-1}\, dz}{2 + \frac{1}{2}\left(z + z^{-1}\right)} \\ &= -i \int \frac{2}{z^2 + 4z + 1} dz. \end{aligned} \tag{30.5}$$

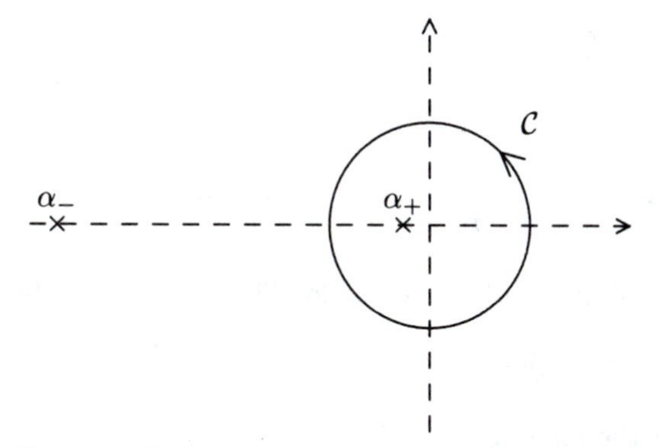

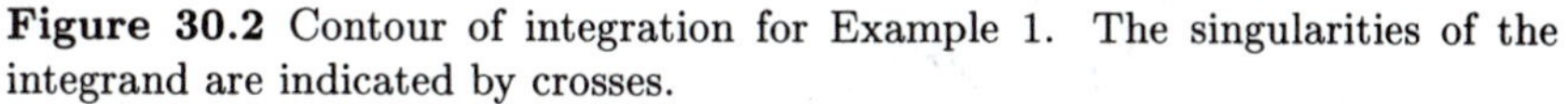
Figure 30.2 Contour of integration for Example 1. The singularities of the integrand are indicated by crosses.

Now we must determine what the new range of integration is. As θ varies from 0 to 2π, z traces out a circle of radius one in the clockwise sense (see Figure 30.2). Hence, we can use the Cauchy residue theorem. If we write I as

$$I = -i \int_{\mathcal{C}} \frac{2}{(z-\alpha_+)(z-\alpha_-)} dz, \tag{30.6}$$

where $\mathcal{C}$ represents the contour in Figure 30.2, and $\alpha_\pm = -2 \pm \sqrt{3}$, then we see that only α_+ is inside of the contour $\mathcal{C}$ (the $\alpha_\pm$ are indicated on Figure 30.2). The Laurent series of the integrand about the point α_+ is

$$\frac{2}{(z-\alpha_+)(z-\alpha_-)} = \left(\frac{2}{\alpha_+ - \alpha_-}\right) \frac{1}{z-\alpha_+} + h(z), \tag{30.7}$$

where $h(z) = \dfrac{2}{(\alpha_- - \alpha_+)(z-\alpha_-)}$ is analytic at $z = \alpha_+$. Hence, the residue of the integrand at α_+ is equal to $2/(\alpha_+ - \alpha_-)$. Using this in (30.1) results in

$$\begin{aligned} I &= -i \times 2\pi i \sum \text{residues of integrand} \\ &= 2\pi \left(\frac{2}{\alpha_+ - \alpha_-}\right) = 2\pi \frac{2}{\left(-2+\sqrt{3}\right) - \left(-2-\sqrt{3}\right)} = \frac{2\pi}{\sqrt{3}}. \end{aligned} \tag{30.8}$$

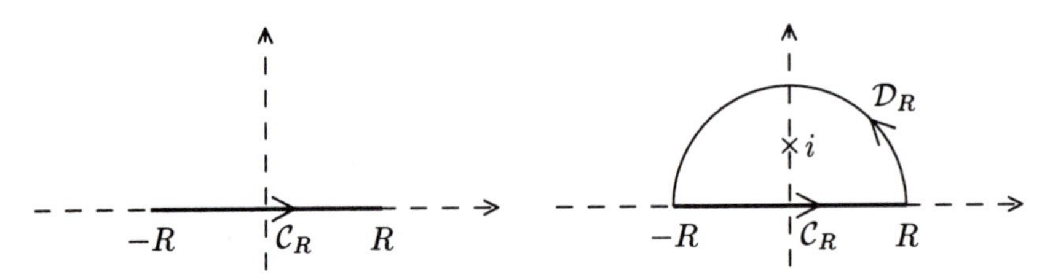

Figure 30.3 (a) Integration contour for (30.9); (b) integration contour for (30.10).

Example 2

Consider the integral

$$J = \int_{-\infty}^{\infty} \frac{e^{ix}}{x^2+1}\,dx = \lim_{R\to\infty} \int_{\mathcal{C}_R} \frac{e^{iz}}{z^2+1}\,dz, \tag{30.9}$$

where the contour $\mathcal{C}_R$ is shown in Figure 30.3.a. To have a closed contour, we define the new integral

$$K = \left(\int_{\mathcal{C}_R} + \int_{\mathcal{D}_R}\right) \frac{e^{iz}}{z^2+1}\,dz, \tag{30.10}$$

where the contour $\mathcal{D}_R$ is shown in Figure 30.3.b. The only isolated singularity inside of the combined contour $\mathcal{C}_R \cup \mathcal{D}_R$ is at $z = i$, and the residue at this pole is $e^{-1}/2i$. This residue is found by using (30.3) with $n = 1$:

$$a_{-1} = \frac{1}{0!} \lim_{z\to i} \left((z-i)\frac{e^{iz}}{z^2+1}\right) = \lim_{z\to i}\left(\frac{e^{iz}}{z+i}\right) = \frac{e^{-1}}{2i}. \tag{30.11}$$

That is, $\dfrac{e^{iz}}{z^2+1} = \dfrac{1}{z-i}\left(\dfrac{e^{-1}}{2i}\right) + h(z)$ where $h(z) = \dfrac{2e^{1+iz} + i(z+i)}{2e(z^2+1)}$ is analytic at $z = i$. Therefore, using the residue theorem, we have $K = 2\pi i\left(e^{-1}/2i\right) = \pi e^{-1}$.

Now, how does K (what we have calculated) relate to J (what we want to find)? Observe that

$$\begin{aligned}
\left|\int_{\mathcal{D}_R} \frac{e^{iz}}{z^2+1}\,dz\right| &\le \int_{\mathcal{D}_R} \left|\frac{e^{iz}}{z^2+1}\,dz\right| \\
&\le \int_{\mathcal{D}_R} \frac{\left|e^{iz}\right|}{\left|z^2+1\right|}\,|dz| \\
&\le \left(\max_{z=Re^{i\theta}} \frac{1}{\left|z^2+1\right|}\right)\int_{\mathcal{D}_R} \left|e^{iz}\right|\,|dz| \\
&\le \left(\max_{z=Re^{i\theta}} \frac{1}{\left|z^2+1\right|}\right)\pi \le \left(\frac{1}{\left|R^2-1\right|}\right)\pi,
\end{aligned} \tag{30.12}$$

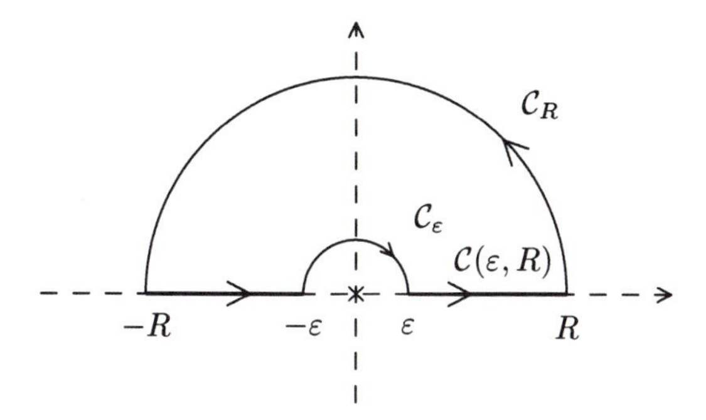

Figure 30.4 Integration contour for (30.14).

where we have used Jordan's lemma (see Notes, below). Hence, as $R \to \infty$, the integral $\int_{\mathcal{D}_R}$ vanishes and we have

$$J = \lim_{R\to\infty} \int_{\mathcal{C}_R} = \lim_{R\to\infty} \left(K - \int_{\mathcal{D}_R} \right) = K = \pi e^{-1}, \tag{30.13}$$

where the integrand is not displayed when it is understood to be the same as in (30.9). Using (30.9) in (30.13), and taking the real and imaginary parts, results in the two integrals

$$\int_{-\infty}^{\infty} \frac{\sin x}{x^2+1}\, dx = 0, \qquad \int_{-\infty}^{\infty} \frac{\cos x}{x^2+1}\, dx = \pi e^{-1}.$$

The first of these two integrals could have been obtained immediately by symmetry considerations.

Example 3

Consider the integral $I = \displaystyle\int_{-\infty}^{\infty} \frac{\sin x}{x}\, dx$. If we define $J = \displaystyle\int_{-\infty}^{\infty} \frac{e^{iz}}{z}\, dz$, then $I = \operatorname{Im} J$. Now consider the following contour integral (see Figure 30.4):

$$K = \lim_{\substack{R\to\infty \\ \varepsilon\to\infty}} \left(\int_{\mathcal{C}(\varepsilon,R)} + \int_{\mathcal{C}_\varepsilon} + \int_{\mathcal{C}_R} \right) \frac{e^{iz}}{z}\, dz. \tag{30.14}$$

Here, $\mathcal{C}_R$ is a semicircular arc of radius R, $\mathcal{C}_\varepsilon$ is a semicircular arc of radius ε, and $\mathcal{C}(\varepsilon, R)$ is the union of two intervals: $\mathcal{C} = [-R, -\varepsilon] \cup [\varepsilon, R]$. Since there are no singularities of the integrand within the contour $\mathcal{C}(\varepsilon, R) \cup \mathcal{C}_\varepsilon \cup \mathcal{C}_R$, we find that $K = 0$.

Now we will determine each of the integrals in (30.14). We immediately recognize that $J = \displaystyle\lim_{\substack{R\to\infty \\ \varepsilon\to\infty}} \int_{\mathcal{C}(\varepsilon,R)} \frac{e^{iz}}{z}\, dz$. By Jordan's lemma, it is easy to see that $\lim_{R\to\infty} \int_{\mathcal{C}_R} = 0$. The integral $\displaystyle\lim_{\varepsilon\to 0} \int_{\mathcal{C}_\varepsilon} \left(e^{iz}/z\right) dz$ can be evaluated

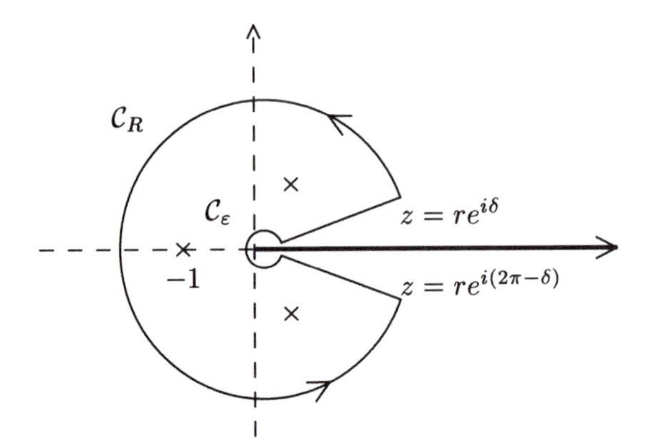

Figure 30.5 Contour used in (30.15). The thick line represents the branch cut, and the crosses indicate the location of the poles.

by using the second theorem in the introductory remarks. We note that the angle is $\phi = -\pi$ (the angle is negative since it is being traversed in the negative direction) and that there is a simple pole at $z = 0$ with residue $a_{-1} = 1$. Hence, $\lim_{\varepsilon\to 0} \int_{\mathcal{C}_\varepsilon} \left(e^{iz}/z\right) dz = i(-\pi)(1) = -i\pi$. Combining everything we have, we find

$$J = \lim_{\substack{R\to\infty \\ \varepsilon\to 0}} \int_{\mathcal{C}(\varepsilon,R)} = K - \lim_{\substack{R\to\infty \\ \varepsilon\to 0}} \left(\int_{\mathcal{C}_\varepsilon} + \int_{\mathcal{C}_R} \right) = 0 - (0 - i\pi) = i\pi,$$

where the integrand is not displayed when it is understood to be the same as in (30.14). Taking the imaginary parts of this expression, we determine that $I = \operatorname{Im} J = \pi$.

Example 4

Consider the integral $I = \displaystyle\int_0^\infty \frac{dx}{x^3+1}$. We first consider, instead, the integral

$$K = \int_{\mathcal{C}} \frac{\log z}{z^3+1}\, dz, \tag{30.15}$$

and will then determine I from K. For K the integration contour is given by $\mathcal{C} = \mathcal{C}_R \cup \mathcal{C}_\varepsilon \cup \mathcal{C}_{\text{ud}} \cup \mathcal{C}_{\text{ld}}$, where "ud" and "ld" stand for the upper diagonal branch and lower diagonal branch, respectively, in Figure 30.5. Since K involves a logarithm, a branch cut in the complex plane is needed; Figure 30.5 shows the location of the branch cut. With this branch cut we have $\log z = \log|z| + i \arg z$ where $0 < \arg z < 2\pi$.

The only poles in $\mathcal{C}$ are at the the cube roots of -1, $\{z_1, z_2, z_3\}$, and it is easy to derive that

$$\begin{aligned}\int_{\mathcal{C}} \frac{\log z}{z^3+1}\,dz &= 2\pi i \sum_{j=1}^{3} \text{residue at } z_j \\ &= -\frac{4\pi^2 i\sqrt{3}}{9} = \int_{\mathcal{C}_R} + \int_{\mathcal{C}_\varepsilon} + \int_{\mathcal{C}_{\mathrm{ud}}} + \int_{\mathcal{C}_{\mathrm{ld}}},\end{aligned} \tag{30.16}$$

where the integrand is not displayed when it is understood to be the same as in (30.15).

In the limits $R \to \infty$ and $\varepsilon \to 0$ the integrals along the two circular arcs vanish, observe:

- $\int_{\mathcal{C}_R} = O\left(\frac{\log R}{R^3}\right) O(R) \to 0$ as $R \to \infty$;
- $\int_{\mathcal{C}_\varepsilon} = O(\log \varepsilon) O(\varepsilon) \to 0$ as $\varepsilon \to 0$.

Taking the limits $R \to \infty$ and $\varepsilon \to 0$ in (30.16) results in

$$\begin{aligned}-\frac{4\pi^2 i\sqrt{3}}{9} &= \lim_{\substack{R\to\infty \\ \varepsilon\to 0}} \left\{\int_{\mathcal{C}_R} + \int_{\mathcal{C}_\varepsilon} + \int_{\mathcal{C}_{\mathrm{ud}}} + \int_{\mathcal{C}_{\mathrm{ld}}}\right\} \\ &= \lim_{\substack{R\to\infty \\ \varepsilon\to 0}} \left\{\int_{\mathcal{C}_{\mathrm{ud}}} + \int_{\mathcal{C}_{\mathrm{ld}}}\right\} \\ &= \int_0^\infty \frac{\log\left(re^{i\delta}\right)}{1+\left(re^{i\delta}\right)^3} e^{i\delta}\,dr + \int_\infty^0 \frac{\log\left(re^{i(2\pi-\delta)}\right)}{1+\left(re^{i(2\pi-\delta)}\right)^3} e^{i(2\pi-\delta)}\,dr.\end{aligned}$$

If we now take the limit $\delta \to 0$ in this last integral, we obtain

$$\begin{aligned}-\frac{4\pi^2 i\sqrt{3}}{9} &= \lim_{\delta\to 0} \left\{\int_0^\infty \frac{\log\left(re^{i\delta}\right)}{1+\left(re^{i\delta}\right)^3} e^{i\delta}\,dr + \int_\infty^0 \frac{\log\left(re^{i(2\pi-\delta)}\right)}{1+\left(re^{i(2\pi-\delta)}\right)^3} e^{i(2\pi-\delta)}\,dr\right\} \\ &= \int_0^\infty \frac{\log r}{1+r^3}\,dr + \int_\infty^0 \frac{\log r + 2\pi i}{1+r^3}\,dr \\ &= \int_\infty^0 \frac{2\pi i}{1+r^3}\,dr = -2\pi i I.\end{aligned}$$

From this we conclude that $I = 2\pi/3\sqrt{3}$.

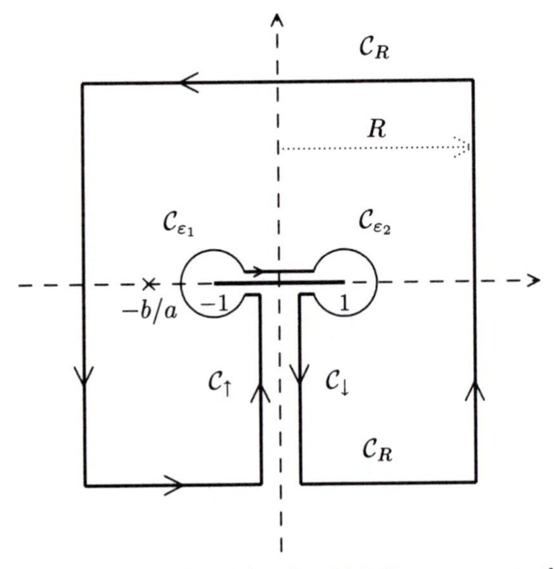

Figure 30.6 Contour used in (30.17). The thick line represents the branch cut, and the cross indicates the location of the pole.

Example 5

Consider the integral $I = \displaystyle\int_{-1}^{1} \frac{dx}{(ax+b)\sqrt{1-x^2}}$. We first consider the integral

$$K = \int_{\mathcal{C}} \frac{dz}{(az+b)\sqrt{z^2-1}} \tag{30.17}$$

and will determine I from K. In K, the contour $\mathcal{C}$ is given by $\mathcal{C} = \mathcal{C}_R \cup \mathcal{C}_\uparrow \cup \mathcal{C}_{\varepsilon_1} \cup \mathcal{C}_\rightarrow \cup \mathcal{C}_{\varepsilon_2} \cup \mathcal{C}_\leftarrow \cup \mathcal{C}_\downarrow$, where the different components should be clear from Figure 30.6. (Note that $\mathcal{C}_\leftarrow$ is along the bottom of the cut, and has two components.) Since K involves a square root, branch cuts are present and we must define how the square root is to be evaluated.

We define the function $(z^2-1)^{1/2}$ to be real and positive for $\operatorname{Re} z > 1$, and we take $z+1 = r_1 e^{i\theta_1}$ and $z - 1 = r_2 e^{i\theta_2}$ with $0 < \theta_i < 2\pi$; see Figure 30.7. With these definitions, along the real axis between -1 and 1 we have $(z^2-1)^{1/2} = \sqrt{1-x^2}e^{i(\theta_1+\theta_2)/2}$. On the top of the cut (i.e., along $\mathcal{C}_\rightarrow$) $\theta_1 = 0$ and $\theta_2 = \pi$; on the bottom of the cut (i.e., along $\mathcal{C}_\leftarrow$) $\theta_1 = 2\pi$ and $\theta_2 = \pi$.

The only pole in $\mathcal{C}$ is at $z = -b/a$, so that

$$K = \int_{\mathcal{C}} \frac{dz}{(az+b)\sqrt{z^2-1}} = 2\pi i \left(\text{residue at } z = -\frac{b}{a}\right) = -\frac{2\pi i}{\sqrt{b^2-a^2}}. \tag{30.18}$$

In the limits $R \to \infty$ and $\varepsilon_i \to 0$ the integrals along the bounding box and the two circular arcs vanish. Observe:

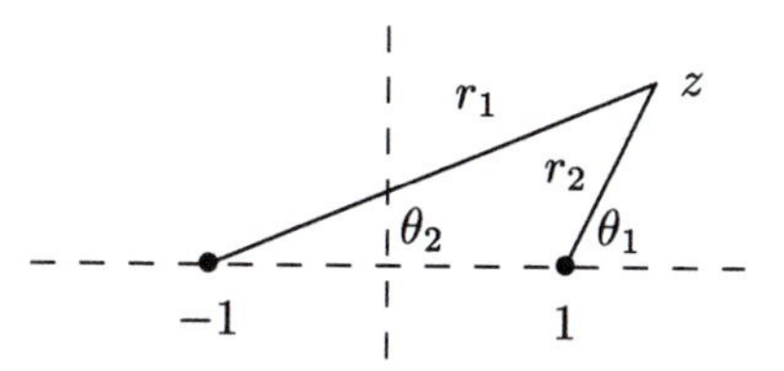

Figure 30.7 How the square root function is defined in (30.17).

- $\displaystyle\int_{C_R} = O\left(\frac{1}{R^2}\right) O(R) \to 0$ as $R \to \infty$;
- $\displaystyle\int_{C_{\varepsilon_i}} = O\left(\frac{1}{\sqrt{\varepsilon_i}}\right) O(\varepsilon_i) \to 0$ as $\varepsilon_i \to 0$.

If we allow the two vertical contours to come together, then the integrals along $\mathcal{C}_\uparrow$ and $\mathcal{C}_\downarrow$ cancel, since the integrand is continuous between these contours and the contours are going in opposite directions. We presume that this limit has been taken, so that $\mathcal{C}_\leftarrow$ is now continuous along the bottom of the cut.

Taking the $R \to \infty$ and $\varepsilon_i \to 0$ limits in (30.17), using (30.18), and disregarding the $\mathcal{C}_\uparrow$ and $\mathcal{C}_\downarrow$ integrals results in

$$\begin{aligned}
-\frac{2\pi i}{\sqrt{b^2 - a^2}} &= \lim_{\varepsilon_i \to 0} \left\{ \int_{\mathcal{C}_\leftarrow} + \int_{\mathcal{C}_\rightarrow} \right\} \frac{dz}{(az+b)\sqrt{z^2-1}} \\
&= \int_{-1}^{1} \frac{dx}{(ax+b)\sqrt{1-x^2}e^{\pi i/2}} + \int_{1}^{-1} \frac{dx}{(ax+b)\sqrt{1-x^2}e^{3\pi i/2}} \\
&= \left(\int_{-1}^{1} \frac{dx}{(ax+b)\sqrt{1-x^2}} \right)(-2i) = -2iI.
\end{aligned}$$

We conclude that $I = \pi/\sqrt{b^2 - a^2}$.

Example 6

The Laplace transform of the function $f(t)$ is the function $F(s) = \mathcal{L}[f(t)] = \int_0^\infty e^{-st} f(t)\, dt$. As a rule, $F(s)$ will converge for some region in the complex s-plane, typically $\operatorname{Re} s > \alpha$, where α is some real constant. The inverse Laplace transform is given by

$$f(t) = \mathcal{L}^{-1}[F(s)] = \frac{1}{2\pi i} \int_{\mathcal{C}} e^{ts} F(s)\, ds,$$

where the contour of integration is the vertical line $z = \sigma + iy$ (with $-\infty < y < \infty$), and σ is "to the right" of any singularities of $F(t)$. Often, the contour in this integral can be made into a closed contour by closing the contour to the left; this results in the Bromwich contour, $\mathcal{C} \cup \mathcal{C}_R$, shown in

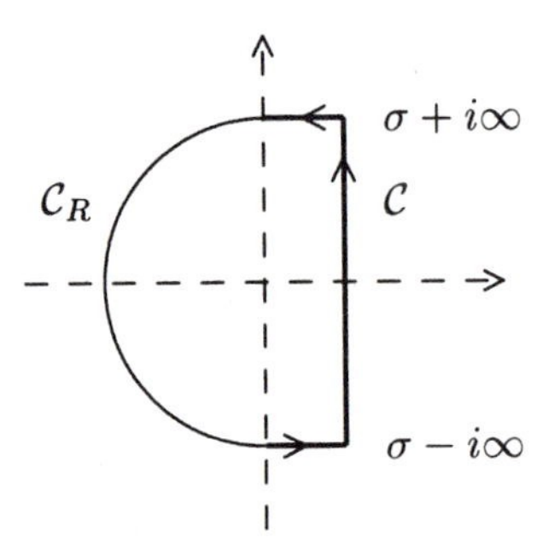

Figure 30.8 Bromwich contour used in inverse Laplace transforms.

Figure 30.8. With a closed contour, of course, Cauchy's theorem can be used.

Consider the inverse Laplace transform of the function $F(s) = 1/(s^2 + 4)$. This function has singularities at $s = \pm 2i$, so we must take $\sigma > 0$. Defining a new integral with the Bromwich contour

$$K(t) = \left(\int_{\mathcal{C}_R} + \int_{\mathcal{C}}\right) \frac{e^{ts}}{s^2+4}\, ds,$$

we note that the only poles are at $s = \pm 2i$. Therefore, the residue theorem yields

$$K(t) = 2\pi i \left[(\text{residue at } s = 2i) + (\text{residue at } s = -2i) \right] = \frac{1}{2} \sin 2t.$$

The difference between $f(t)$ (what we want) and $K(t)$ (what we have calculated) is the integral $\int_{\mathcal{C}_R}$. As we now show, this integral vanishes as $R \to \infty$.

Let $x = \operatorname{Re} s$ along $\mathcal{C}_R$. Then, for positive t, we have $|e^{ts}| < e^{tx} < e^{t\sigma}$ along $\mathcal{C}_R$. Hence,

$$\int_{\mathcal{C}_R} \frac{e^{ts}}{s^2+4}\, ds = O\left(\frac{e^{t\sigma}}{R^2}\right) O(R) \to 0 \qquad \text{as} \qquad R \to \infty,$$

since σ is fixed. We obtain the final result $f(t) = \frac{1}{2} \sin 2t$.

Notes

[1] Jordan's lemma states: If C_R is the contour $z = Re^{i\theta}$, $0 \le \theta \le \pi$, then $\int_{C_R} \left|e^{iz}\right| |dz| < \pi$ (see Levinson and Redheffer [7]). This is often useful for estimating integrals.

[2] The integral in Example 1 had the form $I = \int_0^{2\pi} f(\cos\theta, \sin\theta)\, d\theta$. Integrals of this form can be recast as contour integrals by setting $z = e^{i\theta}$. As θ varies from 0 to 2π, z traverses a unit circle C counterclockwise. Hence,

$$I = \int_C f\left(\frac{z^2+1}{2z}, \frac{z^2-1}{2iz}\right)\frac{dz}{z}.$$

[3] Loop integrals are contour integrals in which the path of integration is given by a closed loop. For example, Hankel's representation of the Gamma function when $\operatorname{Re} x > 0$ is given by

$$\Gamma(x) = -\frac{1}{2i\sin\pi x}\int_C (-t)^{x-1}e^{-t}\, dt$$

where the contour C lies in the complex plane cut along the positive real axis, starting at ∞, going around the origin once counterclockwise, and ending at ∞ again.

[4] In the above theorems, the contours were sometimes described as being "traversed in the positive sense." Consider a closed region surrounding the contour $C(t)$. Consider the vector giving the direction of increasing t; if a counterclockwise rotation of $\pi/2$ causes the vector to point into the interior of the closed region, then the contour C is being traversed in the positive sense and the region is positively oriented.

[5] Plaisted [9] shows that the following problem involving the evaluation of a contour integral is NP-hard:

> Given integers b and N and a set of k sparse polynomials with integer coefficients $\{p_i(z)\}$, it is NP-hard to determine whether the following contour integral is zero:
>
> $$\int_C \frac{\prod_{j=1}^k p_j(z)}{z^N - 1}\left(1 - \frac{1}{z^{kN}}\right)\frac{z^N - b^N}{z - b}\, dz$$
>
> where C is any contour including the origin in its interior.

See also Garey and Johnson [5].

[6] Gluchoff [6] presents the following interpretation of a contour integral:

$$\int_C f(z)\, dz = L(C) \operatorname*{av}_{z\in C}\left[f(z)T(z)\right] \tag{30.19}$$

where $L(C)$ is the length of C, $T(z)$ is the unit tangent to C at z, and "av" denotes the averaging function.

For example, for the integral $I = \int_{|z|=R} dz/z$ we identify $L(C) = 2\pi R$ and $T(z) = iz/R$ so that (30.19) becomes

$$\int_{|z|=R}\frac{dz}{z} = 2\pi R \operatorname*{av}_{|z|=R}\left[\frac{1}{z}\frac{iz}{R}\right] = 2\pi i \operatorname*{av}_{|z|=R}[1] = 2\pi i.$$

[7] Several of the references describe quadrature rules for numerically approximating a contour integral.

[8] The numerical integrator in the language Mathematica allows an arbitrary path in the complex place to be specified. For example, consider the integral in (30.5). The integration contour may be deformed to the square with vertices at $\{\pm 1, \pm i\}$ without changing the value of the integral. If Mathematica is input "`NIntegrate[-2I/(z^2+4z+1),z,1,I,-1,-I,1]`", then the result is `3.6276 + 0. I`, which is the correct numerical value. (Here I is used to represent i.)

[9] The theory of multidimensional residues is based on the general Stokes formula and its corollary, the Cauchy–Poincaré integral theorem. Serious topological difficulties arise for analytic functions of several complex variables because the role of the singular point is now played by surfaces in $\mathbf{C}^n$. These surfaces generally have a complicated structure requiring tools such as algebraic topology.

References

[1] B. Braden, "Polya's Geometric Picture of Complex Contour Integrals," *Math. Mag.*, **60**, No. 5, 1987, pages 321–327.

[2] P. J. Davis and P. Rabinowitz, *Methods of Numerical Integration*, Second Edition, Academic Press, Orlando, Florida, 1984, pages 168–171.

[3] D. Elliot and J. D. Donaldson, "On Quadrature rules for Ordinary and Cauchy Principal Value Integrals over Contours," *SIAM J. Numer. Anal.*, **14**, No. 6, December 1977, pages 1078–1087.

[4] R. J. Fornaro, "Numerical Evaluation of Integrals Around Simple Closed Contours," *SIAM J. Numer. Anal.*, **1**, No. 4, 1973, pages 623–634.

[5] M. R. Garey and D. S. Johnson, *Computers and Intractability*, W. H. Freeman and Co., New York, 1979, page 252.

[6] A. Gluchoff, "A Simple Interpretation of the Complex Contour Integral," *Amer. Math. Monthly*, **98**, No. 7, August–September 1991, pages 641–644.

[7] N. Levinson and R. M. Redheffer, *Complex Variables*, Holden–Day, Inc., San Francisco, 1970, Chapter 5, pages 259–332.

[8] J. N. Lyness and L. M. Delves, "On Numerical Integration Round a Closed Contour," *Math. of Comp.*, **21**, 1967, pages 561–577.

[9] D. A. Plaisted, "Some Polynomial and Integer Divisibility Problems Are NP-Hard," in *Proc. 17th Ann. Symp. on Found. of Comp. Sci.*, IEEE, Long Beach, CA, 1976, pages 264–267.

31. Convolution Techniques

Applicable to Some one dimensional integrals that contain a product of terms.

Yields

An exact evaluation in terms of other integrals.

Idea

If the original integral can be written as a convolution, then the value of the integral may be determined by a sequence of integrals.

Procedure

Given the functions $f(x)$ and $g(x)$, and an integral operator $I[\,]$, define the functions $F(\zeta) = I[f(x)]$ and $G(\zeta) = I[g(x)]$. For many common integral operators there is a relation that relates an integral of f and g, of a specific form, to an integral involving F and G. These relations are often of the form

$$A\left[F(\zeta)G(\zeta)\right] = \int J[f(a(x,\zeta))), g(b(x,\zeta)), \zeta]h(\zeta)\,d\zeta \tag{31.1}$$

where $A[\,]$ is an integral operator. Such a relation is known as a convolution theorem.

Convolution theorems may sometimes be used to simplify integrals. Given an integral in the form of the right-hand side of (31.1), it may be easier to evaluate the left-hand side of (31.1). In the following examples, upper case letters denote transforms of lower case letters.

Example 1

For the Mellin transform, defined by

$$F(s) = \mathcal{M}[f(x)] = \int_0^\infty t^x f(t)\,dt,$$

the convolution theorem is

$$\int_0^\infty f\left(\frac{x}{u}\right) g(u)\frac{du}{u} = \frac{1}{2\pi i}\int_{c-i\infty}^{c+i\infty} F(s)G(s)x^{-s}\,ds$$

where the integral on the right-hand side is a Bromwich integral.

Example 2

For the Laplace transform pair, defined by

$$F(s) = \mathcal{L}[f(t)] = \int_0^\infty e^{-ts}f(t)\,dt,$$

$$f(t) = \mathcal{L}^{-1}[F(s)] = \frac{1}{2\pi i}\int_{c-i\infty}^{c+i\infty} e^{ts}F(s)\,ds,$$

the convolution theorem is

$$\int_0^t f(\tau)g(t-\tau)\,d\tau = \mathcal{L}^{-1}\left[F(s)G(s)\right].$$

Example 3

For the Fourier transform pair, defined by

$$F(t) = \mathcal{F}[f(x)] = \frac{1}{\sqrt{2\pi}} \int_{-\infty}^{\infty} e^{-itx} f(x)\,dx,$$
$$f(x) = \mathcal{F}^{-1}[F(t)] = \frac{1}{\sqrt{2\pi}} \int_{-\infty}^{\infty} e^{itx} F(t)\,dt,$$

the convolution theorem is

$$\frac{1}{\sqrt{2\pi}} \int_{-\infty}^{\infty} f(x) g(x-\xi)\,d\xi = \mathcal{F}^{-1}[F(t)G(t)].$$

Notes

[1] Bouwkamp [2] uses the convolution theorem for Fourier transforms to show that

$$\int_0^{2\pi} \int_0^{\infty} \frac{re^{-\alpha r}\,dr\,d\theta}{\sqrt{r^2 + D^2 - 2rD\cos\theta}} = \frac{2\pi}{\alpha}\left[1 - \alpha D K_0\left(\tfrac{1}{2}\alpha D\right) I_1\left(\tfrac{1}{2}\alpha D\right)\right].$$

References

[1] A. Apelblat, "Repeating Use of Integral Transforms—A New Method for Evaluation of Some Infinite Integrals," *IMA J. Appl. Mathematics*, **27**, 1981, pages 481–496.

[2] C. J. Bouwkamp, "A Double Integral," Problem 71–23 in *SIAM Review*, **14**, No. 3, July 1972, pages 505–506.

[3] E. Butkov, *Mathematical Physics*, Addison–Wesley Publishing Co., Reading, MA, 1968.

[4] N. T. Khai and S. B. Yakubovich, "Some Two-Dimensional Integral Transformations of Convolution Type," *Dokl. Akad. Nauk BSSR*, **34**, No. 5, 1990, pages 396–398.

32. Differentiation and Integration

Applicable to Definite and indefinite integrals.

Yields

An alternative representation of the integral.

Idea

By differentiating or integrating an integral with respect to a parameter, the integral may be more tractable. This parameter may have to be introduced into the original integral.

Procedure

Given an integral, try to differentiate or integrate that integral with respect to a parameter appearing in the integral. If there are no parameters appearing in the integral, insert one and then perform the differentiation or integration.

After the integration is performed then either an integration or a differentiation remains.

Example 1

Given the integral $u(x) = \int_0^\infty e^{-t^2} \cos xt \, dt$, differentiation shows that $u(x)$ satisfies the differential equation $u' + \frac{1}{2}xu = 0$. (This was obtained by integration by parts.) This differential equation is separable (see Zwillinger [3]) and the solution is found to be $u(x) = Ce^{-x^2/4}$, for some constant C. From the defining integral, $u(0) = \int_0^\infty e^{-t^2} \, dt = \frac{1}{2}\sqrt{\pi}$, so we have the final answer

$$\int_0^\infty e^{-t^2} \cos xt \, dt = \tfrac{1}{2}\sqrt{\pi} e^{-x^2/4}. \tag{32.1}$$

Example 2

Given the integral

$$I(p) = \int_0^\infty x \sin(x) e^{-px^2} \, dx$$

we might introduce a parameter a and consider instead

$$J(a,p) = \int_0^\infty x \sin(ax) e^{-px^2} \, dx. \tag{32.2}$$

Note that $I(p) = J(1,p)$. Both sides of (32.2) may be integrated with respect to a to determine

$$\begin{aligned} \int^a J(a,p) \, da &= \int^a \int_0^\infty x \sin(ax) e^{-px^2} \, dx \, da \\ &= -\int_0^\infty \cos(ax) e^{-px^2} \, dx. \end{aligned} \tag{32.3.a–b}$$

This last integral may be evaluated by a simple change of variable and using (32.1). The result is

$$\int^a J(a,p) \, da = \frac{1}{2}\sqrt{\frac{\pi}{p}} e^{-a^2/4p}. \tag{32.4}$$

Differentiating (32.4) with respect to a results in

$$J(a,p) = -\frac{a}{4}\sqrt{\frac{\pi}{p^3}} e^{-a^2/4p}$$

and so

$$I(p) = J(1,p) = -\frac{1}{4}\sqrt{\frac{\pi}{p^3}} e^{-1/4p}.$$

Example 3

Given the integral

$$K(b,q) = \int_0^\infty \frac{\sin(bx)}{x} e^{-qx^2}\, dx \tag{32.5}$$

we differentiate with respect to b to obtain $\frac{d}{db}K(b,q) = \int_0^\infty \cos(bx)e^{-qx^2}$, which is the same as (32.3.b). Therefore

$$\frac{d}{db}K(b,q) = \frac{1}{2}\sqrt{\frac{\pi}{q}}e^{-b^2/4q}. \tag{32.6}$$

From (32.5) we note that $K(0,q) = 0$. Hence, we need to solve the differential equation in (32.6) with this initial condition. The solution is

$$\begin{aligned} K(b,q) &= \int_0^b \left(\frac{1}{2}\sqrt{\frac{\pi}{q}}e^{-b^2/4q}\right) db \\ &= \sqrt{\pi}\int_0^{b/2\sqrt{q}} e^{-z^2}\, dz \\ &= \frac{\pi}{2}\operatorname{erf}\left(\frac{b}{2\sqrt{q}}\right) \end{aligned} \tag{32.7.a–c}$$

where we have recognized (32.7.b) as being an error function (see page 172).

Notes

[1] As another example of the technique illustrated in Example 3, consider the integral

$$F(y) = \int_0^\infty \frac{\log(1+x^2y^2)}{1+x^2}\, dx. \tag{32.8}$$

Differentiating with respect to y produces

$$\begin{aligned} F'(y) &= \int_0^\infty \frac{2x^2y}{1+xy^2}\frac{dx}{1+x^2} \\ &= \frac{2y}{y^2-1}\int_0^\infty \left(\frac{1}{1+x^2} - \frac{1}{1+x^2y^2}\right) dx \\ &= \frac{\pi}{1+y} \end{aligned} \tag{32.9}$$

if $y > 0$. From (32.8) we recognize that $F(0) = 0$. Therefore, we can integrate (32.9) to determine

$$F(y) = \int_0^y \frac{\pi}{1+y}\, dy = \pi \log(1+y).$$

[2] Squire [2] starts with the identity $\int_0^1 x^{t-1}\, dt = t^{-1}$, and integrates with respect to t (from $t=1$ to $t=a$) to obtain

$$\int_0^1 \frac{x^{a-1}-1}{\log x}\, dx = \log a.$$

[3] Many other clever manipulations based on integration and differentiation can be conceived. For example (again from Squire [2]), the identity $x^{-1} = \int_0^\infty e^{-xt}\,dt$ can be used in the following way:

$$\int_0^\infty \frac{\sin x}{x}\,dx = \int_0^\infty \sin x \int_0^\infty e^{-xt}\,dt\,dx = \int_0^\infty \frac{dt}{1+t^2} = \frac{\pi}{2}.$$

References

[1] J. Mathews and R. L. Walker, *Mathematical Methods of Physics*, W. A. Benjamin, Inc., NY, 1965, pages 58–59.

[2] W. Squire, *Integration for Engineers and Scientists*, American Elsevier Publishing Company, New York, 1970, pages 84–85.

[3] D. Zwillinger, *Handbook of Differential Equations*, Academic Press, New York, Second Edition, 1992.

33. Dilogarithms

Applicable to Integrals of a special form.

Yields

An exact solution in terms of dilogarithms.

Idea

Integrals of the form $\int^x P(x, \sqrt{R}) \log Q(x, \sqrt{R})\,dx$, where $P(\,,\,)$ and $Q(\,,\,)$ are rational functions and $R = A^2 + Bx + Cx^2$, can be transformed to a canonical form.

Procedure

Given the integral

$$I = \int^x P(x, \sqrt{R}) \log Q(x, \sqrt{R})\,dx, \tag{33.1}$$

make the change of variable $\sqrt{R} = A + xt$. This transformation results in

$$\begin{aligned}
t &= \frac{\sqrt{A^2 + Bx + Cx^2} - A}{x}\\
x &= \frac{2At - B}{C - t^2}\\
dx &= \frac{2}{(C-t^2)^2}\left[AC - Bt + At^2\right]dt\\
\sqrt{R} &= \frac{AC - Bt + At^2}{C - t^2}
\end{aligned}$$

so that (33.1) becomes

$$I = \int^t \overline{P}(t) \log \overline{Q}(t)\, dt \tag{33.2}$$

where $\overline{P}$ and $\overline{Q}$ are rational functions of t. In principle, partial fraction expansion can be used on $\overline{P}$ to obtain

$$\begin{aligned} \overline{P} &= \widetilde{P}(t) + \sum_{m,n} \frac{\alpha_{mn}}{(\beta_{mn} + t)^n} \\ \overline{Q} &= K \prod_{m,n} (\gamma_{mn} + t)^n \end{aligned} \tag{33.3}$$

where $\widetilde{P}(t)$ is a polynomial in t. Using (33.3) in (33.2), and expanding, leads to many integrals. Those integrals that have the form $\int \widetilde{P}(t) \log(\gamma + t)\, dt$ or the form $\displaystyle\int \frac{\alpha_{mn}}{(\beta_{mn} + t)^n} \log(\gamma + t)\, dt$, for $n \neq -1$, can be evaluated by integrating by parts. The only integral that cannot be evaluated in this manner is

$$\int \frac{\log(\alpha + t)}{\gamma + t}\, dt = \begin{cases} \log(\alpha - \gamma) \log\left(\dfrac{\gamma + t}{\gamma}\right) - \mathrm{Li}_2\left(\dfrac{\gamma + t}{\gamma - \alpha}\right) + C, & \text{if } \alpha \neq \gamma, \\ \frac{1}{2} \log^2(\gamma + t) + C, & \text{if } \alpha = \gamma, \end{cases} \tag{33.4}$$

where $\mathrm{Li}_2\,()$ is the *dilogarithm function* and C is an arbitrary constant. The dilogarithm function is defined by

$$\mathrm{Li}_2\,(x) = -\int_0^x \frac{\log(1 - x)}{x}\, dx. \tag{33.5}$$

Hence, the integral in (33.1) can always be integrated analytically in terms of the dilogarithm and elementary functions. A few integrals involving dilogarithms are shown in Table 33.

Example

The integral

$$I = \int^x \frac{1}{x\sqrt{1 + x^2}} \log\left(x + \sqrt{1 + x^2}\right) dx \tag{33.6}$$

becomes under the change of variable $\sqrt{R} = \sqrt{1 + x^2} = 1 + xt$ (which is the same as $x = 2t/(1 - t^2)$)

$$\begin{aligned} I &= \int^t \frac{1}{t} \left[\log(1 + t) - \log(1 - t)\right] dt, \\ &= \mathrm{Li}_2\,(t) - \mathrm{Li}_2\,(-t), \\ &= \mathrm{Li}_2\left(\frac{\sqrt{1 + x^2} - 1}{x}\right) - \mathrm{Li}_2\left(\frac{1 - \sqrt{1 + x^2}}{x}\right). \end{aligned} \tag{33.7}$$

Table 33. Some integrals involving dilogarithms.

$$\int_0^x \frac{\log(1+ax^n)}{x}\,dx = -\frac{1}{n}\mathrm{Li}_2\left(-ax^n\right)$$

$$\int_0^x \frac{\log(a+bt)}{c+et}\,dx = \frac{1}{2e}\log^2\left(\frac{b}{e}(c+et)\right) - \frac{1}{2e}\log^2\left(\frac{bc}{e}\right) + \frac{1}{e}\mathrm{Li}_2\left(\frac{bc-ae}{b(c+et)}\right) - \frac{1}{e}\mathrm{Li}_2\left(\frac{bc-ae}{bc}\right)$$

$$\int_0^x \frac{\log x}{\sqrt{1+x^2}}\,dx = \frac{1}{2}\mathrm{Li}_2\left(\left(\sqrt{1+x^2}-x\right)^2\right) + \frac{1}{2}\log^2\left(\frac{\sqrt{1+x^2}+x}{2}\right)$$

$$\int_0^x \frac{\log(1+x^2)}{\sqrt{1-x}}\,dx = \frac{1}{4}\mathrm{Li}_2\left(-x^2\right) + \frac{1}{2}\mathrm{Li}_2\left(\frac{2x}{1+x^2}\right) - \mathrm{Li}_2\left(x\right) + \frac{1}{4}\log^2(1+x^2) - \log(1-x)\log(1+x^2)$$

$$\int_0^x \frac{\log x \log(x-1)}{x}\,dx = \mathrm{Li}_3(x) - \log x \mathrm{Li}_2(x)$$

$$\int_0^1 x^{a-1}\mathrm{Li}_2\left(x\right)\,dx = \frac{\pi^2}{6a} - \frac{\Psi(1+a)+\gamma}{a^2}$$

Notes

[1] Tables of the dilogarithm may be found in Lewin [2].

[2] An extension of the dilogarithm function of one argument is the dilogarithm function of two arguments:

$$\mathrm{Li}_2\left(x,\theta\right) = -\frac{1}{2}\int_0^x \frac{\log\sqrt{1-2x\cos\theta+x^2}}{x}\,dx$$

where, of course, $\mathrm{Li}_2(x,0) = \mathrm{Li}_2(x)$ for $-1 \le x \le 1$. The higher order logarithmic functions are defined by $\mathrm{Li}_n(x) = \int_0^x \mathrm{Li}_{n-1}(x)/x\,dx$. All of these functions have been well studied; recurrence relations and other formulae have been determined for each (see Lewin [2].)

References

[1] M. Abramowitz and I. A. Stegun, *Handbook of Mathematical Functions*, National Bureau of Standards, Washington, DC, 1964, Section 27.7, page 1004.

[2] L. Lewin, *Dilogarithms and Associated Functions*, MacDonald & Co., London, 1958.

[3] L. Lewin, *Polylogarithms and Associated Functions*, North–Holland Publishing Co., New York, 1981.

34. Elliptic Integrals

Applicable to Integrals of a special form.

Yields
An exact solution in terms of elliptic functions.

Idea
Integrals of the form $\int_x R(x, \sqrt{T(x)})\,dx$ where $R(\,,\,)$ is a rational function of its arguments and $T(x)$ is a third of fourth order polynomial can be transformed to a canonical form.

Procedure
Given an integral of the form

$$\int_x R(x, \sqrt{T(x)})\,dx \tag{34.1}$$

where

$$R(x, \sqrt{T}) = \frac{f_1(x) + f_2(x)\sqrt{T}}{f_3(x) + f_4(x)\sqrt{T}},$$

$T = a_4x^4 + a_3x^3 + a_2x^2 + a_1x + a_0$, and each $f_i(x)$ represents a polynomial in x, rewrite the integrand as

$$\begin{aligned} R(x, \sqrt{T}) &= \frac{f_1(x) + f_2(x)\sqrt{T}}{f_3(x) + f_4(x)\sqrt{T}} \left[\frac{f_3(x) - f_4(x)\sqrt{T}}{f_3(x) - f_4(x)\sqrt{T}}\right] \left[\frac{\sqrt{T}}{\sqrt{T}}\right] \\ &= \frac{f_5(x) + f_6(x)\sqrt{T}}{f_7(x)\sqrt{T}} \\ &= R_1(x) + \frac{R_2(x)}{\sqrt{T}}, \end{aligned} \tag{34.2}$$

where $R_1(x)$ and $R_2(x)$ are rational functions of x. Clearly, the integral $\int R_1(x)\,dx$ can be evaluated in terms of logarithms and arc-tangents (see page 183).

We can always write T in the form (see Whittaker and Watson [10]) $T = \left[A_1(x-\alpha)^2 + B_1(x-\beta)^2\right]\left[A_2(x-\alpha)^2 + B_2(x-\beta)^2\right]$. Then, changing variables by $t = (x-\alpha)/(x-\beta)$ results in

$$\frac{dx}{\sqrt{T}} = \pm\frac{(\alpha-\beta)^{-1}\,dt}{\sqrt{S}},$$

where $S := \sqrt{(A_1t^2 + B_1)(A_2t^2 + B_2)}$. Hence, we find

$$\int \frac{R_2(x)}{\sqrt{T}}\,dx = \int \frac{R_3(t)}{\sqrt{S}}\,dt.$$

where R_3 is a rational function. We can now write $R_3(t) = R_4(t^2) + tR_5(t^2)$, where R_4 and R_5 are also rational functions (by, say, $R_4(t^2) \equiv (R_3(t) + R_3(-t))/2$ and $R_5(t^2) \equiv (R_3(t) - R_3(-t))/2t$). Clearly, the integral $\int tR_4(t^2)/\sqrt{S}\,dt$ can be evaluated in terms of logarithms and arctangents by introducing the variable $z = t^2$ (see page 183). By a partial fraction decomposition, we can write

$$\frac{R_4(t^2)}{\sqrt{S}} = \sum_s \alpha_s \frac{t^{2s}}{\sqrt{S}} + \sum_{i,s} \beta_{i,s} \frac{1}{(1+n_i t^2)^s \sqrt{S}} \tag{34.3}$$

for some constants $\{\alpha_s\}$, $\{\beta_{i,s}\}$, and $\{n_i\}$. By integrations by parts, it turns out that we only need to be able to evaluate integrals of three canonical forms to determine (34.3) completely. Thus, knowledge of these canonical forms allow us to determine (34.1) completely. These three canonical forms are, in Legendre's notation:

- Elliptic integrals of the first kind (Iyanaga and Kawada [8], page 1452):

$$F(\phi,k) = \int_0^{\phi} \frac{d\theta}{\sqrt{1-k^2\sin^2\theta}} = \int_0^{\sin\phi} \frac{dt}{\sqrt{(1-t^2)(1-k^2t^2)}}. \tag{34.4}$$

- Elliptic integrals of the second kind (Iyanaga and Kawada [8], page 1452):

$$E(\phi,k) = \int_0^{\phi} \sqrt{1-k^2\sin^2\theta}\,d\theta = \int_0^{\sin\phi} \sqrt{\frac{1-k^2t^2}{1-t^2}}\,dt. \tag{34.5}$$

- Elliptic integrals of the third kind (Iyanaga and Kawada [8], page 1452):

$$\begin{aligned}\Pi(\phi,n,k) &= \int_0^{\phi} \frac{d\theta}{\left(1+n\sin^2\theta\right)\sqrt{1-k^2\sin^2\theta}} \\ &= \int_0^{\sin\phi} \frac{dt}{\left(1+nt^2\right)\sqrt{(1-t^2)(1-k^2t^2)}}.\end{aligned} \tag{34.6}$$

Other Manipulations

There are many ways in which to manipulate an integral to obtain an elliptic integral of the first, second, or third kinds. For example, by a partial fraction decomposition of (34.2), we can write

$$\frac{R_2(x)}{\sqrt{T}} = \sum_s A_s \frac{x^s}{\sqrt{T}} + \sum_{i,s} B_{i,s} \frac{1}{(x - C_i)^s \sqrt{T}},$$

for some constants $\{A_s\}$, $\{B_{i,s}\}$, and $\{C_i\}$. By writing

$$\begin{aligned} T &= a_4 x^4 + a_3 x^3 + a_2 x^2 + a_1 x + a_0 \\ &= b_4(x-c)^4 + b_3(x-c)^3 + b_2(x-c)^2 + b_1(x-c) + b_0, \end{aligned} \tag{34.7}$$

we see that we only need to be able to evaluate integrals of the form

$$I_s = \int^x \frac{x^s}{\sqrt{T}}\, dx \qquad \text{and} \qquad J_s = \int^x \frac{dx}{(x-c)^s \sqrt{T}} \tag{34.8}$$

in order to evaluate (34.1). The recurrence relations

$$\begin{aligned} (\mathrm{s}+2)\mathrm{a}_4\mathrm{I}_{\mathrm{s}+3} + \tfrac{1}{2}a_3(2s+3)I_{s+2} + a_2(s+1)I_{s+1} + \tfrac{1}{2}a_1(2s+1)I_s& \\ + s a_0 I_{s-1} = x^s\sqrt{T} \qquad & \text{for } s = 0, 1, 2, \ldots \\ (2-s)b_4 J_{s-3} + \tfrac{1}{2}b_3(3-2s)J_{s-2} + b_2(1-s)J_{s-1} + \tfrac{1}{2}b_3(1-2s)J_s& \\ - s b_0 J_{s+1} = (x-c)^s\sqrt{T} \qquad & \text{for } s = 1, 2, 3, \ldots \end{aligned} \tag{34.9}$$

allow some manipulation of the $\{I_n\}$ and the $\{J_n\}$.

There are several different methods for calculating I_0, depending on the exact form of $T(x)$. For example, if $T(x) = Q_1(x)Q_2(x)$, where

$$\begin{aligned} Q_1(x) &= ax^2 + bx + c, \\ Q_2(x) &= dx^2 + ex + f, \end{aligned} \tag{34.10}$$

then the change of variable $z = \sqrt{Q_1(x)/Q_2(x)}$ allows I_0 to be written as

$$\begin{aligned} I_0 &= \int^x \frac{dx}{\sqrt{Q_1(x)Q_2(x)}} \\ &= \pm \int^z \frac{dz}{\sqrt{(e^2 - 4df)z^4 + 2(2af + 2cd - be)z^2 + (b^2 - 4ac)}}. \end{aligned} \tag{34.11.a–b}$$

Note that the radical in (34.11.b) is the discriminant of $Q_2 z^2 - Q_1$ (i.e., what would appear in the radical if the quadratic formula were used to solve $Q_2 z^2 - Q_1 = 0$ for the variable x).

Example 1

Suppose we have the integral

$$I = \int^x \frac{dx}{\sqrt{x(3x^2+2x+1)}}.$$

Since this has the form of (34.10), with $Q_1(x) = x$ and $Q_2(x) = 3x^2+2x+1$ we make the change of variable

$$t = \sqrt{\frac{Q_1(x)}{Q_2(x)}} = \sqrt{\frac{x}{3x^2+2x+1}}.$$

This leads to (using (34.11)) $I = \pm\int 2/\sqrt{1-4t^2-8t^4}\,dt$. Defining $b^2 = (1+\sqrt{3})/4$ and $-a^2 = (1-\sqrt{3})/4$, we observe that I can be written in the form

$$I = \pm\frac{1}{\sqrt{2}}\int^t \frac{dt}{\sqrt{(t^2+a^2)(t^2-b^2)}}. \tag{34.12}$$

This standard form can be found in Abramowitz and Stegun [1], 17.4.49. If the lower limit on the integral in (34.12) is taken to be b, then the integral is equal to

$$\begin{aligned} I &= \pm\frac{1}{\sqrt{2}}\frac{1}{\sqrt{a^2+b^2}}F\left(\sin^{-1}t, \frac{a}{\sqrt{a^2+b^2}}\right) \\ &= \pm\frac{1}{\sqrt{2}}\frac{1}{\sqrt{a^2+b^2}}\,\mathrm{nc}^{-1}\left(\frac{t}{b}\,\middle|\,\frac{a^2}{a^2+b^2}\right), \end{aligned}$$

where nc is a Jacobian elliptic function (see Table 34).

Example 2

Given the integral

$$I = \int_1^x \frac{dx}{\sqrt{(5x^2-4x-1)\,(12x^2-4x-1)}},$$

we change variables by $y = 1/x$ to obtain

$$I = \int_1^{1/x} \frac{-dy}{\sqrt{(5-4y-y^2)\,(12-4y-y^2)}}.$$

The change of variable $t = \frac{1}{3}(y+2)$ results in

$$\begin{aligned} I &= -\frac{1}{4}\int_1^{(1+2x)/3x} \frac{dt}{\sqrt{(1-t^2)\left(1-\frac{9}{16}t^2\right)}}, \\ &= -\frac{1}{4}\left[F\left(\frac{1+2x}{3x}, \frac{3}{4}\right) - F\left(1, \frac{3}{4}\right)\right]. \end{aligned}$$

Example 3

Given the integral

$$I = \int_0^2 \frac{dx}{\sqrt{(2x - x^2)(3x^2 + 4)}},$$

we change variables by $t = (2 - 3x)/(6 + 3x)$ to obtain

$$I = \sqrt{6} \int_0^{\frac{1}{3}} \frac{dt}{\sqrt{(1 - 9t^2)(1 + 3t^2)}}.$$

Introducing $v^2 = 1 - 9t^2$, we find

$$\begin{aligned} I &= \frac{1}{\sqrt{2}} \int_0^1 \frac{dv}{\sqrt{(1 - v^2)(1 - \frac{1}{4}v^2)}} \\ &= \frac{1}{\sqrt{2}} K\left(\frac{1}{2}\right) \end{aligned}$$

Table 34. Some relationships between elliptic integrals, elliptic functions in Legendre notation, and Jacobian elliptic functions (from Abramowitz and Stegun [1], page 566). In the following, $x = \sin\phi$.

If $a > b$ and $k^2 = (a^2 - b^2)/a^2$, then

$$F(\phi, k) = a \int_0^x \frac{dt}{\sqrt{(t^2 + a^2)(t^2 + b^2)}} = \operatorname{sc}^{-1}\left(\frac{x}{b} \middle| k^2\right)$$

with ϕ defined by $\tan\phi = \dfrac{x}{b}$;

$$F(\phi, k) = a \int_x^\infty \frac{dt}{\sqrt{(t^2 + a^2)(t^2 + b^2)}} = \operatorname{cs}^{-1}\left(\frac{x}{a} \middle| k^2\right)$$

with ϕ defined by $\tan\phi = \dfrac{a}{x}$;

$$F(\phi, k) = a \int_b^x \frac{dt}{\sqrt{(a^2 - t^2)(t^2 - b^2)}} = \operatorname{nd}^{-1}\left(\frac{x}{b} \middle| k^2\right)$$

with ϕ defined by $\sin^2\phi = \dfrac{a^2(x^2 - b^2)}{x^2(a^2 - b^2)}$;

$$F(\phi, k) = a \int_x^a \frac{dt}{\sqrt{(a^2 - t^2)(t^2 - b^2)}} = \operatorname{dn}^{-1}\left(\frac{x}{a} \middle| k^2\right)$$

with ϕ defined by $\sin^2\phi = \dfrac{a^2 - x^2}{a^2 - b^2}$.

If $a > b$ and $k^2 = b^2/a^2$, then

$$F(\phi, k) = a\int_0^x \frac{dt}{\sqrt{(a^2 - t^2)(b^2 - t^2)}} = \mathrm{sn}^{-1}\left(\frac{x}{b}\middle| k^2\right)$$

with ϕ defined by $\sin\phi = \dfrac{x}{b}$;

$$F(\phi, k) = a\int_x^b \frac{dt}{\sqrt{(a^2 - t^2)(b^2 - t^2)}} = \mathrm{cd}^{-1}\left(\frac{x}{b}\middle| k^2\right)$$

with ϕ defined by $\sin^2\phi = \dfrac{a^2(b^2 - x^2)}{b^2(a^2 - x^2)}$;

$$F(\phi, k) = a\int_a^x \frac{dt}{\sqrt{(t^2 - a^2)(t^2 - b^2)}} = \mathrm{dc}^{-1}\left(\frac{x}{a}\middle| k^2\right)$$

with ϕ defined by $\sin^2\phi = \dfrac{x^2 - a^2}{x^2 - b^2}$;

$$F(\phi, k) = a\int_x^\infty \frac{dt}{\sqrt{(t^2 - a^2)(t^2 - b^2)}} = \mathrm{ns}^{-1}\left(\frac{x}{a}\middle| k^2\right)$$

with ϕ defined by $\sin\phi = \dfrac{a}{x}$.

If $k^2 = a^2/(a^2 + b^2)$, then

$$F(\phi, k) = \sqrt{a^2 + b^2}\int_b^x \frac{dt}{\sqrt{(t^2 + a^2)(t^2 - b^2)}} = \mathrm{nc}^{-1}\left(\frac{x}{b}\middle| k^2\right)$$

with ϕ defined by $\cos\phi = \dfrac{b}{x}$;

$$F(\phi, k) = \sqrt{a^2 + b^2}\int_x^\infty \frac{dt}{\sqrt{(t^2 + a^2)(t^2 - b^2)}} = \mathrm{ds}^{-1}\left(\frac{x}{\sqrt{a^2 + b^2}}\middle| k^2\right)$$

with ϕ defined by $\sin^2\phi = \dfrac{a^2 + b^2}{a^2 + x^2}$.

If $k^2 = b^2/(a^2 + b^2)$, then

$$F(\phi, k) = \sqrt{a^2 + b^2}\int_0^x \frac{dt}{\sqrt{(t^2 + a^2)(b^2 - t^2)}} = \mathrm{sd}^{-1}\left(\frac{x\sqrt{a^2 + b^2}}{ab}\middle| k^2\right)$$

with ϕ defined by $\sin^2\phi = \dfrac{x^2(a^2 + b^2)}{b^2(a^2 + x^2)}$;

$$F(\phi, k) = \sqrt{a^2 + b^2}\int_x^b \frac{dt}{\sqrt{(t^2 + a^2)(b^2 - t^2)}} = \mathrm{cn}^{-1}\left(\frac{x}{b}\middle| k^2\right)$$

with ϕ defined by $\cos\phi = \dfrac{x}{b}$.

Notes

[1] For the elliptic integrals in Legendre's notation, k is called the *parameter* or *modulus*, $k' = \sqrt{1-k^2}$ is called the *complementary modulus*, ϕ is called the *amplitude*, and for $k = \sin\alpha$ we define α to be the *modular angle*. In other representations of the elliptic functions, the variables $m = k^2$ and $m_1 = 1 - m = k'^2 = 1 - k^2$ are sometimes used.

[2] The elliptic integrals of the first and second kind are said to be complete when the amplitude is $\phi = \pi/2$, and so $x = 1$. The following special notation is then used:

$$K = K(k) = F\left(\frac{\pi}{2}, k\right) = \int_0^{\pi/2} \frac{d\theta}{\sqrt{1 - k^2 \sin^2\theta}},$$
$$E = E(k) = E\left(\frac{\pi}{2}, k\right) = \int_0^{\pi/2} \sqrt{1 - k^2 \sin^2\theta}\, d\theta.$$

The complementary values are defined by $K' = K'(k) = K(k') = K(\sqrt{1-k^2})$ and $E' = E'(k) = E(k') = E(\sqrt{1-k^2})$. These values are related by Legendre's relation: $EK' + E'K - KK' = \pi/2$.

[3] The 9 Jacobian elliptic functions, {cd, cs, dc, ds, nc, nd, ns, sc, sd}, can be defined in terms of the three "basic" Jacobian elliptic functions: sn, cn, and dn. We have the standard relationships (from Abramowitz and Stegun [1], 16.3)

$$\begin{aligned} \operatorname{cd} u &= \frac{\operatorname{cn} u}{\operatorname{dn} u}, & \operatorname{dc} u &= \frac{\operatorname{dn} u}{\operatorname{cn} u}, & \operatorname{ns} u &= \frac{1}{\operatorname{sn} u}, \\ \operatorname{sd} u &= \frac{\operatorname{sn} u}{\operatorname{dn} u}, & \operatorname{nc} u &= \frac{1}{\operatorname{cn} u}, & \operatorname{ds} u &= \frac{\operatorname{dn} u}{\operatorname{sn} u}, \\ \operatorname{nd} u &= \frac{1}{\operatorname{dn} u}, & \operatorname{sc} u &= \frac{\operatorname{sn} u}{\operatorname{cn} u}, & \operatorname{cs} u &= \frac{\operatorname{cn} u}{\operatorname{sn} u}. \end{aligned}$$

The "basic" functions may be calculated from

$$\begin{aligned} \operatorname{sn} u &= \operatorname{sn}(u|k^2) = \sin\phi \\ \operatorname{cn} u &= \operatorname{cn}(u|k^2) = \cos\phi = \sqrt{1 - \operatorname{sn}^2 u} \\ \operatorname{dn} u &= \operatorname{dn}(u|k^2) = \sqrt{1 - k^2 \operatorname{sn}^2 u} \end{aligned} \tag{34.13}$$

where u is determined by an elliptic function of the first kind: $u = F(\phi, k)$. (That is, sn is the inverse function to F. Observe: $\operatorname{sn}(F(\phi,k)|k^2) = \sin\phi$.) Tables of the Jacobian elliptic functions may be found in Abramowitz and Stegun [1]. A superscript of -1 on any one of these twelve functions denotes the inverse function.

Table 34 relates the Legendre elliptic functions to the Jacobian elliptic functions.

[4] The differential equation $(y')^2 = 1 - y^2$ with $y(0) = 0$ has the solution $y(x) = \sin(x)$, with a period of

$$P = 2\int_{-1}^{1} \frac{dx}{\sqrt{1-x^2}} = 4\int_{0}^{1} \frac{dx}{\sqrt{1-x^2}} = 2\pi.$$

Analogously, the differential equation $(y')^2 = (1-y^2)(1-k^2y^2)$ with $y(0) = 0$ has the solution $y(x) = \operatorname{sn}(x) = \operatorname{sn}(x|k^2)$, with a period of

$$P = 4\int_{0}^{1} \frac{dx}{\sqrt{(1-x^2)(1-k^2x^2)}} = 4K(k).$$

[5] The Jacobian elliptic function $\operatorname{sn}(u)$ satisfies (see (34.13))

$$\operatorname{sn}(u) = \operatorname{sn}(u|k^2) = \operatorname{sn}\left(\int_0^{\phi} \frac{d\theta}{\sqrt{1-k^2\sin^2\theta}} \,\middle|\, k^2\right) = \sin\phi.$$

Under the transformation $t = \sin\theta$ and $x = \sin\phi$ this becomes (see (34.4))

$$\operatorname{sn}\left(\int_0^{x} \frac{dt}{\sqrt{(1-t^2)(1-k^2t^2)}} \,\middle|\, k^2\right) = x.$$

Since $\operatorname{sn}(z|0) = \sin z$, the limit of $k \to 0$ produces

$$\sin\left(\int_0^{x} \frac{dt}{\sqrt{1-t^2}}\right) = x \qquad \text{or} \qquad \sin\left(\sin^{-1} x\right) = x.$$

[6] Technically, any doubly periodic meromorphic function is called an elliptic function. All of the usual elliptic functions are doubly periodic; for example, $\operatorname{sn} x = \operatorname{sn}(x + 2iK') = \operatorname{sn}(x + 4K)$.

[7] Carlson [4]–[7] has introduced a new notation for the elliptic functions that preserves certain symmetries. Define four functions:

$$\begin{aligned}
R_F(x,y,z) &= \frac{1}{2}\int_0^{\infty} \frac{dt}{\sqrt{(t+x)(t+y)(t+z)}} \\
R_J(x,y,z,w) &= \frac{3}{2}\int_0^{\infty} \frac{dt}{\sqrt{(t+x)(t+y)(t+z)}(t+w)} \\
R_C(x,y) &= R_F(x,y,y) \\
R_D(x,y,z) &= R_J(x,y,z,z).
\end{aligned}$$

Each function has the value unity when all of its arguments are unity, and R_C and R_J are interpreted as Cauchy principal values when the last argument is negative. These functions may be calculated by the numerical routines presented in Carlson [5]–[6].

The relationship between Legendre's notation and Carlson's notation is as follows:

$$
\begin{aligned}
R_F(x,y,z) &= \frac{1}{\sqrt{z-x}} F\left(\cos^{-1}\sqrt{\frac{x}{z}}, \sqrt{\frac{z-x}{z-y}}\right) \\
F(\phi,k) &= \sin\phi R_F(\cos^2\phi, 1-k^2\sin^2\phi, 1) \\
E(\phi,) &= \sin\phi R_F(\cos^2\phi, 1-k^2\sin^2\phi, 1) \\
&\quad - \tfrac{1}{3}k^2\sin^3\phi R_D(\cos^2\phi, 1-k^2\sin^2\phi, 1) \\
\Pi(\phi,n,k) &= \sin\phi R_F(\cos^2\phi, 1-k^2\sin^2\phi, 1) \\
&\quad + \tfrac{1}{3}n\sin^3\phi R_J(\cos^2\phi, 1-k^2\sin^2\phi, 1, 1-n\sin^2\phi)
\end{aligned}
$$

From the above we find $E(k) = R_F(0,1-k^2,1) - \frac{1}{3}k^2 R_D(0,1-k^2,1)$ and $K(k) = R_F(0,1-k^2,1)$. Use of Carlson's notation for elliptical integrals dramatically reduces the number of separate cases that need to be tabulated in integral tables.

[8] The Weierstrass $\mathcal{P}$ function is an elliptic function defined by

$$
\mathcal{P}(u) = \mathcal{P}(u;\omega_1,\omega_2) = \frac{1}{u^2} + \sum_{\substack{n=\dots,-1,0,1,\dots \\ m=\dots,-1,0,1,\dots \\ \Omega=2m\omega_1+2n\omega_2}}{}' \left(\frac{1}{(u-\Omega)^2} - \frac{1}{\Omega^2}\right)
$$

where the prime means that the term $n = m = 0$ is not present in the sum. The function $z = \mathcal{P}(u)$ is the inverse function of the elliptic integral (see Iyanaga and Kawada [8], page 483)

$$
u = \int_{-\infty}^{z} \frac{dz}{\sqrt{4z^3 - g_2 z - g_3}}. \tag{34.14}
$$

[9] Weierstrass showed that any elliptic integral could be transformed into terms of the form of (34.14) by a linear fractional transformation.

It is not hard to see that an integral of the form $I = \int R_1(x, \sqrt{\phi(x)})\,dx$, where R_1 is a rational function and $\phi(x)$ is a quartic polynomial, can be mapped to the form of (34.14). With a bilinear transformation, $x = \dfrac{\alpha y + \beta}{\gamma y + \delta}$ (with $\begin{vmatrix} \alpha & \beta \\ \gamma & \delta \end{vmatrix} \neq 0$), I takes the form $I = \int R_2(y, \sqrt{\psi(y)})\,dy$, where R_2 is a rational function and $\psi(y)$ is a quartic polynomial. By an appropriate choice of coefficients in the transformation, ψ can have the form $\psi(y) = 4y^3 - g_2 y - g_3$.

To understand this transformation, let c be a root of $\phi(x)$, and presume that $\phi(x)$ does not have multiple roots. Then the transformation $x = c + 1/z$ results in

$$
\phi(x) = \frac{\phi'(c)z^3 + \frac{1}{2}\phi''(c)z^2 + \dots}{z^4}.
$$

That is, it yields a new elliptic integral that only involves a cubic polynomial. If we further let $z = ay + b$, then, for a suitable choice of b the coefficient of y^2 in the cubic polynomial will vanish. Finally, by choice of a we can force the coefficient of y^3 to be 4. For more details, see Akhiezer [2], pages 15–16.

[10] Integrals of the form $I = \int R(x, \sqrt{\Xi(x)})\, dx$, where R is a rational function and Ξ is polynomial of degree $n > 4$ are called hyperelliptic integrals. If $n = 2p + 2$ (where p is an integer) then, by rational transformations, one can obtain an equivalent integral in which the polynomial is of degree $2p+1$ (see the previous note). See Byrd and Friedman [3] for details.

[11] Consider an ellipse described by the parametric equations $x = a\cos\theta$ and $y = b\sin\theta$, with $b > a > 0$. The length of arc from $\theta = 0$ to $\theta = \phi$ is given by $s = b\int_0^{\phi} \sqrt{1 - k^2 \sin^2\theta}\, d\theta$, where $k^2 = (b^2 - a^2)/b^2$. It is because of this application that the integrals in this section are called elliptic integrals.

[12] Since there are many representations of elliptic functions, and many of them do not indicate the full functional dependence, symbolic and numerical tables of elliptic functions should be used carefully.

References

[1] M. Abramowitz and I. A. Stegun, *Handbook of Mathematical Functions*, National Bureau of Standards, Washington, DC, 1964, Chapter 17, pages 587–626.

[2] N. I. Akhiezer, *Elements of the Theory of Elliptic Functions*, Translations of Mathematical Monographs, Volume 79, Amer. Math. Soc., Providence, Rhode Island, 1990.

[3] P. F. Byrd and M. D. Friedman, *Handbook of Elliptic Integrals for Engineers and Physicists*, Springer–Verlag, New York, 1954, pages 252–271.

[4] B. C. Carlson, "Elliptic Integrals of the First Kind," *SIAM J. Math. Anal.*, **8**, No. 2, April 1977, pages 231–242.

[5] B. C. Carlson, "A Table of Elliptic Integrals of the Second Kind," *Math. of Comp.*, **49**, 1987, pages 595–606 (Supplement, *ibid.*, S13–S17).

[6] B. C. Carlson, "A Table of Elliptic Integrals of the Third Kind," *Math. of Comp.*, **51**, 1988, pages 267–280 (Supplement, *ibid.*, S1–S5).

[7] B. C. Carlson, "A Table of Elliptic Integrals: Cubic Cases," *Math. of Comp.*, **53**, No. 187, July 1989, pages 327–333.

[8] S. Iyanaga and Y. Kawada, *Encyclopedic Dictionary of Mathematics*, MIT Press, Cambridge, MA, 1980.

[9] D. F. Lauden, *Elliptic Functions and Applications*, Springer–Verlag, New York, 1989.

[10] E. T. Whittaker and G. N. Watson, *A Course of Modern Analysis*, Cambridge University Press, New York, 1962.

35. Frullanian Integrals

Applicable to Integrals of a special form.

Yields

An analytic expression for the integral.

Idea

A convergent integral can sometimes be written as the difference of two integrals that each diverge. If these two integrals diverge in the same way, then the difference may be evaluated by certain limiting processes.

Procedure

A special case will illustrate the general procedure. Consider the convergent integral

$$I = \int_0^\infty \frac{\sin^3 x}{x^2}\,dx = \frac{1}{4}\int_0^\infty \frac{3\sin x - \sin 3x}{x^2}\,dx.$$

It is improper to write this integral as

$$I = \frac{1}{4}\int_0^\infty \frac{3\sin x}{x^2}\,dx - \frac{1}{4}\int_0^\infty \frac{\sin 3x}{x^2}\,dx,$$

since both of these integrals diverge. It is proper, however, to write

$$I = \frac{1}{4}\lim_{\delta\to 0}\left(\int_\delta^\infty \frac{3\sin x}{x^2}\,dx - \int_\delta^\infty \frac{\sin 3x}{x^2}\,dx\right),$$

which can be written as (using $y = 3x$ in the second integral)

$$\begin{aligned} I &= \frac{3}{4}\lim_{\delta\to 0}\left(\int_\delta^\infty \frac{\sin x}{x^2}\,dx - \int_{3\delta}^\infty \frac{\sin y}{y^2}\,dy\right) \\ &= \frac{3}{4}\lim_{\delta\to 0}\int_\delta^{3\delta} \frac{\sin x}{x^2}\,dx \\ &= \frac{3}{4}\lim_{\delta\to 0}\int_\delta^{3\delta} \frac{1}{x}\,dx \\ &= \tfrac{3}{4}\log 3, \end{aligned} \tag{35.1}$$

since $\dfrac{\sin x}{x^2} \to \dfrac{1}{x}$ as $x \to 0$.

Example

The above procedure can be used to derive the general rule

$$\int_0^\infty \frac{f(ax) - f(bx)}{x}\,dx = [f(\infty) - f(0)]\log\frac{a}{b}. \tag{35.2}$$

The integral $J = \displaystyle\int_0^\infty \frac{\tanh ax - \tanh bx}{x}\,dx$ is in the form of (35.2), with $f(z) = \tanh z$. Since $\tanh(\infty) = 1$ and $\tanh(0) = 0$ we find that $J = \log(a/b)$.

Notes

[1] A modification of the formula in (35.2) is (see Ostrowski [3])

$$\int_0^\infty \frac{f(ax) - f(bx)}{x}\, dx = (M[f] - m[f]) \log \frac{a}{b},$$

where $M[f] = \lim_{t\to\infty} \frac{1}{t} \int_0^t f(z)\, dz$ and $m[f] = \lim_{t\to 0} t \int_t^1 f(z)\, dz$.

[2] A generalization of the formula in (35.2) is

$$\int_0^\infty [\dot{u} f(u) - \dot{v} f(v)]\, dx = M[xf(x)] \log \frac{u(\infty)}{v(\infty)} - m[xf(x)] \log \frac{u(0)}{v(0)}$$

when u and v are positive absolutely continuous functions and the limits involving u/v are positive, and $M[xf(x)]$ and $m[xf(x)]$ exist.

[3] A different generalization of the formula in (35.2) is

$$\begin{aligned} &\int_0^\infty \frac{f(ax^p) - f(bx^q)}{x}\, dx \\ &= [f(\infty) - f(0)] \left[\frac{\log a}{p} - \frac{\log b}{q}\right] + \frac{p-q}{pq} \int_0^\infty f(x) \log x\, dx. \end{aligned}$$

This formula can be extended (see Hardy [2]) to

$$\begin{aligned} &\int_0^\infty [f(ax^p) - f(bx^q)] \log^N x\, dx \\ &= \frac{1}{N+1} \sum_{n=0}^{N+1} (-1)^n \binom{N+1}{n} \left[\frac{\log^n a}{p^{N+1}} - \frac{\log^n b}{q^{N+1}}\right] \int_0^\infty f(x) \log^{N-n+1} x\, dx. \end{aligned}$$

References

[1] E. B. Elliot, "On Some (General) Classes of Multiple Definite Integrals," *Proc. London Math. Soc.*, **8**, 1877, pages 35–47 and 146–158.

[2] G. H. Hardy, "A Generalization of Frullani's Integral," *Messenger Math.*, **34**, 1905, pages 11–18.

[3] A. Ostrowski, "On Some Generalizations of the Cauchy–Frullani Integral," *Proc. Natl. Acad. Sci. USA*, **35**, 1949, pages 612–616.

[4] W. Squire, *Integration for Engineers and Scientists*, American Elsevier Publishing Company, New York, 1970, pages 99–105.

36. Functional Equations

Applicable to Definite integrals.

Yields

Sometimes an integral can be formulated as the solution to a functional equation. (That is, an algebraic equation relating the unknown function at different values of the dependent variable.)

Idea

By manipulating an integral into a functional equation, it may be possible to evaluate the integral or to obtain information about the integral.

Procedure

There are many types of functional equations that may exist. There are no general rules on how best to proceed.

Example

Consider the integral

$$I(a) = \int_0^\pi \log\left(1 + 2a\cos x + a^2\right)\,dx. \tag{36.1}$$

Breaking up the region of integration, $\int_0^\pi = \int_0^{\pi/2} + \int_{\pi/2}^\pi$, shows that $I(a) = I(-a)$. Therefore:

$$\begin{aligned} 2I(a) &= \int_0^\pi \left[\log\left(1 + 2a\cos x + a^2\right) + \log\left(1 - 2a\cos x + a^2\right)\right]\,dx \\ &= \int_0^\pi \log\left(1 + a^4 + 2a^2(1 - 2\cos^2 x)\right)\,dx. \end{aligned} \tag{36.2}$$

By performing many manipulations, such as

$$\int_{\pi/2}^\pi \log\left(1 + 2a\cos x + a^2\right)\,dx = 2\int_{\pi/2}^{3\pi/4} \log\left(1 + 2a\sin x + a^2\right)\,dx,$$

the last integral in (36.2) can be shown to be equal to

$$2I(a) = \int_0^\pi \log\left(1 + 2a^2\sin^2 x + a^4\right)\,dx = I(a^2).$$

Hence, our functional equation is $2I(a) = I(a^2)$. This is equivalent to

$$I(a) = \frac{1}{2^n} I\left(a^{2^n}\right). \tag{36.3}$$

Now note that $I(0) = 0$. When $a^2 < 1$, both terms on the right-hand side of (36.3) are approaching zero as $n \to \infty$. Therefore, we conclude that $I(a) = 0$ for $a^2 < 1$.

Notes

[1] The above example is from Squire [2]. Squire shows that the complete evaluation of (36.1) is given by $I(a) = \begin{cases} 0 & \text{for } a^2 \leq 1 \\ \pi \log a^2 & \text{for } a^2 > 1 \end{cases}$.

[2] In Book [1] the problem was to show that

$$\int_0^\infty \left\{1 - e^{-q(t)}\right\} \frac{dt}{t^{3/2}} = \pi, \qquad \text{where} \qquad q(t) = \frac{1}{\pi} \int_0^\infty \log(1+st) \frac{ds}{1+s^2}.$$

The first solution given in Book [1] started from the two functional equations for $q(t)$:

$$q(t) = q\left(\frac{1}{t}\right) + \frac{1}{2} \log t \qquad \text{and} \qquad q(t) + q(-t) = \log(1 - it),$$

where each functional equation has a different domain of applicability.

[3] Consider the integral $I = \int_0^{\pi/2} \sin^2 x \, dx$. This integral can be manipulated into $I = \int_0^{\pi/2} \cos^2 x \, dx$. Adding these two representations of I results in $2I = \int_0^{\pi/2} (\cos^2 x + \sin^2 x) \, dx = \int_0^{\pi/2} dx = \frac{\pi}{2}$, so that $I = \frac{\pi}{4}$.

[4] Consider the integral

$$I = \int_0^\infty \frac{\log x}{1+x^2} \, dx.$$

Using the change of variable $z = 1/x$ we obtain

$$I = -\int_0^\infty \frac{\log z}{1+z^2} \, dz = -I.$$

We conclude that $I = 0$.

[5] Consider the integral

$$I = \int_0^{\pi/2} \frac{\sin^n x}{\sin^n x + \cos^n x} \, dx.$$

Using $z = \pi/2 - x$ we find

$$I = \int_0^{\pi/2} \frac{\cos^n z}{\sin^n z + \cos^n z} \, dz.$$

Adding these two expressions for I results in $2I = \int_0^{\pi/2} dy = \pi/2$. Hence, $I = \pi/4$ (independent of n).

References

[1] D. L. Book, Problem #6575, *Amer. Math. Monthly*, **97**, No. 6, June–July 1990, pages 537–540.

[2] W. Squire, *Integration for Engineers and Scientists*, American Elsevier Publishing Company, New York, 1970, pages 85–87.

37. Integration by Parts

Applicable to Single and multiple integrals.

Yields

A reformulation of the integral.

Idea

There is a simple integration by parts formula; it enables many integrals to be evaluated exactly.

Procedure

The single integral $I = \int_a^b f(x)\,dx$ may often be cast into the form

$$I = \int_a^b u(x)\,dv(x). \tag{37.1}$$

If this is the case, then (37.1) may be evaluated by the integration by parts formula to obtain

$$I = u(x)v(x)\Big|_a^b - \int_a^b v(x)du(x). \tag{37.2}$$

Example 1

If we have the integral

$$J = \int_0^y x\cos x\,dx$$

we recognize that $u = x$ and $v = \sin x$ (since $dv = \cos x\,dx$). Hence, (37.2) may be used to obtain

$$J = x\sin x\Big|_0^y - \int_0^y \sin x\,dx \tag{37.3}$$

since $du = dx$. The last integral appearing in (37.3) is elementary and so

$$\begin{aligned} J &= x\sin x\Big|_0^y + \cos x\Big|_0^y \\ &= y\sin y + \cos y - 1. \end{aligned} \tag{37.4}$$

Example 2

The Gamma function is defined by the integral $\Gamma(z) = \int_0^\infty t^{z-1}e^{-t}\,dt$. If we choose $x = e^{-t}$ and $u = t^z/z$ (so that $du = t^{z-1}\,dt$), then this integral has the form of (37.1). Hence, from (37.2) we have

$$\begin{aligned}\Gamma(z) &= e^{-t}\frac{t^z}{z}\bigg|_0^\infty - \int_0^\infty \frac{t^z}{z}\left(-e^{-t}\right)dt \\ &= \frac{1}{z}\int_0^\infty t^z\left(-e^{-t}\right)dt \\ &= \frac{1}{z}\Gamma(z+1)\end{aligned}$$

or $\Gamma(z+1) = z\Gamma(z)$. Since we can easily determine that $\Gamma(1) = 1$, we conclude that, when n is a positive integer, $\Gamma(n+1) = n\cdot(n-1)\cdot\ldots\cdot 2\cdot 1 = n!$. Hence, the Gamma function is the generalization of the factorial function.

Notes

[1] The integration by parts formula may be re-applied to (37.2). For example, if f_n stands for the n-th derivative of f and g_n stands for the n-th integral of g then (see Brown [1])

$$\int_a^b fg\,dx = fg_1\bigg|_a^b - f_1g_2\bigg|_a^b + f_2g_3\bigg|_a^b - f_3g_4\bigg|_a^b + \ldots. \tag{37.5}$$

[2] Green's theorem is essentially a multidimensional generalization of the usual integration by parts formula, since it relates the value of an integral to the values of some functions on the boundary of the region. See the section on line and surface integrals (page 164) for more details.

[3] Henrici [4] uses Cauchy's theorem to relate contour integrals to area integrals. For example, for the region R bounded by the curve Γ we have

$$\begin{aligned}\int_\Gamma f(x)\,dz &= 2i\iint_R \frac{\partial f(z)}{\partial \overline{z}}\,dx\,dy, \\ \int_\Gamma f(x)\,\overline{dz} &= -2i\iint_R \frac{\partial f(z)}{\partial z}\,dx\,dy.\end{aligned}$$

[4] Integration by parts may also be used to obtain an asymptotic expansion of an integral, see page 215.

References

[1] J. W. Brown, "An Extension of Integration by Parts," *Amer. Math. Monthly*, **67**, No. 4, April 1960, page 372.

[2] P. S. Bullen, "A Survey of Integration by Parts for Perron Integrals," *J. Austral. Math. Soc. Ser. A*, **40**, No. 3, 1986, pages 343–363.

[3] U. Das, and A. G. Das, "Integration by Parts for Some General Integrals," *Bull. Austral. Math. Soc.*, **37**, No. 1, 1988, pages 1–15.

[4] P. Henrici, *Applied and Computational Complex Analysis*, Volume 3, John Wiley & Sons, New York, 1986, pages 289–290.

[5] W. Kaplan, *Advanced Calculus*, Addison–Wesley Publishing Co., Reading, MA, 1952.

[6] G. B. Thomas, Jr., and R. L. Finney, *Calculus and Analytic Geometry*, 7th Edition, Addison–Wesley Publishing Co., Reading, MA, 1988.

38. Line and Surface Integrals

Applicable to Line and surface integrals.

Yields

A reformulation as an ordinary integral.

Idea

Using a parameterization, line integrals and surface integrals can be written as ordinary integrals.

Procedure: Line Integrals

A line integral is an integral whose path of integration is a path in n-dimensional space. For example, in two dimensions, if $f(x, y)$ is continuous on the curve $\mathcal{C}$, then the integrals $\int_{\mathcal{C}} f(x,y)\,dx$ and $\int_{\mathcal{C}} f(x,y)\,dy$ both exist. Here, $\mathcal{C}$ is either continuous, or piece-wise continuous (in which case the above integrals are interpreted to be the sum of many integrals, each one of which has a smooth contour).

Line integrals can be evaluated by reducing them to ordinary integrals. For example, if $f(x, y)$ is continuous on $\mathcal{C}$, and the integration contour is parameterized by $(\phi(t), \psi(t))$ as t varies from a to b, then

$$\int_{\mathcal{C}} f(x,y)\,dx = \int_a^b f\left(\phi(t), \psi(t)\right) \phi'(t)\,dt,$$
$$\int_{\mathcal{C}} f(x,y)\,dy = \int_a^b f\left(\phi(t), \psi(t)\right) \psi'(t)\,dt.$$

In many applications, line integrals appear in the combination

$$J = \int_{\mathcal{C}} P(x,y)\,dx + \int_{\mathcal{C}} Q(x,y)\,dy,$$

which is often abbreviated as

$$J = \int_{\mathcal{C}} P(x,y)\,dx + Q(x,y)\,dy, \tag{38.1}$$

where the parentheses are implicit.

If the vector $\mathbf{u}$ is defined by $\mathbf{u} = P(x,y)\mathbf{i} + Q(x,y)\mathbf{j}$, then the integral in (38.1) can be represented as $J = \int_{\mathcal{C}} u_T\,ds$, where ds is an element of arc-length, and $u_T = \mathbf{u}\cdot\mathbf{T}$ denotes the tangential component of $\mathbf{u}$ (that is, the component of $\mathbf{u}$ in the direction of the unit tangent vector $\mathbf{T}$, the sense given by increasing s). Alternately, if the vector $\mathbf{v}$ is defined by $\mathbf{v} = Q(x,y)\mathbf{i} - P(x,y)\mathbf{j}$, then the integral in (38.1) can be represented as $J = \int_{\mathcal{C}} v_n\,ds$, where $v_n = \mathbf{v}\cdot\mathbf{n}$ denotes the normal component of $\mathbf{v}$ (that is, the component of $\mathbf{v}$ in the direction of the unit normal vector $\mathbf{n}$ which is 90° behind $\mathbf{T}$).

Path Independence

Let X, Y, Z be continuous in a domain D of space. The line integral $I = \int X\,dx + Y\,dy + Z\,dz$ will be independent of the path in D

- if and only if $\mathbf{u} = (X,Y,Z)$ is a gradient vector: $\mathbf{u} = \operatorname{grad} F$, where F is defined in D (that is, $F_x = X$, $F_y = Y$, and $F_z = Z$ throughout D),
- if and only if $\int_{\mathcal{C}} X\,dx + Y\,dy + Z\,dz = 0$ on every simple closed curve $\mathcal{C}$ in D.

Green's Theorems

Green's theorem, in its simplest form, relates a two-dimensional line integral to an integral over an area (see Kaplan [1]):

> **Theorem:** (*Green's*) Let D be a domain of the xy plane and let $\mathcal{C}$ be a piece-wise smooth simple closed curve in D whose interior R is also in D. Let $P(x,y)$ and $Q(x,y)$ be functions defined in D and having continuous first partial derivatives in D. Then
>
> $$\oint_{\mathcal{C}} (P\,dx + Q\,dy) = \iint_R \left(\frac{\partial Q}{\partial x} - \frac{\partial P}{\partial y}\right) dx\,dy. \tag{38.2}$$

Green's theorem can be written in the two alternative forms (using $\mathbf{u} = P(x,y)\mathbf{i} + Q(x,y)\mathbf{j}$ and $\mathbf{v} = Q(x,y)\mathbf{i} - P(x,y)\mathbf{j}$, as above):

$$\begin{aligned} \oint_{\mathcal{C}} u_T\,ds &= \iint_R \operatorname{curl}\mathbf{u}\,dx\,dy \\ \oint_{\mathcal{C}} v_n\,ds &= \iint_R \operatorname{div}\mathbf{v}\,dx\,dy. \end{aligned} \tag{38.3a–b}$$

The second relation in (38.3) is also known as Stokes' theorem. This theorem is sometimes stated as

Theorem: (*Stokes*) Let S be a piecewise smooth oriented surface in space, whose boundary $\mathcal{C}$ is a piecewise simple smooth simple closed curve, directed in accordance with the given orientation of S. Let $\mathbf{u} = L\mathbf{i}+M\mathbf{j}+N\mathbf{k}$ be a vector field, with continuous and differentiable components, in a domain D of space including S. Then, $\int_{\mathcal{C}} u_T \, ds = \iint_S (\operatorname{curl} \mathbf{u} \cdot \mathbf{n}) \, d\sigma$, where $\mathbf{n}$ is the chosen unit normal vector on S. That is

$$\int_{\mathcal{C}} L\,dx + M\,dy + N\,dz = \iint_S \left(\frac{\partial N}{\partial y} - \frac{\partial M}{\partial z}\right) dy\,dz + \left(\frac{\partial L}{\partial z} - \frac{\partial N}{\partial x}\right) dz\,dx + \left(\frac{\partial M}{\partial x} - \frac{\partial L}{\partial y}\right) dx\,dy$$

Green's theorem can be extended to multiply connected domains as follows:

Theorem: Let $P(x, y)$ and $Q(x, y)$ be continuous and have continuous derivatives in a domain D of the plane. Let R be a closed region in D whose boundary consists of n distinct simple closed curves $\{\mathcal{C}_1, \mathcal{C}_2, \ldots, \mathcal{C}_n\}$, where $\mathcal{C}_1$ includes $\{\mathcal{C}_2, \ldots, \mathcal{C}_n\}$ in its interior. Then

$$\oint_{\mathcal{C}_1} [P\,dx + Q\,dy] + \oint_{\mathcal{C}_2} [P\,dx + Q\,dy] + \ldots + \oint_{\mathcal{C}_n} [P\,dx + Q\,dy] = \iint_R \left(\frac{\partial Q}{\partial x} - \frac{\partial P}{\partial y}\right) dx\,dy.$$

Specifically, if $\dfrac{\partial Q}{\partial x} = \dfrac{\partial P}{\partial y}$ in D, then

$$\oint_{\mathcal{C}_1} [P\,dx + Q\,dy] + \oint_{\mathcal{C}_2} [P\,dx + Q\,dy] + \ldots + \oint_{\mathcal{C}_n} [P\,dx + Q\,dy] = 0.$$

Procedure: Surface Integrals

If a surface S is given in the form $z = f(x, y)$ for (x, y) in R_{xy}, with normal vector $\mathbf{n}$, then the surface integral

$$\iint_S L\,dy\,dz + M\,dz\,dx + N\,dx\,dy = \pm \iint_{R_{xy}} \left(-L\frac{\partial f}{\partial x} - M\frac{\partial f}{\partial y} + N\right) dx\,dy,$$

with the $+$ sign when $\mathbf{n}$ is the upper normal, and the $-$ sign when $\mathbf{n}$ is the lower normal. If we define $\mathbf{v} = L\mathbf{i} + M\mathbf{j} + N\mathbf{k}$ then we may also write

$$\iint_S L\,dy\,dz + M\,dz\,dx + N\,dx\,dy = \iint_S \mathbf{v} \cdot \mathbf{n}\,d\sigma$$

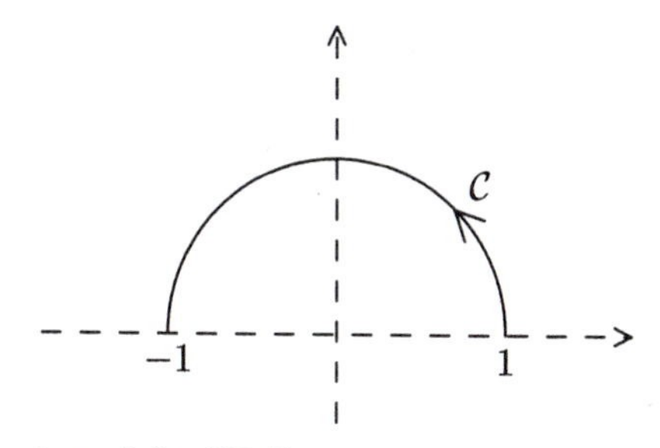

Figure 38. The contour C for (38.4).

where $d\sigma$ is an element of surface area. Here, $\mathbf{n} = \pm(-f_x\mathbf{i} - f_y\mathbf{j} + \mathbf{k})/\sqrt{1 + f_x^2 + f_y^2}$, with the + or − sign used according to whether $\mathbf{n}$ is the upper normal or lower normal.

The generalization of (38.3.a) to 3 dimensions is known as the divergence theorem, or as Gauss' theorem:

> **Theorem:** (*Divergence*) Let $\mathbf{v} = L\mathbf{i} + M\mathbf{j} + N\mathbf{k}$ be a vector field in a domain D of space. Let L, M, and N be continuous and have continuous derivatives in D. Let S be a piecewise smooth surface in D that forms the complete boundary of a bounded closed region R in D. Let $\mathbf{n}$ be the outer normal of S with respect to R. Then
>
> $$\iint_S v_n \, d\sigma = \iiint_R \operatorname{div} \mathbf{v} \, dx \, dy \, dz;$$
>
> that is
>
> $$\iint_S L \, dy \, dz + M \, dz \, dx + N \, dx \, dy$$
>
> $$= \iiint_R \left(\frac{\partial L}{\partial x} + \frac{\partial M}{\partial y} + \frac{\partial N}{\partial z} \right) dx \, dy \, dz.$$

Example 1

Consider the integral

$$I = \int_C \left(x^3 - y^3\right) dy \tag{38.4}$$

where C is the semicircle $y = \sqrt{1 - x^2}$ shown in Figure 38. The contour C can be represented parametrically by $x = \cos t$ and $y = \sin t$ for $0 \le t \le \pi$. Hence, the integral can be evaluated as

$$I = \int_0^\pi \left(\cos^3 t - \sin^3 t\right) \cos t \, dt = \frac{3\pi}{8}. \tag{38.5}$$

Alternatively, the integral could have been evaluated by using the x-parameterization throughout

$$I = \int_1^{-1} \left(x^3 - (1-x^2)^{3/2}\right)\left(\frac{-x}{\sqrt{1-x^2}}\right)\,dx.$$

This integral, which looks more awkward, is equivalent to (38.5) under the substitution $x = \cos t$.

Example 2

Here are a few examples of Green's theorem:

(A) Consider the integral $K = \oint_{\mathcal{C}} \left[\left(y^2 + \sin x^2\right)\,dx + \left(\cos y^2 - x\right)\,dy\right]$, where $\mathcal{C}$ is the boundary of the unit square ($R := \{0 \le x \le 1, 0 \le y \le 1\}$). A direct evaluation of this integral by parameterizing $\mathcal{C}$ is quite difficult. For example, using x as the parameter on the bottom piece of $\mathcal{C}$, $\{y = 0,\ 0 \le x \le 1\}$, necessitates the evaluation of the integral $\int_0^1 \sin x^2\,dx$, which is not elementary. However, using Green's theorem in (38.2) (with $P = y^2 + \sin x^2$ and $Q = \cos y^2 - x$) we may write this integral as $K = \iint_R (-1-2y)\,dx\,dy$. As a set of iterated integrals, we readily find that $K = -2$.

(B) Let $\mathcal{C}$ be the circle $x^2 + y^2 = 1$. Then, using (38.2)

$$\oint_{\mathcal{C}} \left[4xy^3\,dx + 6x^2y^2\,dy\right] = \iint_R \left(12xy^2 - 12xy^2\right)\,dx\,dy = 0.$$

(C) Let $\mathcal{C}$ be the ellipse $x^2 + 4y^2 = 4$. Then, using (38.2)

$$\begin{aligned}\oint_{\mathcal{C}} \left[(2x-y)\,dx + (x+3y)\,dy\right] &= \iint_R (1+1)\,dx\,dy \\ &= 2(\text{area of ellipse}) = 4\pi.\end{aligned}$$

Notes

[1] When the contour of integration in a line integral is closed, then we often represent the integral by the symbol $\oint$, rather than the usual $\int$.

[2] When the contour in a two-dimensional line integral is closed, and the integrand is analytic, then contour integration techniques may be used (see page 129). Sometimes a two-dimensional line integral can be extended to a closed contour and be evaluated in this manner. (Note, however, that the integrand in (38.4) is not analytic.)

[3] The evaluation of the integral in Example 2.C required that an area be known. Using Green's theorems, we can write the following integral expressions for the area bounded by the contour $\mathcal{C}$:

$$\text{area} = \oint_{\mathcal{C}} x\,dy = -\oint_{\mathcal{C}} y\,dx.$$

[4] If D is a three-dimensional domain with boundary B, let dV represent the volume element of D, let ds represent the surface element of B, and let $d\mathbf{S} = \mathbf{n}dS$, where $\mathbf{n}$ is the outer normal vector of the surface B. Then Gauss's formulas are (see Iyanaga and Kawada [3], page 1400)

$$\iiint_D \nabla \mathbf{A}\, dV = \iint_B d\mathbf{S} \cdot \mathbf{A} = \iint_B (\mathbf{n} \cdot \mathbf{A})\, dS$$

$$\iiint_D \nabla \times \mathbf{A}\, dV = \iint_B d\mathbf{S} \times \mathbf{A} = \iint_B (\mathbf{n} \times \mathbf{A})\, dS$$

$$\iiint_D \nabla\phi\, dV = \iint_B \phi\, d\mathbf{S}$$

where ϕ is an arbitrary scalar and $\mathbf{A}$ is an arbitrary vector.

[5] There are also Green's theorems that relate a volume integral to a surface integral. Let V be a volume with surface S, which we assume to be simple and closed. Define n to be the outward normal to S. Let ϕ and ψ be scalar functions which, together with $\nabla^2\phi$ and $\nabla^2\psi$, are defined in V and on S. Then (see Gradshteyn and Ryzhik [2], page 1089)

(A) Green's first theorem states that

$$\int_S \phi \frac{\partial\psi}{\partial n}\, dS = \int_V \left(\phi\nabla^2\psi + \nabla\phi \cdot \nabla\psi\right)\, dV.$$

(B) Green's second theorem states that

$$\int_S \left(\phi\frac{\partial\psi}{\partial n} - \psi\frac{\partial\phi}{\partial n}\right)\, dS = \int_V \left(\phi\nabla^2\psi - \psi\nabla^2\phi\right)\, dV.$$

References

[1] W. Kaplan, *Advanced Calculus*, Addison–Wesley Publishing Co., Reading, MA, 1952.

[2] I. S. Gradshteyn and I. M. Ryzhik, *Tables of Integrals, Series, and Products*, Academic Press, New York, 1980.

[3] S. Iyanaga and Y. Kawada, *Encyclopedic Dictionary of Mathematics*, MIT Press, Cambridge, MA, 1980.

[4] G. B. Thomas, Jr., and R. L. Finney, *Calculus and Analytic Geometry*, 7th Edition, Addison–Wesley Publishing Co., Reading, MA, 1988.

39. Look Up Technique

Applicable to Integrals of certain forms.

Yields

An exact evaluation, an approximate evaluation, or a numerical technique.

Idea

Many integrals have been named and well studied. If a given integral can be transformed to a known form, then information about the evaluation may be obtained from the appropriate reference.

Procedure

Compare the integral of interest with the lists on the following pages. If the integral of interest appears, see the reference cited for that integral.

There are four lists of integrals, those with no parameters (i.e., constants), those with one parameter, those with two parameters, and those with three or more parameters.

Notes

[1] The integrals in this section cannot be evaluated, in closed form, in terms of elementary functions (see page 77).

[2] Some of the integrals are only defined for some values of the parameters; these restrictions are not listed in the tables. For example, the gamma function $\Gamma(x)$ is not defined when x is a negative integer.

[3] Realize that the same integral may look different when written in different variables. A transformation of your integral may be required to make it look like one of the forms listed.

[4] In this section, the references follow the listings of integrals.

[5] If the integral desired is not of a common form, then it will not appear in this section. However, it might be tabulated in one of the tables of integrals, see page 190.

Constants Defined by Integrals

Catalan's constant (see Lewin [21], page 34)

$$G = \mathrm{Ti}_2(1) = \int_0^1 \frac{\tan^{-1} t}{t}\, dt = \sum_{k=0}^{\infty} \frac{(-1)^k}{(2k+1)^2} \simeq 0.91596559.$$

Euler's constant (see Gradshteyn and Ryzhik [15], 8.367.4, page 946)

$$\gamma = -\int_0^\infty e^{-t} \log t\, dt \simeq 0.577215.$$

Unnamed – related to random permutations (see Goh and Schmutz [14])

$$Z = \int_0^\infty \log\log\left(\frac{e}{1-e^{-t}}\right) dt \simeq 1.11786.$$

Integrals with One Parameter

Airy function (see Spanier and Oldham [26], 56:3:1, page 555)

$$\mathrm{Ai}(x) = \frac{1}{\pi}\int_0^\infty \cos\left(\frac{1}{3}t^3 + xt\right) dt.$$

Related to Airy's function (see Abramowitz and Stegun [1], 10.4.42, page 448)

$$\mathrm{Gi}(z) = \frac{1}{\pi}\int_0^\infty \sin\left(\frac{1}{3}t^3 + zt\right) dt.$$

Related to Airy's function (see Abramowitz and Stegun [1], 10.4.44, page 448)

$$\mathrm{Hi}(z) = \frac{1}{\pi}\int_0^\infty \exp\left(\frac{1}{3}t^3 + zt\right) dt.$$

Bairy function (see Abramowitz and Stegun [1], 10.4)

$$\mathrm{Bi}(x) = \frac{1}{\pi}\int_0^\infty \cos\left(\frac{t^3}{3} + xt\right) \exp\left(-\frac{t^3}{3} + xt\right) dt.$$

Binet integrals (see Van Der Laan and Temme [19], page 122)

$$S(z) = 2\int_0^\infty \frac{\tan^{-1}(t/z)}{e^{2\pi t} - 1}\, dt.$$

Bloch–Gruneisen integral (see Deutsch [9])

$$z(x) = \int_0^x \frac{t^5}{(e^t - 1)(1 - e^{-t})}\, dt.$$

Clausen's integral (see Lewin [21], Chapter 4, pages 91–105)

$$\mathrm{Cl}_2(\theta) = -\int_0^\theta \log\left(2\sin\frac{t}{2}\right)\,dt.$$

Cosine integral (see Abramowitz and Stegun [1], 5.2.2, page 231)

$$\mathrm{Ci}(z) = \gamma + \log z + \int_0^z \frac{\cos t - 1}{t}\,dt.$$

Auxiliary cosine integral (see Spanier and Oldham [26], 38:13:2, page 371)

$$\mathrm{gi}(x) = \int_0^\infty \frac{te^{-xt}}{t^2+1}\,dt.$$

Hyperbolic cosine integral (see Abramowitz and Stegun [1], 5.2.4, page 231)

$$\mathrm{Chi}(z) = \gamma + \log z + \int_0^z \frac{\cosh t - 1}{t}\,dt.$$

Dawson's integral (see Spanier and Oldham [26], Chapter 42, pages 405–410)

$$\mathrm{daw}(x) = \int_0^\infty e^{t^2-x^2}\,dt.$$

Dilogarithm (see Lewin [21])

$$\mathrm{Li}_2(x) = -\int_1^x \frac{\log t}{t-1}\,dt.$$

Error function (see Abramowitz and Stegun [1], 7.1.1, page 297)

$$\mathrm{erf}(z) = \frac{2}{\sqrt{\pi}}\int_0^z e^{-t^2}\,dt.$$

Complementary error function (see Abramowitz and Stegun [1], 7.1.2, page 297)

$$\mathrm{erfc}(z) = \frac{2}{\sqrt{\pi}}\int_z^\infty e^{-t^2}\,dt.$$

Complete elliptic integral of the first kind (see page 154)

$$K(k) = \int_0^{\pi/2} \frac{d\theta}{\sqrt{1-k^2\sin^2\theta}}.$$

Complete elliptic integral of the second kind (see page 154)

$$E(k) = \int_0^{\pi/2} \sqrt{1-k^2\sin^2\theta}\,d\theta.$$

Exponential integral (see Abramowitz and Stegun [1], 5.1.1, page 228)

$$E_1(z) = \int_z^\infty \frac{e^{-t}}{t}\,dt.$$

Exponential integral (see Abramowitz and Stegun [1], 5.1.2, page 228)

$$\mathrm{Ei}(x) = -\!\!\!\!\!\!\int_{-x}^{\infty} \frac{e^{-t}}{t} dt.$$

Fresnel integral (see Abramowitz and Stegun [1], 7.3.1, page 300)

$$C(z) = \int_0^z \cos\left(\frac{\pi}{2}t^2\right) dt.$$

Fresnel integral (see Abramowitz and Stegun [1], 7.3.2, page 300)

$$S(z) = \int_0^z \sin\left(\frac{\pi}{2}t^2\right) dt.$$

Gamma function (see Abramowitz and Stegun [1], 6.1.1, page 255)

$$\Gamma(z) = \int_0^\infty t^{z-1}e^{-t}\, dt.$$

Product of Gamma functions (see Abramowitz and Stegun [1], 6.1.17, page 256)

$$\Gamma(z)\Gamma(1-z) = \int_{-\infty}^{\infty} \frac{t^{z-1}}{t+1}\, dt.$$

Inverse tangent integral (see Lewin [21], Chapter 2, pages 33–60)

$$\mathrm{Ti}_2(x) = \int_0^x \frac{\tan^{-1}(t)}{t}\, dt.$$

Lebesgue constants (see Wong [27], page 40)

$$L_n = \frac{2}{\pi}\int_0^{\pi/2} \frac{|\sin(2n+1)\,t|}{\sin t}\, dt.$$

Legendre's Chi function (see Lewin [21], page 17)

$$\chi_2(x) = \frac{1}{2}\int_0^x \log\left(\frac{1+t}{1-t}\right)\frac{dt}{t}.$$

Logarithmic integral (see Abramowitz and Stegun [1], 5.1.3, page 228)

$$\mathrm{li}(x) = -\!\!\!\!\!\!\int_0^x \frac{dt}{\log t}.$$

Psi (Digamma) function (see Abramowitz and Stegun [1], 6.3.21, page 259)

$$\psi(z) = \int_0^\infty \left[\frac{e^{-t}}{t} - \frac{e^{-zt}}{1-e^{-t}}\right] dt.$$

Phi function (normal probability function) (see Abramowitz and Stegun [1], 26.2.2, page 931)

$$\Phi(x) = \frac{1}{\sqrt{2\pi}}\int_{-\infty}^x e^{-t^2/2}\, dt.$$

Riemann's zeta function (see Abramowitz and Stegun [1], 27.1.3, page 998)

$$\zeta(n+1) = \frac{1}{n!}\int_0^\infty \frac{t^n\,dt}{e^t-1} = \sum_{k=1}^{\infty}\frac{1}{k^{n+1}}.$$

Sine integral (see Abramowitz and Stegun [1], 5.2.1, page 231)

$$\mathrm{Si}(z) = \int_0^z \frac{\sin t}{t}dt.$$

Auxiliary sine integral (see Spanier and Oldham [26], 38:13:1, page 370)

$$\mathrm{fi}(x) = \int_0^\infty \frac{\sin x}{t+x}\,dt.$$

Hyperbolic sine integral (see Abramowitz and Stegun [1], 5.2.3, page 231)

$$\mathrm{Shi}(z) = \int_0^z \frac{\sinh t}{t}dt.$$

Trilogarithm (see Lewin [21], Chapter 6, pages 136–168)

$$\mathrm{Li}_3(z) = \int_0^z \frac{\mathrm{Li}_2(t)}{t}\,dt.$$

Integrals with Two Parameters

α-order Green's function (see Iyanaga and Kawada [16], page 165B)

$$G_\alpha(x) = \frac{1}{2}\int_0^\infty \frac{e^{-\alpha t}}{(2\pi t)^{d/2}}\exp\left(-\frac{x^2}{2t}\right)\,dt.$$

Anger function (see Abramowitz and Stegun [1], 12.3.1, page 498)

$$\mathbf{J}_\nu(z) = \frac{1}{\pi}\int_0^\pi \cos(\nu\theta - z\sin\theta)\,d\theta.$$

Bessel function (see Abramowitz and Stegun [1], 9.1.22, page 360)

$$J_\nu(z) = \frac{1}{\pi}\int_0^\pi \cos(z\sin\theta - \nu\theta)\,d\theta - \frac{\sin\nu\pi}{\pi}\int_0^\infty e^{-z\sinh t - \nu t}\,dt.$$

Bessel function (see Abramowitz and Stegun [1], 9.1.22, page 360)

$$Y_\nu(z) = \frac{1}{\pi}\int_0^\pi \sin(z\sin\theta - \nu\theta)\,d\theta - \frac{1}{\pi}\int_0^\infty \left\{e^{\nu t} + e^{-\nu t}\cos(\nu\pi)\right\}e^{-z\sinh t - \nu t}\,dt.$$

Beta function (see Abramowitz and Stegun [1], 6.2.1, page 258)

$$B(z,w) = \int_0^1 t^{z-1}(1-t)^{w-1}\,dt.$$

Bickley function (see Amos [2])

$$\mathrm{Ki}_n(x) = \begin{cases} \int_x^\infty \mathrm{Ki}_{n-1}(t)\,dt & \text{for } n = 1, 2, \dots \\ K_0(x) & \text{for } n = 0 \end{cases}.$$

Debye function (see Abramowitz and Stegun [1], 27.1.1, page 998)

$$z(x, n) = \int_0^x \frac{t^n\,dt}{e^t - 1}.$$

Dnestrovskii function of index q (see Robinson [25])

$$F_q(z) = \frac{1}{\Gamma(q)} \int_0^\infty \frac{t^{q-1} e^{-t}}{t + z}\,dt.$$

Elliptic integral of the first kind (see page 147)

$$F(\phi, k) = \int_0^{\sin\phi} \frac{dt}{\sqrt{(1 - t^2)(1 - k^2 t^2)}}.$$

Elliptic integral of the second kind (see page 147)

$$E(\phi, k) = \int_0^{\sin\phi} \sqrt{\frac{1 - k^2 t^2}{1 - t^2}}\,dt.$$

Repeated integrals of the error function (see Abramowitz and Stegun [1], 7.2.3, page 299)

$$\mathrm{i}^n \operatorname{erfc}(z) = \frac{2}{\sqrt{\pi}} \int_z^\infty \frac{(t - z)^n}{n!} e^{-t^2}\,dt.$$

Exponential integral (see Abramowitz and Stegun [1], 5.1.4, page 228)

$$E_n(z) = \int_1^\infty \frac{e^{-zt}}{t^n}\,dt.$$

Generalized exponential integral (see Chiccoli *et al.* [7])

$$E_v(z) = \int_1^\infty \frac{e^{-zt}}{t^v}\,dt.$$

Fermi–Dirac integral (see Fullerton and Rinker [11])

$$F_\nu(\alpha) = \int_0^\infty \frac{t^\nu}{1 + e^{t-\alpha}}\,dt.$$

Logarithmic Fermi–Dirac integral (see Fullerton and Rinker [11])

$$G_\nu(\alpha) = \int_0^\infty \frac{t^\nu \log\left(1 + e^{\alpha - t}\right)}{1 + e^{t-\alpha}}\,dt.$$

Generalized Fresnel integral (see Abramowitz and Stegun [1], 6.5.7, page 262)

$$C(x, a) = \int_x^\infty t^{a-1} \cos t\,dt.$$

Generalized Fresnel integral (see Abramowitz and Stegun [1], 6.5.8, page 262)

$$S(x,a) = \int_x^\infty t^{a-1} \sin t \, dt.$$

Hubbell rectangular-source integral (see Gabutti *et al.* [12])

$$f(a,b) = \int_0^b \tan^{-1}\left(\frac{1}{\sqrt{1+x^2}}\right) \frac{dx}{\sqrt{1+x^2}}.$$

Incomplete Gamma function (see Abramowitz and Stegun [1], 6.5.3, page 260)

$$\Gamma(a,x) = \int_x^\infty t^{a-1} e^{-t} \, dt.$$

Generalized inverse tangent integral (see Lewin [21], Chapter 3, pages 61–90)

$$\mathrm{Ti}_2(x,a) = \int_0^x \frac{\tan^{-1}(t)}{t+a} \, dt.$$

Hurwitz function (see Spanier and Oldham [26], 64:3:1, page 655)

$$\zeta(\nu;u) = \frac{1}{\Gamma(\nu)} \int_0^\infty \frac{t^{\nu-1} e^{-ut}}{1-e^{-t}} \, dt.$$

Repeated integrals of K_0 (see Abramowitz and Stegun [1], 11.2.10, page 483)

$$\mathrm{Ki}_r(z) = \int_z^\infty \frac{e^{-z\cosh t}}{\cosh^r t} \, dt.$$

Kummer's function (see Lewin [21], page 178)

$$\Lambda_n(z) = \int_0^z \frac{\log^{n-1}|t|}{1+t} \, dt.$$

Legendre function of the first kind (see Spanier and Oldham [26], 59:3:1, page 583)

$$P_\nu(x) = \frac{1}{\pi} \int_0^\infty \left[x + \sqrt{x^2-1}\cos t\right]^\nu \, dt.$$

Legendre function of the second kind (see Spanier and Oldham [26], 59:3:2, page 583)

$$Q_\nu(x) = \int_0^\infty \left[x + \sqrt{x^2-1}\cosh t\right]^{-\nu-1} \, dt.$$

Pearcey integral (see Kaminski [18])

$$P(x,y) = \int_{-\infty}^\infty \exp\left\{i\left(\frac{t^4}{4} + x\frac{t^2}{2} + yt\right)\right\} \, dt.$$

Polygamma function (see Abramowitz and Stegun [1], 6.4, page 260)

$$\Psi^{(n)}(x) = (-1)^{n+1} \int_0^\infty \frac{t^n e^{-xt}}{1-t} \, dt.$$

Log-sine integral of order n (see Lewin [21], page 243)

$$\mathrm{Ls}_n(\theta) = -\int_0^\theta \log^{n-1}\left|2\sin\frac{\theta}{2}\right| d\theta.$$

Extended log-sine integral of the third order of argument θ and parameter α (see Lewin [21], page 243)

$$\mathrm{Ls}_3(\theta,\alpha) = -\int_0^\theta \log\left|2\sin\frac{\theta}{2}\right| \log\left|2\sin\left(\frac{\theta}{2}+\frac{\alpha}{2}\right)\right| d\theta.$$

Sievert integral (see Abramowitz and Stegun [1], 27.4, page 1000)

$$z(x,\theta) = \int_0^\theta e^{-x\sec\phi}\, d\phi.$$

Polylogarithm of order n (see Lewin [21], page 169)

$$\mathrm{Li}_n(z) = \int_0^z \frac{\mathrm{Li}_{n-1}(t)}{t}\, dt.$$

Struve function (see Abramowitz and Stegun [1], 12.1.7, page 496)

$$\mathbf{H}_\nu(z) = \frac{2\left(\frac{z}{2}\right)^\nu}{\sqrt{\pi}\,\Gamma\left(\nu+\frac{1}{2}\right)} \int_0^{\pi/2} \sin(z\cos\theta)\sin^{2\nu}\theta\, d\theta.$$

Modified Struve function (see Abramowitz and Stegun [1], 12.2.2, page 498)

$$\mathbf{L}_\nu(z) = \frac{2\left(\frac{z}{2}\right)^\nu}{\sqrt{\pi}\,\Gamma\left(\nu+\frac{1}{2}\right)} \int_0^{\pi/2} \sinh(z\cos\theta)\sin^{2\nu}\theta\, d\theta.$$

Voigt function (see Lether and Wenston [20])

$$V(x,y) = \frac{y}{\pi}\int_{-\infty}^{\infty} \frac{e^{-t^2}}{(x-t)^2+y^2}\, dt.$$

Weber function (see Abramowitz and Stegun [1], 12.3.3, page 498)

$$\mathbf{E}_\nu(z) = \frac{1}{\pi}\int_0^\pi \sin(\nu\theta - z\sin\theta)\, d\theta.$$

Unnamed integral (see Bonham [5])

$$f_{2n}(x) = (-1)^n \int_0^1 P_{2n}(t)\sin xt\, dt.$$

Unnamed integral (see Bonham [5])

$$f_{2n+1}(x) = (-1)^{n+1} \int_0^1 P_{2n+1}(t)\cos xt\, dt.$$

Unnamed integral (see Chahine and Narasimha [6])

$$z(x,u) = \int_0^\infty t^n e^{-(t-u)^2 - x/t}\, dt.$$

Unnamed integral (see Glasser [13])

$$I_\nu(x) = \int_0^1 t^\nu (1-t)^\nu |\sin xt|\, dt.$$

Unnamed integral (see Nagarja [23])

$$z(x,p) = \int_0^p \frac{e^{-t^2}}{t+x}\, dt.$$

Integrals with Three or More Parameters

Incomplete Beta function (see Abramowitz and Stegun [1], 6.6.1, page 263)

$$B_x(a,b) = \int_0^x t^{a-1}(1-t)^{b-1}\, dt.$$

Elliptic integral of the third kind (see page 147)

$$\Pi(\phi, n, k) = \int_0^{\sin\phi} \frac{dt}{\left(1+nt^2\right)\sqrt{(1-t^2)(1-k^2t^2)}}.$$

Double Fermi–Dirac integral (see Fullerton and Rinker [11])

$$H_{\nu,\mu}(\alpha) = \int_0^\infty dt \frac{t^\nu}{1+e^{t-\alpha}} \int_t^\infty ds \frac{s^\nu}{1+e^{s-\alpha}}.$$

Generalized Fermi–Dirac integral (see Pichon [24])

$$F_k(\eta,\theta) = \int_0^\infty \frac{t^k\sqrt{1+\frac{1}{2}\theta t}}{1+e^{t-\eta}}\, dt.$$

Gauss function (see Spanier and Oldham [26], 60:3:1, page 600)

$$F(a,b,c;x) = \frac{\Gamma(c)}{\Gamma(b)\Gamma(c-b)} \int_0^1 \frac{t^{b-1}\, dt}{(1-t)^{1+b-c}(1-xt)^a}.$$

Incomplete hyperelliptic integral (see Loiseau, Codaccioni, and Caboz [22])

$$H(a,X;\lambda_2,\lambda_n) = \int_0^X \frac{dx}{\sqrt{a-\lambda_2 x^2 - \lambda_n x^n}}.$$

Hypergeometric function (see Abramowitz and Stegun [1], 15.3.1, page 558)

$$F(a,b,c;z) = \frac{\Gamma(c)}{\Gamma(b)\Gamma(c-b)}\int_0^1 t^{b-1}(1-t)^{c-b-1}(1-tz)^{-a}\,dt.$$

Confluent hypergeometric function (see Abramowitz and Stegun [1], 13.2.1, page 505)

$$M(a,b,c) = \frac{\Gamma(b)}{\Gamma(b-a)\Gamma(a)}\int_0^1 e^{zt}t^{a-1}(1-t)^{b-a-1}\,dt.$$

Confluent hypergeometric function (see Abramowitz and Stegun [1], 13.2.5, page 505)

$$U(a,b,c) = \frac{1}{\Gamma(a)}\int_0^\infty e^{-zt}t^{a-1}(1+t)^{b-a-1}\,dt.$$

Lerch's function (see Spanier and Oldham [26], 64:12:2, page 661)

$$\Phi(x;\nu;u) = \frac{1}{\Gamma(\nu)}\int_0^\infty \frac{t^{\nu-1}e^{-ut}}{1-xe^{-t}}\,dt.$$

Generalized log-sine integral of order n and index m (see Lewin [21], page 243)

$$\mathrm{Ls}_n^{(m)}(\theta) = -\int_0^\theta \theta^m \log^{n-m-1}\left|2\sin\frac{\theta}{2}\right|\,d\theta.$$

Selberg Integral (see Aomoto [3])

$$S(\alpha,\beta,\gamma) = \int_{[0,1]^{N-2}} \prod_{j=3}^{N} x_j^\alpha(1-x_j)^\beta \prod_{3\le i<j\le N} |x_i-x_j|^\gamma\,dx_3\wedge\ldots\wedge dx_N\wedge dx_N.$$

Complex Selberg Integral (see Aomoto [3])

$$S_c(\alpha,\beta,\gamma) = \left(\frac{i}{2}\right)^N \int_{[0,1]^{2(N-2)}} \prod_{j=3}^{N} |z_j|^{2\alpha}|z_j-1|^{2\beta} \prod_{3\le i<j\le N} |z_i-z_j|^{2\gamma}$$
$$dz_3\wedge d\overline{z}_3\wedge\ldots\wedge dz_N\wedge d\overline{z}_N.$$

Shkarofsky functions (see Robinson [25])

$$\mathcal{F}_q(z,a) = -i\int_0^\infty \frac{dt}{(1-it)^q}\exp\left(izt-\frac{at^2}{1-it}\right).$$

Wu's integral (see Fettis [10])

$$J_{p,\gamma}(\xi) = \int_0^\infty e^{-(1/2)(z+\xi z\gamma)}z^p\,dz.$$

Unnamed integral (see Anderson and Macomber [4])

$$H_n(p,q) = \int_0^\infty dt\,t^{n-2}\exp\left(-\frac{1}{2}(t-p)^2-\frac{q}{t}\right).$$

Unnamed integral (see Cole [8])

$$I(y,z,s) = \int_0^\infty \exp[-s(t + y\cos t - z\sin t)]\,dt.$$

Unnamed integral (see Kölbig [17])

$$z(\nu,\lambda,m) = \int_0^1 x^{\nu-1}(1-x)^{-\lambda}\log^m x\,dx.$$

References

[1] M. Abramowitz and I. A. Stegun, *Handbook of Mathematical Functions*, National Bureau of Standards, Washington, DC, 1964.

[2] D. E. Amos, "Algorithm 609: A Portable FORTRAN Subroutine for the Bickley Functions $\mathrm{Ki}_n(x)$," *ACM Trans. Math. Software*, **9**, No. 4, December 1983, pages 480–493.

[3] K. Aomoto, "On the Complex Selberg Integral," *Quart. J. Math. Oxford*, **38**, No. 2, 1987, pages 385–399.

[4] D. G. Anderson and H. K. Macomber, "Evaluation of $\frac{1}{\sqrt{2\pi}}\int_0^\infty du\, u^{n-2} \exp\left\{-\frac{1}{2}(u-p)^2 - \frac{q}{u}\right\}$," *J. Math. and Physics*, **45**, 1966, pages 109–120.

[5] R. A. Bonham, "On Some Properties of the Integrals $\int_0^1 P_{2n}(t)\sin xt\,dt$ and $\int_0^1 P_{2n+1}(t)\cos xt\,dt$," *J. Math. and Physics*, **45**, 1966, pages 331–334.

[6] M. T. Chahine and R. Narasimha, "The Integral $\int_0^\infty v^n \exp\left[-(v-u)^2 - x/v\right] dv$," *J. Math. and Physics*, **43**, 1964, pages 163–168.

[7] C. Chiccoli, S. Lorenzutta, and G. Maino, "Recent Results for Generalized Exponential Integrals," *Comp. & Maths. with Appls.*, **19**, No. 5, 1990, pages 21–29.

[8] R. J. Cole, "Two Series Representations of the Integral $\int_0^\infty \exp[-s(\psi+y\cos\psi - z\sin\psi)]\,d\psi$," *J. Comput. Physics*, **44**, 1981, pages 388–396.

[9] M. Deutsch, "An Accurate Analytic Representation for the Bloch–Gruneisen Integral," *J. Phys. A: Math. Gen.*, **20**, 1987, pages L811-L813.

[10] H. E. Fettis, "On the Calculation of Wu's Integral," *J. Comput. Physics*, **53**, 1984, pages 197–204.

[11] I. W. Fullerton and G. A. Rinker, "Generalized Fermi–Direc Integrals – FD, FDG, FDH," *Comput. Physics Comm.*, **39**, 1986, pages 181–185.

[12] B. Gabutti, S. L. Kalla, and J. H. Hubbell, "Some Expressions Related to the Hubbell Rectangular-Source Integral," *J. Comput. Appl. Math.*, **37**, 1991, pages 273–285.

[13] M. L. Glasser, "A Note on the Integral $\int_0^1 s^\nu(1-s)^\nu|\sin xs|\,ds$ with an Application to Schlomilch Series," *J. Math. and Physics*, **43**, 1964, pages 158–162.

[14] W. M. Y. Goh and E. Schmutz, "The Expected Order of a Random Permutation," *Bull. London Math. Soc.*, **23**, 1991, pages 34–42.

[15] I. S. Gradshteyn and I. M. Ryzhik, *Tables of Integrals, Series, and Products*, Academic Press, New York, 1980.

[16] S. Iyanaga and Y. Kawada, *Encyclopedic Dictionary of Mathematics*, MIT Press, Cambridge, MA, 1980.

[17] K. S. Kölbig, "On the Integral $\int_0^1 x^{\nu-1}(1-x)^{-\lambda} \log^m x\, dx$," *J. Comput. Appl. Math.*, **18**, No. 3, 1987, pages 369–394.

[18] D. Kaminski, "Asymptotic Expansion of the Pearcey Integral Near the Caustic," *SIAM J. Math. Anal.*, **20**, No. 4, July 1989, pages 987–1005.

[19] C. G. van der Laan and N. M. Temme, *Calculation of Special Functions: The Gamma function, the Exponential Integrals and Error-like Functions*, Centrum voor Wiskunde en Informatica, Amsterdam, 1984.

[20] F. G. Lether and P. R. Wenston, "The Numerical Computation of the Voigt Function by the Corrected Midpoint Quadrature Rule for $(-\infty, \infty)$," *J. Comput. Appl. Math.*, **34**, 1991, pages 75–92.

[21] L. Lewin, *Dilogarithms and Associated Functions*, MacDonald & Co., London, 1958.

[22] J. F. Loiseau, J. P. Codaccioni, and R. Caboz, "Incomplete Hyperelliptic Integrals and Hypergeometric Series," *Math. of Comp.*, **53**, No. 187, July 1989, pages 335–342.

[23] K. S. Nagarja, "Concerning the Value of $\int_0^p \frac{e^{-u^2}}{u+x}\, du$," *J. Math. and Physics*, **44**, 1965, pages 182–188.

[24] B. Pichon, "Numerical Calculation of the Generalized Fermi–Direc Integrals," *Comput. Physics Comm.*, **55**, 1989, pages 127–136.

[25] P. A. Robinson, "Relativistic Plasma Dispersion Functions: Series, Integrals, and Approximations," *J. Math. Physics*, **28**, No. 5, 1987, pages 1203–1205.

[26] J. Spanier and K. B. Oldham, *An Atlas of Functions*, Hemisphere Publishing corporation, New York, 1987.

[27] R. Wong, *Asymptotic Approximation of Integrals*, Academic Press, New York, 1989.

40. Special Integration Techniques

Applicable to There are many specialized integration techniques, each of which works on a special class of integrals.

Integrands involving solutions of second order differential equations

Consider the integral

$$I = \int f\left(y_1(x), y_2(x)\right)\, dx \tag{40.1}$$

where y_1 and y_2 are linearly independent solutions to the ordinary differential equation: $y'' = Q(x)y$. (Note that the Wronskian of y_1 and y_2 is a constant; $W(y_1, y_2) = c \neq 0$.) For this method to work, we require that f be homogeneous of degree -2, that is $f(ay_1, ay_2) = a^{-2} f(y_1, y_2)$. In this case we find

$$I = \int f\left(y_1(x), y_2(x)\right)\, dx = \int f\left(1, \frac{y_2}{y_1}\right) \frac{W(y_1, y_2)}{c} \frac{dx}{y_1^2}$$
$$= \frac{1}{c} \int f(1, u)\, du \tag{40.2}$$

where $u = y_2/y_1$.

Example 1

The following example is from Ashbaugh [3]. The integral

$$J = \int \frac{\sin^2 x}{\cos^4 x + \cos x \sin^3 x}\, dx$$

can be evaluated by identifying $y_1 = \cos x$ and $y_2 = \sin x$ as being solutions to $y'' + y = 0$. Hence, we write J as

$$\begin{aligned} J = \int \frac{y_2^2}{y_1^4 + y_1 y_2^3}\, dx &= \int \frac{y_2^2}{y_1^4 + y_1 y_2^3} \frac{y_1 y_2' - y_1' y_2}{1}\, dx \\ &= \int \frac{u^2}{1+u^3}\, du \\ &= \tfrac{1}{3} \log\left|1 + u^3\right| + C \\ &= \tfrac{1}{3} \log\left|1 + \left(\frac{y_2}{y_1}\right)^3\right| + C \\ &= \tfrac{1}{3} \log\left|1 + \left(\frac{\cos x}{\sin x}\right)^3\right| + C \\ &= \tfrac{1}{3} \log\left|1 + \tan^3 x\right| + C \end{aligned}$$

where C is an arbitrary constant.

Example 2

The following example is from Ashbaugh [3]. The integral

$$K = \int_0^\infty \frac{dx}{(\mathrm{Ai}(x) - i\,\mathrm{Bi}(x))^2}$$

(where Ai (Bi) is the Airy (Bairy) function) has the form of (40.1), with $y_1(x) = \mathrm{Ai}(x) - i\,\mathrm{Bi}(x)$ and $y_2(x) = \mathrm{Bi}(x)$. We can write K as (recall that

Ai and Bi satisfy Airy's equation: $y'' = xy$)

$$K = \int_0^\infty \frac{dx}{(\mathrm{Ai}(x) - i\,\mathrm{Bi}(x))^2} = \int_0^\infty \frac{y_1 y_2' - y_1' y_2}{\pi^{-1}} \frac{dx}{y_1^2}$$
$$= \pi \int_{x=0}^{x=\infty} 1\, du$$
$$= \pi \left. (u) \right|_{x=0}^{x=\infty}$$
$$= \pi \left. \left(\frac{\mathrm{Bi}(x)}{\mathrm{Ai}(x) - i\,\mathrm{Bi}(x)} \right) \right|_{x=0}^{x=\infty}$$
$$= \pi \frac{i - \sqrt{3}}{4}.$$

Integration of Rational Functions

A rational function can always be integrated into a sum of rational functions and logarithmic (or arctangent) functions. Using $D(X)$ and $N(x)$ to represent polynomials, a rational function can be written using partial fractions as

$$f(x) = \frac{N(x)}{D(x)} = (\text{polynomial in } x) + \sum_i \frac{N(r_i)}{(x - r_i)\prod_{j \neq i}(r_i - r_j)}$$

where the $\{r_i\}$ are the roots of $D(x) = 0$ (assumed here to be distinct). This integrand can be directly integrated to obtain

$$\begin{aligned} \int f(x)\, dx &= \int (\text{polynomial in } x)\, dx + \int \sum_i \frac{N(r_i)}{(x - r_i)\prod_{j \neq i}(r_i - r_j)}\, dx \\ &= (\text{different polynomial in } x) + \sum_i \frac{N(r_i) \log(x - r_i)}{\prod_{j \neq i}(r_i - r_j)}. \end{aligned} \tag{40.3}$$

This is formally correct, even when the roots are complex. Since logarithms and arctangents are related by $\log(u - iv) = 2i \tan^{-1}(u/v)$, the expression in (40.3) is always applicable.

Example 3

$$\begin{aligned} \int \frac{dx}{x^4 - 1} &= \int \frac{dx}{(x-1)(x+1)(x-i)(x+i)} \\ &= \tfrac{1}{4} \log(x - 1) + \tfrac{1}{4} \log(x + 1) + \tfrac{1}{4i} \log(x - i) + \tfrac{1}{4i} \log(x + i) \\ &= \frac{1}{4} \log \frac{x - 1}{x + 1} - \frac{1}{2} \tan^{-1} x. \end{aligned}$$

For the case when some of the roots are repeated, see Stroud [6].

Use of Infinite Series

Sometime an integral may be evaluated by expanding the integrand in a series, integrating term by term, and then re-summing the result. This technique may also be used to obtain an asymptotic expansion, see Wong [7] for details.

Example 4

Consider the integral $I = \int_0^\infty \frac{\sin ax}{e^x - 1}\,dx$. Formally expanding the denominator, and interchanging integration and summation results in

$$\begin{aligned} I &= \sum_{n=1}^{\infty} \int_0^\infty e^{-nx} \sin ax\,dx \\ &= \sum_{n=1}^{\infty} \frac{a}{a^2 + n^2} \\ &= \frac{\pi}{2}\left(\cot a\pi - \frac{1}{a\pi}\right) \end{aligned}$$

where we used Jolley [4] to recognize the cotangent sum.

Example 5

Consider the integral

$$I = \int_0^\infty e^{-\alpha x^2} J_1(\beta x)\,dx \tag{40.4}$$

where J_1 is a Bessel function. From Abramowitz and Stegun [1] 9.1.10, we note that

$$J_1(\beta x) = \frac{\beta x}{2} \sum_{k=0}^{\infty} \frac{\left(-\frac{1}{4}\beta^2 x^2\right)^k}{k!(k+1)!}. \tag{40.5}$$

Using (40.5) in (40.4), and interchanging orders of integration (see page 109), results in

$$\begin{aligned} I &= \frac{\beta}{2} \sum_{k=0}^{\infty} \frac{1}{k!(k+1)!} \left(-\frac{\beta^2}{4}\right)^k \int_0^\infty e^{-\alpha x^2} x^{2k-1}\,dx \\ &= \frac{\beta}{2} \sum_{k=0}^{\infty} \frac{1}{k!(k+1)!} \left(-\frac{\beta^2}{4}\right)^k \frac{k!}{2\alpha^{k+1}} \\ &= -\frac{1}{\beta} \sum_{k=0}^{\infty} \left(-\frac{\beta^2}{4\alpha}\right)^{k+1} \frac{1}{(k+1)!} \\ &= \frac{1}{\beta}\left(1 - e^{-\beta^2/4\alpha}\right) \end{aligned}$$

where we had to recognize the series for the exponential function.

Series Transpositions

Squire [5] presents an elegant technique using series expansions and re-summing. An example will demonstrate the general ideas.

Consider the integral $I = \int_0^\infty \frac{\sin x}{x}\, dx$. The infinite integration region can be written as an infinite sum of finite integration regions, and then each finite region of integration can have the variables changed:

$$
\begin{aligned}
I &= \sum_{k=0}^{\infty} \int_{k\pi/2}^{(k+1)\pi/2} \frac{\sin x}{x}\, dx = \sum_{k=0}^{\infty} \int_0^{\pi/2} \sin\left(\frac{k\pi}{2} + u\right) \frac{du}{\frac{1}{2}k\pi + u} \\
&= \sum_{k=0}^{\infty} \int_0^{\pi/2} \left(\frac{\sin u \cos \frac{k\pi}{2}}{\frac{1}{2}k\pi + u} + \frac{\cos u \sin \frac{k\pi}{2}}{\frac{1}{2}k\pi + u} \right) du \\
&= \sum_{k=0}^{\infty} \int_0^{\pi/2} \left(\frac{\sin u \cos \frac{k\pi}{2}}{\frac{1}{2}k\pi + u} + \frac{\sin u \sin \frac{k\pi}{2}}{\frac{1}{2}(k+1)\pi - u} \right) du \\
&= \int_0^{\pi/2} du\, \sin u \sum_{k=0}^{\infty} \left(\frac{\cos \frac{k\pi}{2}}{\frac{1}{2}k\pi + u} + \frac{\sin \frac{k\pi}{2}}{\frac{1}{2}(k+1)\pi - u} \right) \\
&= \int_0^{\pi/2} du\, \sin u \sum_{k=0}^{\infty} \left(\left[\frac{1}{u} - \frac{1}{\pi + u} + \dots \right] + \left[\frac{1}{\pi - u} - \frac{1}{2\pi - u} + \dots \right] \right) \\
&= \int_0^{\pi/2} du\, \sin u \left(\frac{1}{\sin u} \right) = \int_0^{\pi/2} du = \frac{\pi}{2}
\end{aligned}
$$

where the changes of variables used were $u = x - k\pi/2$ and then $u = \pi/2 - u$. The final summation was found in Jolley [4], summation #769.

Note

[1] Arora *et al.* [2] notice that the two properties

(A) $\int_0^a f(x)\, dx = \int_0^{a/2} f(x)\, dx + \int_0^{a/2} f(a-x)\, dx$

(B) $\int_0^a f(x)\, dx = \int_0^a f(a-x)\, dx$,

if used cleverly, can evaluate some sophisticated integrals. They evaluate the integral $\int_0^\infty \frac{\log x}{1+x^3}\, dx = -\frac{\pi}{4}$ to demonstrate this.

References

[1] M. Abramowitz and I. A. Stegun, *Handbook of Mathematical Functions*, National Bureau of Standards, Washington, DC, 1964.

[2] A. K. Arora, S. K Goel, and D. M. Rodriguez, "Special Integration Techniques for Trigonometric Integrals," *Amer. Math. Monthly*, February 1988, Vol95, No. 2, pages 126–130.

[3] M. S. Ashbaugh, "On Integrals of Combination of Solutions of Second-Order Differential Equations," *J. Phys. A: Math. Gen.*, **19**, 1986, pages 3701–3703.

[4] L. B. W. Jolley, *Summation of Series*, Second Edition, Dover Publications, Inc., New York, 1961.

[5] W. Squire, *Integration for Engineers and Scientists*, American Elsevier Publishing Company, New York, 1970, page 92.

[6] A. H. Stroud, *Approximate Calculation of Multiple Integrals*, Prentice–Hall Inc., Englewood Cliffs, NJ, 1971, Section 2.4, pages 49–52.

[7] R. Wong, *Asymptotic Approximation of Integrals*, Academic Press, New York, 1989, Chapter 4, pages 195–240.

41. Stochastic Integration

Applicable to Integrals in which the measure involves Brownian noise.

Yields

Information on how to evaluate Ito and Stratonovich integrals.

Idea

Different types of stochastic integrals exist, depending on how points are chosen in a limiting process.

Background

Suppose that $w(t)$ is the Wiener process and $G = G(t) = G(t, w(t))$ is an arbitrary function of $w(t)$ and the time t. (See the Notes section for information about the Wiener process.) The stochastic integral

$$I = \int_{t_0}^{t} G(s)\, dw(s) \tag{41.1}$$

is defined as a kind of Riemann–Stieltjes integral. That is, first divide the interval $[t_0, t]$ into n sub-intervals: $t_0 \leq t_1 \leq t_2 \leq \cdots \leq t_{n-1} \leq t_n = t$. Then choose the points $\{\tau_i\}$, for $i = 1, 2, \ldots, n$, such that τ_i lies in the i-th sub-interval: $t_{i-1} \leq \tau_i \leq t_i$. The stochastic integral I is now defined as a limit of partial sums, $I = \lim_{n\to\infty} S_n$, with

$$S_n = \sum_{i=1}^{n} G(\tau_i, w(\tau_i))[w(t_i) - w(t_{i-1})]. \tag{41.2}$$

Note that this limit *depends on the particular choice of the intermediate points* $\{\tau_i\}$.

Consider, for example, the special case of $G(t) = w(t)$. Then we find the expectation of S_n to be

$$
\begin{aligned}
E[S_n] &= E\left[\sum_{i=1}^{n} w(\tau_i)[w(t_i) - w(t_{i-1})]\right] \\
&= \sum_{i=1}^{n}[\min(\tau_i, t_i) - \min(\tau_i, t_{i-1})] \\
&= \sum_{i=1}^{n}(\tau_i - t_{i-1}).
\end{aligned}
$$

If, for example, we take $\tau_i = \alpha t_i + (1-\alpha)t_{i-1}$, where $0 < \alpha < 1$, then $E[S_n]$ becomes

$$
E[S_n] = \sum_{i=1}^{n}(t_i - t_{i-1})\alpha = (t - t_0)\alpha. \tag{41.3}
$$

Clearly, the value of S_n (and hence I), in this example, depends on α.

For consistency, some specific choice must be made for the points $\{\tau_i\}$. For the Ito stochastic integral we choose $\tau_i = t_{i-1}$ (i.e., $\alpha = 0$ in the above). We show this by use of the notation $I\!\!\int$. That is

$$
I\!\!\int_{t_0}^{t} G(s, w(s))\, dw(s) = \underset{n\to\infty}{\text{ms-lim}} \left\{ \sum_{i=1}^{n} G(t_{i-1}, w(t_{i-1}))[w(t_i) - w(t_{i-1})] \right\}, \tag{41.4}
$$

where ms-lim refers to the mean square limit.

Example

Suppose we would like to evaluate the Ito stochastic integral

$$I\!\!\!\int_{t_0}^{t} w(s)\, dw(s). \tag{41.5}$$

If we write w_i for $w(t_i)$ then (41.4) becomes (using $G(s) = w(s)$) $I\!\!\!\int_{t_0}^{t} w(s)\, dw(s) = \text{ms-lim}_{n\to\infty} S_n$ with

$$\begin{aligned}
S_n &= \sum_{i=1}^{n} w_{i-1}[w_i - w_{i-1}] \\
&= \sum_{i=1}^{n} w_{i-1}\Delta w_i \\
&= \tfrac{1}{2}\sum_{i=1}^{n} \left[(w_{i-1} + \Delta w_i)^2 - (w_{i-1})^2 - (\Delta w_i)^2\right] \\
&= \tfrac{1}{2}\sum_{i=1}^{n} \left[(w_i)^2 - (w_{i-1})^2\right] - \tfrac{1}{2}\sum_{i=1}^{n}(\Delta w_i)^2 \\
&= \tfrac{1}{2}\left[w^2(t) - w^2(t_0)\right] - \tfrac{1}{2}\sum_{i=1}^{n}(\Delta w_i)^2
\end{aligned}$$

where $\Delta w_i = w_i - w_{i-1}$. Now the mean-square limit of $\frac{1}{2}\sum_{i=1}^{n}(\Delta w_i)^2$ can be shown to be $\frac{1}{2}(t - t_0)$. Hence,

$$\begin{aligned}
I\!\!\!\int_{t_0}^{t} w(s)\, dw(s) &= \underset{n\to\infty}{\text{ms-lim}} \left\{ \tfrac{1}{2}\left[w^2(t) - w^2(t_0)\right] - \tfrac{1}{2}\sum_{i=1}^{n}(\Delta w_i)^2 \right\} \\
&= \tfrac{1}{2}\left[w^2(t) - w^2(t_0) - (t - t_0)\right].
\end{aligned} \tag{41.6}$$

Note that the result is not the same result that we would have obtained by the usual Riemann–Stieltjes integral (in which the last term would be absent). Note that the expectation of (41.6) yields the value 0, which is the same value as given by (41.3).

Notes

[1] The Wiener process is a Gaussian random process that has a fixed mean given by its starting point, $E[w(t)] = w_0 = w(t_0)$, and a variance of $E[(w(t) - w_0)^2] = t - t_0$. From this we can compute that $E[w(t)w(s)] = \min(t, s)$. The sample paths of $w(t)$ are continuous, but not differentiable.

[2] We define the Stratonovich stochastic integral (indicated by use of the notation $\fint$) to be (see Schuss [2])

$$\fint_{t_0}^{t} G(w(s), x)\, dw(s) = \operatorname*{ms-lim}_{n \to \infty} \left\{ \sum_{i=1}^{n} G\left(\frac{w(t) + w(t_{i-1})}{2}, t_{i-1} \right) [w(t_i) - w(t_{i-1})] \right\}.$$

[3] It can be shown that the Stratonovich integral has the usual properties of integrals. In particular, we have the fundamental theorem of integral calculus,

$$\fint_{t_0}^{t} f'(w(s))\, dw(s) = f(w(t)) - f(w(t_0)),$$

integration by parts, etc. Taking the Stratonovich integral of the integrand in (41.5) results in $\fint_{t_0}^{t} w(s)\, dw(s) = \frac{1}{2}\left[w^2(t) - w^2(t_0)\right]$.

[4] For arbitrary functions G, there is no connection between the Ito integral and the Stratonovich integral. However, when $x(t)$ satisfies the stochastic differential equation $dx(t) = a[x(t), t]\, dt + b[x(t), t]\, dw(t)$, it can be shown that (see Gardiner [1])

$$\fint_{t_0}^{t} b[x(s), s]\, dw(s) = \mathcal{I}\!\!\int_{t_0}^{t} b[x(s), s]\, dw(s) + \frac{1}{2} \int_{t_0}^{t} b[x(s), s] \frac{\partial b[x(s), s]}{\partial x}\, ds.$$

This relates, in a way, the Stratonovich integral and the Ito integral.

[5] Stochastic integration can also refer to the (ordinary) integration of random variables. Since the linear operations of integration and expectation commute, the following results are straightforward to derive. Let $\{X(t)\}$ be a continuous parameter stochastic process with finite second moments, whose mean ($m(t) = E[X(t)]$) and covariance ($K(s, t) = \operatorname{Cov}[X(s), X(t)]$) are continuous functions of s and t. Then

$$E\left[\int_a^b X(t)\, dt\right] = \int_a^b m(t)\, dt$$

$$E\left[\left|\int_a^b X(t)\, dt\right|^2\right] = \int_a^b \int_a^b E[X(s)X(t)]\, dt\, ds$$

$$\operatorname{Var}\left[\int_a^b X(t)\, dt\right] = \int_a^b \int_a^b K(s, t)\, dt\, ds.$$

[6] Suppose that $x(t)$ is a random process that satisfies the stochastic differential equation $dx(t) = a[x(t), t]\,dt + b[x(t), t]\,dw(t)$. If we define $\alpha(x, t) = a(x, t) - \frac{1}{2}b(x, t)\partial b(x, t)/\partial x$, then the solution to the stochastic differential equation, $x(t)$, can be shown to satisfy (see Gardiner [1])

$$x(t) = x(t_0) + \int_{t_0}^{t} \alpha[x(s), s]\,ds + S\!\!\int_{t_0}^{t} b[x(s), s]\,dw(s).$$

References

[1] C. W. Gardiner, *Handbook of Stochastic Methods*, Springer–Verlag, New York, Second Edition, 1985.

[2] Z. Schuss, *Theory and Applications of Stochastic Differential Equations*, John Wiley & Sons, New York, 1980.

[3] K. L. Chung and R. J. Williams, *Introduction to Stochastic Integration*, Birkhäuser, Basel, 1990.

42. Tables of Integrals

Applicable to Specific definite and indefinite integrals.

Yields

An exact evaluation.

Idea

The evaluation of many integrals has been tabulated.

Procedure

Compare the integral of interest with the tables in the references. These tables typically include indefinite and definite integration of both elementary and special functions (including, for instance, Bessel functions and hypergeometric functions).

Example

Suppose that we would like to determine the value of the definite integral $I = \int_0^{\pi/2} \tan^a x\,dx$. Some tables have I tabulated more or less as written, some tables give only the indefinite integral, and other tables give information that can be manipulated to yield the value of I. We find:

- Beyer [3] has a recursion for the indefinite integral in number 423:
$$\int (\tan^n ax)\,dx = \frac{\tan^{n-1} ax}{a(n-1)} - \int (\tan^{n-2} ax)\,dx.$$
- Bois [4] lists several recursions for an integrand similar to that in I. One of these (on page 133) is:
$$\int \frac{\sin^m x}{\cos^n x} x\,dx = \frac{\sin^{m-1} x}{(n-1)\cos^{n-1} x} - \frac{m-1}{n-1}\int \frac{\sin^{m-1} x}{\cos^{n-1} x} x\,dx.$$

- Dwight [5] has a recursion for the indefinite integral in number 480.9:

$$\int \tan^n x\,dx = \frac{\tan^{n-1} x}{n-1} - \int \tan^{n-2} x\,dx.$$

- In Gradshteyn and Ryzhik [7], the definite integral appears as number 3.622.1:

$$\int_0^{\pi/2} \operatorname{tg}^{\pm\mu} x\,dx = \frac{\pi}{2}\sec\frac{\mu\pi}{2} \qquad \text{[when]} \qquad |\operatorname{Re}\mu| < 1.$$

- In Gradshteyn and Ryzhik [7], the indefinite integral appears as numbers 2.527.2, 2.527.3, and 2.527.4:

$$\int \operatorname{tg}^p x\,dx = \frac{\operatorname{tg}^{p-1} x}{p-1} - \int \operatorname{tg}^{p-2} x\,dx \qquad \text{[when]} \qquad p \neq 1$$

$$\int \operatorname{tg}^{2n+1} x\,dx = \sum_{k=1}^{n} \frac{(-1)^{k-1}\operatorname{tg}^{2n-2k+2} x}{2n-2k+2} - (-1)^n \ln\cos x$$

$$\int \operatorname{tg}^{2n} x\,dx = \sum_{k=1}^{n} (-1)^{k-1} \frac{\operatorname{tg}^{2n-2k+1} x}{2n-2k+1} - (-1)^n x.$$

- In Gröbner and Hofreiter [8] the definite integral appears as number 331.31:

$$\int_0^{\pi/2} \operatorname{tg}^{\lambda} x\,dx = \frac{\pi}{2\cos\frac{\lambda\pi}{2}} \qquad \text{[when]} \qquad -1 < \lambda < 1.$$

- In Prudnikov, Brychov, and Marichev [12], the integral appears as number 2.5.26.7:

$$\int_0^{\pi/2} \operatorname{tg}^{\mu} x\,dx = \frac{\pi}{2\cos(\mu\pi/2)} \qquad \text{[when]} \qquad |\operatorname{Re}\mu| < 1.$$

- In Prudnikov, Brychov, and Marichev [12], the indefinite integral appears as numbers 1.5.8.1, 1.5.8.2, and 1.5.8.3:

$$\int (\operatorname{tg} x)^p\,dx = \pm\frac{1}{p-1}(\operatorname{tg} x)^{p-1} - \int (\operatorname{tg} x)^{p-2}\,dx$$

$$\int (\operatorname{tg} x)^{2n}\,dx = \mp\sum_{k=1}^{n} (-1)^k \frac{1}{2n-2k+1}(\operatorname{tg} x)^{2n-2k+1} + (-1)^n x$$

$$\int (\operatorname{tg} x)^{2n+1}\,dx = \mp\sum_{k=1}^{n} (-1)^k \frac{1}{2n-2k+2}(\operatorname{tg} x)^{2n-2k+2} \mp (-1)^n \log|\cos x|.$$

Notes

[1] Realize that the same integral may look different when written in different variables. A transformation of your integral may be required to make it look like one of the forms listed.

[2] It is not always clear that having a symbolic evaluation of an integral is more useful than the original integral. For example, it is straightforward to show that

$$\begin{aligned} I(x) = \int_0^x \frac{dt}{1+t^4} &= \frac{1}{4\sqrt{2}} \log\left(\frac{x^2 + x\sqrt{2}+1}{x^2 - x\sqrt{2}+1}\right) \\ &+ \frac{1}{2\sqrt{2}}\left[\tan^{-1}\left(\frac{x}{\sqrt{2}+x}\right) + \tan^{-1}\left(\frac{x}{\sqrt{2}-x}\right)\right]. \end{aligned}$$

Given a specific value of x, numerically determining $I(x)$ by using this formula requires the computation of logarithms and inverse tangents. In some cases it might be easier to approximate $I(x)$ numerically directly from its definition.

[3] It is an unfortunate fact that a not insignificant fraction of the tabulated integrals are in error. See, for example, Klerer and Grossman [9].

A common error is to produce a discontinuous antiderivative when a continuous integral is available. For example, the symbolic computer language REDUCE produces (see page 117):

$$\begin{aligned} I &= \int^x \frac{dx}{2+\cos x} \\ &= -\frac{2\sqrt{3}}{3}\arctan\left(\frac{\sin x}{\cos x + 1}\right) + \frac{2\sqrt{3}}{3}\arctan\left(\frac{\sqrt{3}\sin x}{3(\cos x+1)}\right) + \frac{x}{\sqrt{3}}. \end{aligned}$$

This (correct) antiderivative is continuous, yet Abramowitz and Stegun [1], 4.3.133, report the discontinuous result $\widehat{I} = \frac{2}{\sqrt{3}}\arctan\frac{\tan(x/2)}{\sqrt{3}}$. These antiderivatives agree on the interval $-\pi < x < \pi$, but $\widehat{I}$ is periodic while I is not.

[4] If an integral is recognized to be of a certain form, then appropriate tables may be used. For example, an integral of the form $I = \int_{-\infty}^{\infty} f(x)e^{-xt}\,dx$ represents a Fourier transform of the function $f(x)$. Hence, a table of Fourier transforms (such as Oberhettinger [11]) might be an appropriate place to look for an evaluation of I.

[5] Note that Oberhettinger [11] has tables of Fourier transforms, Fourier sine transforms, and Fourier cosine transforms.

[6] All of the integral evaluations in Gradshteyn and Ryzhik [7] are referenced, so that one level of checking against typographic errors can be performed.

References

[1] M. Abramowitz and I. A. Stegun, *Handbook of Mathematical Functions*, National Bureau of Standards, Washington, DC, 1964.

[2] A. Apelblat, *Table of Definite and Indefinite Integrals*, American Elsevier Publishing Company, New York, 1983.

[3] W. H. Beyer (ed.), *CRC Standard Mathematical Tables and Formulae*, 29th Edition, CRC Press, Boca Raton, Florida, 1991.

[4] G. P. Bois, *Tables of Indefinite Integrals*, Dover Publications, Inc., New York, 1961.

[5] H. B. Dwight, *Tables of Integrals and Other Mathematical Data*, The MacMillan Company, New York, 1957.

[6] Staff of the Bateman Manuscript Project, A. Erdélyi (ed.), *Tables of Integral Transforms*, in 3 volumes, McGraw–Hill Book Company, New York, 1954.

[7] I. S. Gradshteyn and I. M. Ryzhik, *Tables of Integrals, Series, and Products*, Academic Press, New York, 1980.

[8] W. Gröbner and N. Hofreiter, *Integralyafel*, Springer–Verlag, New York, 1949.

[9] M. Klerer and F. Grossman, "Error Rates in Tables of Indefinite Integrals," *Indust. Math.*, **18**, Part 1, 1968, pages 31–62.

[10] G. F. Miller, *Mathematical Tables: Volume 3. Tables of Generalized Exponential Integrals*, Her Majesty's Stationery Office, London, 1960.

[11] F. Oberhettinger, *Tables of Fourier Transforms and Fourier Transforms of Distributions*, Springer–Verlag, New York, 1990.

[12] A. P. Prudnikov, Yu. A. Brychov, and O. I. Marichev, *Integrals and Series*, Volumes 1, 2, and 3, translated by N. M. Queen, Gordon and Breach, New York, 1990.

[13] A. D. Wheelon, *Tables of Summable Series and Integrals Involving Bessel Functions*, Holden–Day, Inc., San Francisco, 1968.

IV

Approximate Analytical Methods

43. Asymptotic Expansions

Applicable to Definite integrals that depend on a parameter.

Yields

An asymptotic expansion.

Idea

When a parameter in an integral tends to some limit, it may be possible to find an asymptotic expansion of the integral that is valid in that limit.

Procedure

There are several general asymptotic expansion theorems that can be used to determine the asymptotic nature of an integral; we enumerate only a few.

Theorem (Bleistein and Handelsman [2], page 71): Define $I(\lambda) = \int_a^b h(t;\lambda) f(t)\,dt$, where $f^{(n)}(t)$ is continuous for $n = 0, 1, \ldots, N+1$, and $f^{(N+2)}(t)$ is piecewise continuous in the interval $[a, b]$. If

$$\left|h^{(-n-1)}(t;\lambda)\right| \leq \alpha_n(t)\phi_n(\lambda), \quad \text{for } n = 0, 1, \ldots, N+1,$$

where the functions $\{\alpha_n(t)\}$ are continuous on the interval $[a, b]$ and the functions $\{\phi_n(x)\}$ form an auxiliary asymptotic sequence as $\lambda \to \lambda_0$, then

$$I(\lambda) \sim \sum_{n=0}^{N} S_n(\lambda), \qquad \text{as } \lambda \to \lambda_0, \tag{43.1}$$

where the $S_n(\lambda)$ are defined by

$$S_n(\lambda) = (-1)^n \left[f^{(n)}(b) h^{(-n-1)}(b;\lambda) - f^{(n)}(a) h^{(-n-1)}(a;\lambda) \right].$$

Special Case 1

Under appropriate smoothness and boundedness conditions, the integral $I(\lambda) = \int_a^b h(\lambda t) f(t)\, dt$ has the asymptotic expansion

$$I(\lambda) \sim \sum_{n=0}^{N} \frac{(-1)^n}{\lambda^{n+1}} \left[f^{(n)}(b) h^{(-n-1)}(\lambda b) - f^{(n)}(a) h^{(-n-1)}(\lambda a) \right]$$

as $\lambda \to \infty$.

Special Case 2

Under appropriate smoothness and boundedness conditions, the integral $I(\lambda) = \int_a^b e^{i\lambda t} f(t)\, dt$ (which is a special case of Special Case 1), has the asymptotic expansion

$$I(\lambda) \sim \sum_{n=0}^{N} \frac{(-1)^n}{(i\lambda)^{n+1}} \left[f^{(n)}(b) e^{i\lambda b} - f^{(n)}(a) e^{i\lambda a} \right]$$

as $\lambda \to \infty$.

Special Case 3

Under appropriate smoothness and boundedness conditions, the integral $I(\lambda) = \int_a^b e^{-\lambda t} f(t)\, dt$ (which is a special case of Special Case 1), has the asymptotic expansion

$$I(\lambda) \sim \sum_{n=0}^{N} \frac{e^{-\lambda a}}{\lambda^{n+1}} f^{(n)}(a)$$

as $\lambda \to \infty$.

Watson's Lemma (Bleistein and Handelsman [2], page 103, or Wong [5], page 20): If $f(t)$ is locally absolutely integrable on $(0, \infty)$, as $t \to \infty$, $f(t) = O(e^{\alpha t})$ for some real number α, and, as $t \to 0+$, $f(t) \sim \sum_{m=0}^{\infty} c_m t^{a_m}$, where $\operatorname{Re}(a_m)$

increases monotonically to $+\infty$ as $m \to \infty$, and $\mathrm{Re}(a_0) > -1$, then, as $\lambda \to \infty$,

$$\int_0^\infty e^{-\lambda t} f(t)\, dt \sim \sum_{m=0}^{\infty} \frac{c_m \Gamma(a_m + 1)}{\lambda^{a_m+1}}.$$

Theorem (Bleistein and Handelsman [2], page 120): Let $h(t)$ and $f(t)$ be sufficiently smooth functions on the infinite interval $(0, \infty)$ having the asymptotic forms

$$h(t) \sim \exp\left(-dt^{\nu}\right) \sum_{m=0}^{\infty} \sum_{n=0}^{N(m)} c_{mn} t^{-r_m} (\log t)^n,$$

$$f(t) \sim \exp\left(-qt^{\mu}\right) \sum_{m=0}^{\infty} \sum_{n=0}^{\overline{N}(m)} p_{mn} t^{-a_m} (\log t)^n,$$

with some conditions on the range of the parameters appearing in the expansion. Let the Mellin transforms of h and f be denoted by $M[h; z]$ and $M[f; z]$ (see the Notes). If some technical conditions are satisfied, then

$$\int_0^\infty h(\lambda t) f(t)\, dt \sim -\sum_z \operatorname{res}\left(\lambda^{-z} M[h; z]\, M[f; 1-z]\right) \tag{43.2}$$

represents a finite asymptotic expansion as $\lambda \to \infty$ with respect to the asymptotic sequence $\{\lambda^{-a_j} (\log \lambda)^{n_j - m}\}$. The expression in (43.2) represents a sum of the residues over all of the poles in a specific region of the complex plane.

Bleistein and Handelsman [2] simplify the expression appearing in (43.2) in five different cases, depending on the values of the parameters.

Example

If $J(\lambda) = \int_a^b t^\lambda f(t)\, dt$, then $h(t; \lambda) = t^\lambda f(t)$ and so

$$h^{(-n-1)}(t; \lambda) = \frac{t^{\lambda+n+1}}{\prod_{j=0}^{n} (\lambda + j + 1)}$$

for $n = 0, 1, \ldots$ where we have chosen the limit of integration in the repeated integrals of h to be zero. From (43.1) this results in the following asymptotic expansion:

$$J(\lambda) \sim \sum_{n=0}^{N} \frac{t^{\lambda+n+1}}{\prod_{j=0}^{n} (\lambda + j + 1)} \left[f^{(n)}(b) b^{\lambda+n+1} - f^{(n)}(a) a^{\lambda+n+1} \right] \tag{43.3}$$

whenever f satisfies the hypotheses of the theorem. If f were a polynomial, then the expansion in (43.3) would be the exact evaluation for $J(\lambda)$.

Notes

[1] The Mellin transform of the function $f(t)$ is $M[f;z] = \int_0^\infty t^{z-1} f(t)\,dt$ which can be interpreted as the $(z-1)$-st moment of $f(t)$. The so-called bilateral Laplace transform of $g(t)$ is the Mellin transform of $f(t) = g(-\log t)$.

[2] Consider the integral $I = \int_a^b e^{\lambda f(x)} g(x)\,dx$ in the limit $\lambda \to \infty$. An asymptotic expansion methodology for I is given by
(A) the method of steepest descents (page 229) when $f(x)$ is complex,
(B) Laplace's method (page 221) when $f(x)$ is real,
(C) the method of stationary phase (page 226) when $f(x)$ is purely imaginary (no real component).

[3] In asymptotic formulas, it is important to describe fully the region in which a parameter is tending to a limit. Error estimates should also be supplied with an asymptotic formula. For example, the complementary error function has the following expansions as $x \to \infty$:

$$\begin{aligned} \operatorname{erfc}(x) &\sim \qquad e^{-x^2}\sum_{k=0}^{\infty}(-1)^k \frac{(2k-1)!!}{2^{k+1}x^{2k+1}} && \text{for } -\frac{3\pi}{4} < \arg x < \frac{3\pi}{4}, \\ \operatorname{erfc}(x) &\sim \sqrt{\pi}+e^{-x^2}\sum_{k=0}^{\infty}(-1)^k \frac{(2k-1)!!}{2^{k+1}x^{2k+1}} && \text{for } \quad \frac{\pi}{4} < \arg x < \frac{7\pi}{4}. \end{aligned} \tag{43.4a–b}$$

Note that these expansions both apply in the region of overlap, $\pi/4 < \arg x < 3\pi/4$. In this overlap region equation (43.4.a) has a large error while equation (43.4.b) has a small error.

[4] Wong [5] (page 22) has a generalized Watson's lemma:

> Define $I(\lambda) = \int_0^{\infty e^{i\gamma}} f(t)e^{-\lambda t}\,dt$, and assume that $I(\lambda_0)$ exists. If $f(t) \sim \sum_{m=0}^{\infty} c_m t^{a_m}$ as $t \to 0$ along $\arg t = \gamma$, where $\operatorname{Re} a_0 > 0$ and $\operatorname{Re} a_{m+1} > \operatorname{Re} a_m$, then $I(\lambda) \sim \sum_{m=0}^{\infty} \dfrac{c_m \Gamma(a_m+1)}{\lambda^{a_m+1}}$.

[5] Salvy [4] has created a package for automatically determining the asymptotic expansion of some classes of integrals, using the symbolic computer language MAPLE. Salvy gives the following sample outputs from his program:

- $\displaystyle\int_0^\infty \frac{\sin\left(\frac{1}{t}\right)}{t(t+x)}\,dx \sim \frac{\pi}{2x} + \frac{\gamma-1}{x^2} - \frac{\log x}{x^2} - \frac{\pi}{4x^3} + O\left(\frac{1}{x^4}\right)$
- $\displaystyle E_1(x) = \int_1^\infty \frac{e^{-xt}}{t}\,dt \sim \frac{1}{xe^x} - \frac{1}{x^2e^x} + \frac{2}{x^3e^x} - \frac{6}{x^4e^x} + \dots$
- $\displaystyle\int_0^\infty \frac{\sqrt{t}e^{-t}\log t}{1+tx}\,dt \sim -\frac{\sqrt{\pi}(\gamma+2\log 2)}{x} + \frac{\pi\log x}{x^{3/2}} + \frac{\sqrt{\pi}(4-2\gamma-4\log 2)}{x^2} + O\left(\frac{\log x}{x^{5/2}}\right)$

as $x \to \infty$. Here, γ is Euler's constant.

References

[1] C. M. Bender and S. A. Orszag, *Advanced Mathematical Methods for Scientists and Engineers*, McGraw–Hill, New York, 1978.

[2] N. Bleistein and R. A. Handelsman, *Asymptotic Expansions of Integrals*, Dover Publications, Inc., New York, 1986.

[3] F. W. J. Olver, "Uniform, Exponentially Improved, Asymptotic Expansions for the Generalized Exponential Integral," *SIAM J. Math. Anal.*, **22**, No. 5, September 1991, pages 1460–1474.

[4] B. Salvy, "Examples of Automatic Asymptotic Expansions," *SIGSAM Bulletin*, ACM, New York, **25**, No. 2, April 1991, pages 4–17.

[5] R. Wong, *Asymptotic Approximation of Integrals*, Academic Press, New York, 1989.

44. Asymptotic Expansions: Multiple Integrals

Applicable to Multidimensional definite integrals that depend on a parameter.

Yields

An asymptotic expansion.

Idea

When a parameter in an integral tends to some limit, it may be possible to find an asymptotic expansion of the integral that is valid in that limit.

Procedure

It is difficult to state concisely very much about the different asymptotic behaviors that are possible in multiple integrals. In this section we will only focus on the integral

$$I(\lambda) = \int_D e^{\lambda \phi(\mathbf{x})} g_0(\mathbf{x})\, d\mathbf{x} \tag{44.1}$$

when ϕ is a real function and λ is real with $|\lambda| \to \infty$. Integrals of this form are known as integrals of Laplace type. (The other interesting case that occurs in applications is when λ is purely imaginary with $|\lambda| \to \infty$; this leads to integrals of Fourier type.) Here, D is some (not necessarily bounded) domain in n-dimensional $\mathbf{x}$ space.

Laplace Type Integrals

We presume that λ is real and that D is a bounded simply connected domain. The boundary of D, denoted by Γ, is an $(n-1)$-dimensional hypersurface. We assume it can be represented as

$$\Gamma : \mathbf{x}(\boldsymbol{\sigma}), \quad \boldsymbol{\sigma} = (\sigma_1, \ldots, \sigma_{n-1})$$

where $\boldsymbol{\sigma}$ takes values in the set P. We presume that ϕ, g_0, and $x_i(\boldsymbol{\sigma})$ are sufficiently differentiable for what follows.

There are now two cases, depending on where the maximum of ϕ appears in $\overline{D}$.

Maximum on the Boundary

We presume that the maximum of ϕ appears on the boundary $\overline{D}$ at the unique point $\mathbf{x} = \mathbf{x}_0$.

Let $\mathbf{N}$ be the outward normal to Γ, and let $d\Sigma$ be the differential element of "surface area" on Γ. Now define the gradient operation, $\nabla = (\partial_{x_1}, \ldots, \partial_{x_n})$, the functions $\psi(\boldsymbol{\sigma}) = \phi(\mathbf{x}(\boldsymbol{\sigma}))$, $\mathbf{H}_j = g_j \nabla\phi / |\nabla\phi|^2$, and $g_{j+1} = \nabla \cdot \mathbf{H}_j$ (for $j = 0, 1, \ldots$). Then an exact representation of the integral in (44.1) is given by (see Bleistein and Handelsman [2], page 332)

$$I(\lambda) = -\sum_{j=0}^{M-1} (-\lambda)^{-j-1} \int_\Gamma (\mathbf{H}_j \cdot \mathbf{N}) e^{\lambda\phi}\, d\Sigma + \frac{(-1)^M}{\lambda^M} \int_D g_M e^{\lambda\phi}\, d\mathbf{x} \quad (44.2)$$

for $M = 2, 3, \ldots$.

The first terms in this expression are lower dimensional integrals for which asymptotic expressions may be found (recursively, if necessary). The last term can be bounded and will become the "error term."

For the particular case of $n = 2$, we can parameterize the boundary by $\Gamma : (x_1(s), x_2(s))$ where $s = 0$ corresponds to the maximum at $\mathbf{x} = \mathbf{x}_0$. The leading order term in (44.2) (e.g., the $j = 0$ term) can then be written as

$$I(\lambda) \sim e^{\lambda\phi(\mathbf{x}_0)} \sqrt{\frac{2\pi}{\lambda^3 |\Psi''(0)|}} (\mathbf{H}_0 \cdot \mathbf{N}) \Big|_{s=0}, \quad (44.3)$$

if the maximum of Ψ at $s = 0$ is simple (so that $\Psi''(0) < 0$). If $\kappa(0)$ is the curvature of Γ at $\mathbf{x} = \mathbf{x}_0$, then (44.3) can be further simplified to

$$\begin{aligned} I(\lambda) \sim & e^{\lambda\phi(\mathbf{x}_0)} g_0(\mathbf{x}_0) \sqrt{\frac{2\pi}{\lambda^3}} \\ & \times \left| \phi_{x_1x_1}\phi_{x_2}^2 - 2\phi_{x_1x_2}\phi_{x_1}\phi_{x_2} + \phi_{x_2x_2}\phi_{x_1}^2 \mp \kappa |\nabla\phi|^3 \right|^{-1/2} \Bigg|_{\mathbf{x}=\mathbf{x}_0}, \end{aligned} \quad (44.4)$$

where the minus (plus) sign holds when Γ is convex (concave) at $\mathbf{x} = \mathbf{x}_0$.

Maximum Not on the Boundary

We presume that the maximum of ϕ appears in the interior of $\overline{D}$ at the unique point $\mathbf{x} = \mathbf{x}_0$.

Near $\mathbf{x} = \mathbf{x}_0$ we can expand ϕ in the form $\phi(\mathbf{x}) - \phi(\mathbf{x}_0) \approx \frac{1}{2}(\mathbf{x} - \mathbf{x}_0)A(\mathbf{x} - \mathbf{x}_0)^{\mathrm{T}}$, where the matrix $A = (a_{ij})$ is defined by $a_{ij} = \phi_{x_i x_j}(\mathbf{x}_0)$. Let Q be an orthogonal matrix that diagonalizes A, i.e.,

$$Q^{\mathrm{T}} A Q = \begin{pmatrix} \lambda_1 & & 0 \\ & \ddots & \\ 0 & & \lambda_n \end{pmatrix}.$$

Then define the variable $\mathbf{z}$ by $(\mathbf{x} - \mathbf{x}_0) = QR\mathbf{z}^{\mathrm{T}}$ where the matrix $R = (r_{ij})$ is defined by $r_{ij} = \delta_{ij} |\lambda_i|^{-1/2}$. Now the functions $\{h_i\}$ are chosen so that $\zeta_i = h_i(\mathbf{z}) = z_i + o(|\mathbf{z}|)$ (as $|\mathbf{z}| \to 0$) and $\sum_{i=1}^n h_i^2 = 2\left(\phi(\mathbf{x}_0) - \phi(\mathbf{x}(\mathbf{z}))\right)$.

With these functions we define the Jacobian $J(\boldsymbol{\zeta}) = \dfrac{\partial(x_1, \ldots, x_n)}{\partial(\zeta_1, \ldots, \zeta_n)}$ and then $g_0(\mathbf{x}(\boldsymbol{\zeta}))J(\boldsymbol{\zeta}) = G_0(\boldsymbol{\zeta}) = G_0(\mathbf{0}) + \boldsymbol{\zeta} \cdot \mathbf{H}_0$. Then we have the recursive definitions $G_j(\boldsymbol{\zeta}) = G_j(\mathbf{0}) + \boldsymbol{\zeta} \cdot \mathbf{H}_j(\boldsymbol{\zeta})$ and $G_{j+1}(\boldsymbol{\zeta}) = \nabla \cdot \mathbf{H}_j(\boldsymbol{\zeta})$. (Note that there is an ambiguity in the $\{\mathbf{H}_n\}$, this is not important.) Finally, the approximation to $I(\lambda)$ is given by (see Bleistein and Handelsman [2], page 335)

$$I(\lambda) \sim e^{\lambda\phi(\mathbf{x}_0)} \left[\sum_{j=0}^{M-1} \lambda^{-j} G_j(\mathbf{0}) Z[1] + \lambda^{-M} Z\left[G_M(\boldsymbol{\zeta})\right] \right]. \tag{44.5}$$

where $Z\left[k(\boldsymbol{\zeta})\right] = \int_D k(\boldsymbol{\zeta}) \exp(-\frac{1}{2}\lambda\boldsymbol{\zeta} \cdot \boldsymbol{\zeta})\, d\boldsymbol{\zeta}$. Note that (44.5) is not an exact representation, since M exponentially small boundary integrals have been discarded. The leading order term in (44.5) can be written as

$$I(\lambda) \sim \frac{e^{\lambda\phi(\mathbf{x}_0)}}{\sqrt{\left|\det\left(\phi_{x_i x_j}(\mathbf{x}_0)\right)\right|}} \left(\frac{2\pi}{\lambda}\right)^{n/2} g_0(\mathbf{x}_0). \tag{44.6}$$

For the particular case of $n = 2$, the result in (44.6) can be written as

$$I(\lambda) \sim \frac{2\pi}{\lambda} \frac{g_0(\mathbf{x}_0) e^{\lambda\phi(\mathbf{x}_0)}}{\sqrt{\phi_{x_1x_1}(\mathbf{x}_0)\phi_{x_2x_2}(\mathbf{x}_0) - \phi_{x_1x_2}^2(\mathbf{x}_0)}}. \tag{44.7}$$

Example 1

Consider the two-dimensional integral $J(\lambda) = \iint\limits_D e^{\lambda(x-y^2)}\, dx\, dy$ where D is the unit circle. This integral has $g_0 = 1$ and $\phi = x - y^2$. In D, the maximum of ϕ is at $\mathbf{x}_0 = (1, 0)$. Since this a boundary point of D, (44.4) is the appropriate formula to use. The only nonzero terms that appear in (44.4) are: $g_0(\mathbf{x}_0) = 1$, $\phi(\mathbf{x}_0) = 1$, $\phi_x(\mathbf{x}_0) = 1$, $\kappa(\mathbf{x}_0) = 1$, and $|\nabla\phi(\mathbf{x}_0)| = 1$. Using these values results in the approximation $J(\lambda) \sim e^{\lambda}\sqrt{2\pi/\lambda^3}$.

Example 2

Consider the two-dimensional integral $K(\lambda) = \iint_D e^{\lambda(2-x^2-y^2)}\, dx\, dy$ where D is the unit circle. This integral has $g_0 = 1$ and $\phi = 2 - x^2 - y^2$. In D, the maximum of ϕ is at $\mathbf{x}_0 = (0,0)$. Since this an interior point of D, (44.7) is the appropriate formula to use. The only nonzero terms that appear in (44.7) are $g_0(\mathbf{x}_0) = 1$, $\phi(\mathbf{x}_0) = 2$, $\phi_{xx}(\mathbf{x}_0) = -2$, and $\phi_{yy}(\mathbf{x}_0) = -2$. Using these values results in the approximation $K(\lambda) \sim \pi e^{2\lambda}/\lambda$.

For this example, the integral can be computed exactly. We find

$$K(\lambda) = \int_0^{2\pi} \int_0^1 e^{\lambda(2-r^2)} r\, dr\, d\theta = -\frac{\pi}{\lambda} e^{\lambda(2-r^2)} \Big|_0^1 = \pi \frac{e^{2\lambda} - e^{\lambda}}{\lambda}$$
$$\sim \pi \frac{e^{2\lambda}}{\lambda} \qquad \text{as } \lambda \to \infty.$$

Notes

[1] Bleistein and Handelsman [2] also describe asymptotic results for multidimensional Fourier integrals.

[2] Consider the integral $I(\lambda) = \iint_{R^2} f(\mathbf{x}) e^{i\lambda\phi(\mathbf{x})}\, d\mathbf{x}$, where λ is a large parameter. At caustic points (also known as turning points), defined by

$$\nabla\phi(\mathbf{x}_0) = 0 \qquad \text{and} \qquad \det\left(\frac{\partial^2\phi}{\partial x_i \partial x_j}\right) = 0,$$

the classical stationary phase techniques do not apply. For caustic points where the Hessian determinant vanishes, but the Hessian is not identically the zero matrix, there are several canonical forms of physical interest. These include the following possibilities for $\phi(x,y)$:

$$\begin{aligned} \phi(x,y) &= x^3 + y^2 && \text{fold} \\ \phi(x,y) &= x^4 + y^2 && \text{cusp} \\ \phi(x,y) &= x^5 + y^2 && \text{swallowtail} \\ \phi(x,y) &= x^6 + y^2 && \text{butterfly.} \end{aligned}$$

For caustic points where the Hessian is identically the zero matrix, the canonical forms for $\phi(x,y)$ (for co-dimension less than 5) are

$$\begin{aligned} \phi(x,y) &= x^3 - xy^2 && \text{elliptic umbilic} \\ \phi(x,y) &= x^3 + y^3 \quad \text{or} \quad x^3 + xy^2 && \text{hyperbolic umbilic} \\ \phi(x,y) &= x^4 + xy^2 && \text{parabolic umbilic.} \end{aligned}$$

See Gorman and Wells [4] for details.

[3] Brüning and Heintze [3] derive an asymptotic expansion for the integral $\int_{[0,1]^n} g\left(\frac{\mathbf{x}^\alpha}{s}\right) \mathbf{x}^\beta \log^\gamma \mathbf{x} f(\mathbf{x})\, dx$, as $s \to 0^+$.

[4] McClure and Wong [5] derive an asymptotic expansion for the integral $\int_{[0,1]^n} g\left(\frac{\mathbf{x}^\alpha}{s}\right) \mathbf{x}^\beta f(\mathbf{x})\, dx$, as $s \to 0^+$.

References

[1] C. M. Bender and S. A. Orszag, *Advanced Mathematical Methods for Scientists and Engineers*, McGraw–Hill, New York, 1978.

[2] N. Bleistein and R. A. Handelsman, *Asymptotic Expansions of Integrals*, Dover Publications, Inc., New York, 1986, Chapter 8, pages 321–366.

[3] J. Brüning and E. Heintze, "The Minakschisundaram–Pleijel Expansion in the Equivariant Case," *Duke Math. J.*, **51**, 1984, pages 959–980.

[4] A. D. Gorman and R. Wells, "The Asymptotic Expansion of Certain Canonical Integrals," *J. Math. Anal. Appl.*, **152**, 1984, pages 566–584.

[5] J. P. McClure and R. Wong, "Asymptotic Expansion of a Multiple Integral," *SIAM J. Math. Anal.*, **18**, No. 6, November 1987, pages 1630–1637.

[6] F. W. Olver, *Asymptotics and Special Functions*, Academic Press, New York, 1974.

[7] R. Wong, *Asymptotic Approximations of Integrals*, Academic Press, New York, 1989, Chapters 8 and 9, pages 423–515.

45. Continued Fractions

Applicable to Integrals for which a continued fraction can be found.

Yields

A numerical approximation of an integral.

Idea

If a continued fraction can be found for an integral, then it may be used to approximate the value of that integral.

Procedure

A continued fraction is an expression of the form

$$C = \mathop{\Phi}_{i=1}^{\infty} \frac{a_i}{b_i} = \cfrac{a_1}{b_1 + \cfrac{a_2}{b_2 + \cfrac{a_3}{b_3 + \dotsb}}}.$$

Here, a_k is called the k-th partial numerator and b_k is called the k-th partial denominator, for $k = 1, 2, \ldots, \infty$. For typographical convenience, and also to save space, continued fractions are sometimes represented as

$$\frac{a_1|}{|b_1} + \frac{a_2|}{|b_2} + \frac{a_3|}{|b_3} + \cdots \qquad \text{or as} \qquad \frac{a_1}{b_1+}\,\frac{a_2}{b_2+}\,\frac{a_3}{b_3+}\cdots.$$

We define

$$c_k = \mathop{\Phi}_{i=1}^{k} \frac{a_i}{b_i} = \cfrac{a_1}{b_1 + \cfrac{a_2}{b_2 + \cfrac{a_3}{b_3 + \cfrac{\cdots}{\cdots + a_k}}}}$$

to be the k-th convergent of the continued fraction I. The continued fraction is said to converge if the sequence $\{c_n\}$ converges.

In many cases, an integral may be written as a continued fraction. Partial convergents of the continued fraction then yield approximations to the original integral.

Notes

[1] Given $\{a_k\}$ and $\{b_k\}$, define the sequences $\{p_k\}$ and $\{q_k\}$ by the recurrence relations $q_k = b_k q_{k-1} + a_k q_{k-1}$ and $p_k = b_k p_{k-1} + a_k p_{k-1}$ for $k = 1, 2, \ldots$. The initial values are given by $p_{-1} = 1$, $p_0 = 0$, $q_{-1} = 0$, and $q_0 = 1$. Then $c_k = p_k/q_k$ for $k = 1, 2, \ldots$. Note that this relates continued fractions to recurrence relations.

It might be easier to observe that the three-term recurrence relation, $y_n + a_n y_{n+1} + b_n y_{n+1} = 0$, is formally equivalent to the continued fraction

$$\frac{y_n}{y_{n+1}} = -a_n - \frac{b_n}{a_{n+1}-} \frac{b_{n+1}}{a_{n+2}-} \cdots.$$

[2] From a Taylor series in the form $F(z) = \sum_{i=0}^{\infty} d_i z^{-(i+1)}$, a z-fraction, which is a continued fraction of the form $\frac{e_0|}{|z} - \frac{f_1|}{|1} - \frac{e_1|}{|z} - \frac{f_2|}{|1} - \ldots$, may easily be constructed. By defining $F_k(z) = \sum_{i=0}^{\infty} d_{i+k} z^{-(i+1)} = c_k(z)$, where

$$c_k(z) = \frac{d_k|}{|z} - \frac{f_{1,k}|}{|1} - \frac{e_{1,k}|}{|z} - \frac{f_{2,k}|}{|1} - \frac{e_{2,k}|}{|z} - \frac{f_{3,k}|}{|1} - \ldots,$$

we obtain the recurrence relations

$$\begin{aligned} e_{j-1,k+1} + f_{j,k+1} &= f_{j,k} + e_{j,k} \quad \text{for } e \\ f_{j,k+1} e_{j,k+1} &= e_{j,k} f_{j+1,k} \quad \text{for } f \end{aligned}$$

with $e_{0,k} = 0$ and $f_{1,k} = d_{k+1}/d_k$. After determining the $\{e_{j,k}, f_{j,k}\}$, we find $e_j = e_{j,0}$ and $f_j = f_{j,0}$ for $j \geq 1$, and $e_0 = d_0$. This is known as the QD (for quotient difference) algorithm. See van der Laan and Temme [3] for details.

Using this algorithm and the asymptotic formula

$$\operatorname{erfc}(z) = \frac{2}{\sqrt{\pi}} \int_z^{\infty} e^{-t^2}\, dt \sim \frac{2z}{\sqrt{\pi}} e^{-z^2} \left(\frac{1}{2z^2} - \frac{1}{4z^4} + \frac{3}{8z^6} - \ldots \right),$$

we can derive a continued fraction approximation to the erfc function:

$$\frac{\sqrt{\pi}}{z} e^{z^2} \operatorname{erfc}(z) = \frac{1|}{|z^2} + \frac{\frac{1}{2}|}{|1} + \frac{\frac{2}{2}|}{|z^2} + \frac{\frac{3}{2}|}{|1} + \frac{\frac{4}{2}|}{|z^2} + \cdots.$$

References

[1] B. Char, "On Stieltjes Continued Fraction for the Gamma Functions," *Math. of Comp.*, **34**, 1980, pages 547–551.

[2] D. Dijkstra, "A Continued Fraction Expansion for a Generalization of Dawson's Integral," *Math. of Comp.*, **31**, 1977, pages 503–510.

[3] C. G. van der Laan and N. M. Temme, *Calculation of Special Functions: The Gamma function, the Exponential Integrals and Error–like Functions*, Centrum voor Wiskunde en Informatica, Amsterdam, 1984.

46. Integral Inequalities

Idea

Some integrals may be easily bounded by known theorems.

Procedure

Given an integral that is to be bounded, a formula should be located that has the desired form. This is not a straightforward process.

Example

Suppose we would like to bound the integral $I = \displaystyle\int_0^1 \frac{e^{-x}}{\sqrt{1+x^2}}\,dx$. If we write this integral as $I = \int_0^1 f(x)g(x)\,dx$, with $f(x) = e^{-x}$ and $g(x) = 1/\sqrt{1+x^2}$, then we note that both f and g are decreasing functions on the interval $[0,1]$. Hence, Tschebyscheff's inequality can be used to derive a lower bound (see the table at the end of this section for an exact statement of the inequality). We have

$$\begin{aligned} I &\geq \frac{1}{1}\left(\int_0^1 e^{-x}\,dx\right)\left(\int_0^1 \frac{1}{\sqrt{1+x^2}}\,dx\right) \\ &= \left(-e^{-x}\right)\Big|_0^1 \log\left(x+\sqrt{1+x^2}\right)\Big|_0^1 \\ &= \frac{e-1}{e}\log(1+\sqrt{2}) \simeq 0.557. \end{aligned}$$

To obtain an upper bound, we can use Hölder's inequality with $p = q = 2$ (see the table at the end of this section for an exact statement of the inequality). We have

$$
\begin{aligned}
I &\le \sqrt{\int_0^1 e^{-2x}\,dx}\sqrt{\int_0^1 \frac{1}{1+x^2}\,dx} \\
&= \sqrt{\left(-\frac{e^{-2x}}{2}\right)\Big|_0^1}\sqrt{\left(\tan^{-1}x\right)\Big|_0^1} \\
&= \sqrt{\frac{e^2-1}{2e^2}}\sqrt{\frac{\pi}{4}} \simeq 0.583.
\end{aligned}
$$

Hence, we have found a fairly tight bound for I (that is, $0.556 < I < 0.584$), without having to perform much computation.

One Dimensional Inequalities – Named

[1] Carleman's inequality (see Iyanaga and Kawada [9], page 1422)

$$\int_0^\infty \exp\left(\frac{1}{x}\int_0^x \log f(t)\,dt\right)dx < e\int_0^\infty f(x)\,dx$$

when $f(x) > 0$.

[2] Cauchy–Schwartz–Bunyakowsky inequality (see Squire [19], page 21)

$$\left(\int^b f(x)g(x)\,dx\right)^2 \le \left(\int^b f^2(x)\,dx\right)\left(\int^b g^2(x)\,dx\right).$$

Equality occurs only when $f(x) = kg(x)$, with k real.

[3] Hardy's inequality (see Iyanaga and Kawada [9], page 1422)

$$\int_0^\infty \left(\frac{F(x)}{x}\right)^p dx \le \left(\frac{p}{1-p}\right)^p \int_0^\infty f^p(x)\,dx$$

when $p > 1$ and $f(x) > 0$. Equality is achieved only if $f(x) = 0$.

[4] Modified Hardy's inequality (see Izumi and Izumi [8])

$$\int_0^\pi x^{-m}\left(\int_{x/2}^x f(t)\,dt\right)^p dx < \left(\frac{p}{m-1}\right)^p \int_0^\pi x^{-m}\left|f\left(\frac{x}{2}\right) - f(x)\right|^p dx$$

when $m > 1$, $p > 1$, and $f(x) > 0$.

[5] Hardy–Littlewood supremum theorem (see Hardy, Littlewood, and Polya [7], page 298 (#398))

$$\int_0^a \left(\sup_{0\le\xi<x} \frac{1}{x-\xi}\int_\xi^x f(t)\,dt\right)^k dx \le \left(\frac{k}{k-1}\right)^k \int_0^a f^k(t)\,dt$$

if $k > 1$ and $f(x)$ is non-negative and integrable.

[6] Hölder's inequality (see Squire [19], page 21)

$$\left|\int_a^b f(x)g(x)\,dx\right| \le \left(\int_a^b |f(x)|^p\,dx\right)^{1/p} \left(\int_a^b |g(x)|^q\,dx\right)^{1/q}$$

when p and q are positive and $1/p + 1/q = 1$. Equality occurs only when $\alpha|f(x)|^p = \beta|g(x)|^q$, where α and β are positive constants.

[7] Backward Hölder's inequality (see Brown and Shepp [1])

$$\sup_x \int [f(x-y)g(y)]\,dy \le \left(\int |f(x)|^p\,dx\right)^{1/p} \left(\int |g(x)|^q\,dx\right)^{1/q}$$

when f and g have compact support, p and q are positive, and $1/p+1/q = 1$.

[8] Backward Hölder's inequality (see Brown and Shepp [1])

$$\int \sup_y [f(x-y)g(y)]\,dx \ge \left(\int |f(x)|^p\,dx\right)^{1/p} \left(\int |g(x)|^q\,dx\right)^{1/q}$$

when f and g have compact support, p and q are positive, and $1/p+1/q = 1$.

[9] Minkowski's inequality (see Squire [19], page 21)

$$\left(\int_a^b |f(x)+g(x)|^p\,dx\right)^{1/p} \le \left(\int_a^b |f(x)|^p\,dx\right)^{1/p} \left(\int_a^b |g(x)|^p\,dx\right)^{1/p}$$

for $p > 1$. Equality occurs only when $f(x) = kg(x)$, with k non-negative.

[10] Ostrowski inequality (see Gradshteyn and Ryzhik [6], page 1100)

$$\left|\int_a^b f(x)g(x)\,dx\right| \le |f(x)| \max_{a\le\xi\le b} \left|\int_a^\xi g(x)\,dx\right|$$

when $f(x)$ is monotonic decreasing and $f(a)f(b) \ge 0$.

[11] Tschebyscheff inequality (see Squire [19], page 22)

$$\int_a^b f(x)g(x)\,dx \ge \frac{1}{b-a}\left(\int_a^b f(x)\,dx\right)\left(\int_a^b g(x)\,dx\right)$$

when $f(x)$ and $g(x)$ are both increasing or both decreasing functions.

[12] Tschebyscheff inequality (see Squire [19], page 22)

$$\int_a^b f(x)g(x)\,dx \le \frac{1}{b-a}\left(\int_a^b f(x)\,dx\right)\left(\int_a^b g(x)\,dx\right)$$

when $f(x)$ is an increasing function and $g(x)$ is a decreasing function (or vice-versa).

[13] Wirtinger's inequality (see Hardy, Littlewood, and Polya [7], page 185 (#257))

$$\int_0^\pi f^2(x)\,dx \le \int_0^\pi (f')^2(x)\,dx$$

If $f(0) = f(\pi) = 0$ and f' is L^2. Equality is obtained only if $f(x) = C\sin x$.

[14] Generalized Wirtinger's inequality ($1 \le k < \infty$), (see Tananika [20])

$$\left(\int_0^1 |u|^k\,dt\right)^{1/k} \le \sqrt{\frac{k}{\pi}}\,2^{(1-k)/k}(k+2)^{(k-2)/2k}\frac{\Gamma((k+2)/2k)}{\Gamma(1/k)}\left(\int_0^1 |u|^2\,dt\right)^{1/2}.$$

[15] Young's inequality (see Hardy, Littlewood, and Polya [7], page 111 (#156))

$$ab \leq \int_0^a f(x)\,dx + \int_0^b f^{-1}(x)\,dx$$

when $f(x)$ is continuous, strictly monotone increasing in $x \geq 0$, $f(0) = 0$, $a \geq 0$, and $b \geq 0$. Equality occurs only if $b = f(a)$.

One Dimensional Inequalities: Arbitrary Intervals – Unnamed

[16] If $p(x) > 0$ and $\int p(x)\,dx = 1$ (see Hardy, Littlewood, and Polya [7], page 137 (#184)), then (unless f is a constant)

$$\exp\left(\int p(x) \log f(x)\,dx\right) < \int p(x) f(x)\,dx.$$

[17] If $0 < r < s$, $p(x) > 0$, and $\int p(x)\,dx = 1$ (see Hardy, Littlewood, and Polya [7], page 143 (#192)), then (unless f is a constant)

$$\left(\int p(x) f^r(x)\,dx\right)^{1/r} < \left(\int p(x) f^s(x)\,dx\right)^{1/s}.$$

[18] If $0 < a \leq f(x) \leq A < \infty$ and $0 < b \leq g(x) \leq B < \infty$ (see Hardy, Littlewood, and Polya [7], page 166 (#230)), then

$$\left(\int f^2(x)\,dx\right)\left(\int g^2(x)\,dx\right) \leq \left(\frac{1}{2}\left[\sqrt{\frac{AB}{ab}} + \sqrt{\frac{ab}{AB}}\right] \int f(x) g(x)\,dx\right)^2.$$

[19] If a, b, α, β are positive and $f(x)$ is an increasing positive function (see Hardy, Littlewood, and Polya [7], page 297 (#397)), then

$$\int_a^{a+\alpha} f\left(\frac{a}{x}\right)\,dx + \int_b^{b+\beta} f\left(\frac{b}{x}\right)\,dx \leq \int_{a+b}^{a+b+\alpha+\beta} f\left(\frac{a+b}{x}\right)\,dx.$$

[20] If $1 \leq r < p$ and $f(x)$ and $g(x)$ are positive functions in L^p (see Potze and Urbach [21]), then

$$\left|\exp\left(-\int f^r g^{p-r}\,dx\right) - \exp\left(-\int f^{r-1} g^{p+1-r}\,dx\right)\right|$$
$$\leq C_{r,p} \left(\int |f-g|^p\right)^{1/p} dx$$

where $C_{r,p} \geq 0$.

One Dimensional Inequalities: Finite Intervals – Unnamed

[21] If $a \geq 0$, $b \geq 0$, $a \neq 1$, $f(x)$ is non-negative and decreasing, and $f(x) \neq C$ (see Hardy, Littlewood, and Polya [7], page 166 (#229)), then

$$\left(\int_0^1 x^{a+b} f\,dx\right)^2 \leq \left[1 - \left(\frac{a-b}{a+b+1}\right)^2\right] \left(\int_0^1 x^{2a} f\,dx\right) \left(\int_0^1 x^{2b} f\,dx\right).$$

[22] If $f(x)$ has period 2π, $\int_0^{2\pi} f\,dx = 0$, f' is L^2, and $f(x) \neq A\sin x + B\cos x$ (see Hardy, Littlewood, and Polya [7], page 185 (#258)), then

$$\int_0^{2\pi} f^2(x)\,dx < \int_0^{2\pi} (f'(x))^2\,dx.$$

[23] If $0 \leq f' \leq 1$, $0 \leq g(x) < x$, and $k > 1$ (see Hardy, Littlewood, and Polya [7], page 298 (#400)), then

$$\int_0^1 \left(\frac{f(x) - f(g(x))}{x - g(x)}\right)^k dx \leq \frac{kf(1) - f^k(1)}{k-1}.$$

[24] If $0 \leq f' \leq 1$ and $0 \leq g(x) < x$ (see Hardy, Littlewood, and Polya [7], page 298 (#400)), then

$$\int_0^1 \frac{f(x) - f(g(x))}{x - g(x)}\,dx \leq f(1)\,(1 - \log f(1))\,.$$

One Dimensional Inequalities: Infinite Intervals – Unnamed

[25] If $m > 1$, $n > -1$, f is positive (see Hardy, Littlewood, and Polya [7], page 165 (#226)), then

$$\int_0^\infty x^n f^m(x)\,dx$$

$$\leq \frac{m}{n+1} \left(\int_0^\infty x^{m(n+1)/(m-1)} f^m(x)\,dx\right)^{(m-1)/m} \left(\int_0^\infty |f'(x)|^m\,dx\right)^{1/m}.$$

Equality occurs only when $f = B\exp\left(-Cx^{(m+n)/(m-1)}\right)$, where $B \geq 0$ and $C > 0$.

[26] If $a \geq 0$, $b \geq 0$, $a \neq b$, and f is non-negative and decreasing (see Hardy, Littlewood, and Polya [7], page 166 (#228)), then (unless $f(x) = \begin{cases} C & \text{in } (0,\xi) \\ 0 & \text{in } (\xi,\infty) \end{cases}$ with $C > 0$)

$$\left(\int_0^\infty x^{a+b} f\,dx\right)^2 \leq \left[1 - \left(\frac{a-b}{a+b+1}\right)^2\right] \left(\int_0^\infty x^{2a} f\,dx\right) \left(\int_0^\infty x^{2b} f\,dx\right).$$

[27] If f and f'' are in $L^2[0,\infty]$ (see Hardy, Littlewood, and Polya [7], page 187 (#259)), then

$$\left(\int_0^\infty (f'(x))^2\,dx\right)^2 \leq 4\left(\int_0^\infty f^2(x)\,dx\right)\left(\int_0^\infty (f''(x))^2\,dx\right).$$

Equality occurs only when $f(x) = Ae^{-Bx/2}\sin\left(Bx\sin\frac{\pi}{3} - \frac{\pi}{3}\right)$

[28] If f and f'' are in $L^2[0,\infty]$ (see Hardy, Littlewood, and Polya [7], page 188 (#260)), then

$$\int_0^\infty \left(f^2(x) - (f'(x))^2 + (f''(x))^2\right)\,dx \geq 0.$$

Equality occurs only when $f(x) = Ae^{-Bx/2}\sin\left(Bx\sin\frac{\pi}{3} - \frac{\pi}{3}\right)$

[29] If f and f'' are in $L^2[-\infty,\infty]$ (see Hardy, Littlewood, and Polya [7], page 193 (#261)), then (unless $f(x) = 0$)

$$\left(\int_{-\infty}^\infty (f'(x))^2\,dx\right)^2 < \left(\int_{-\infty}^\infty f^2(x)\,dx\right)\left(\int_{-\infty}^\infty (f''(x))^2\,dx\right).$$

[30] If $p > 1$ and $f(x) \geq 0$ (see Hardy, Littlewood, and Polya [7], page 240 (#327)), then (unless $f(x) = 0$)

$$\int_0^\infty \left(\frac{1}{x}\int_0^x f(t)\,dt\right)^p dx < \left(\frac{p}{p-1}\right)^p \int_0^\infty f^p(x)\,dx.$$

[31] If $p > 1$, $0 \leq \alpha < 1/p$, and $p \leq q \leq p/(1-\alpha p)$ (see Hardy, Littlewood, and Polya [7], page 298 (#402)), then

$$\int_0^\infty x^{-(p-q+pq\alpha)/p}\left(\frac{1}{\Gamma(\alpha)}\int_0^x f(y)(x-y)^{\alpha-1}\,dy\right)^q \leq K\left(\int_0^\infty f^p(x)\,dx\right)^{q/p}.$$

This result is also true if $\alpha \geq 1/p$, $p > 1$, and $p \leq q$. In both cases $K = K(p,q,\alpha) > 0$.

[32] Under some continuity requirements, with $\alpha > 0$ (see Mingarelli [10])

$$\int_{-\infty}^\infty |f(x)|^2 e^{\alpha x^2}\,dx \leq \frac{1}{2\alpha}\int_{-\infty}^\infty |f'(x)|^2 e^{\alpha x^2}\,dx.$$

Two Dimensional Inequalities

[33] If $p > 1$, $q > 1$, $p^{-1}+q^{-1} \geq 1$, $\lambda = 2-p^{-1}-q^{-1}$, $h < 1-p^{-1}$, $k < 1-q^{-1}$, $h+k \geq 0$, and $h+k > 0$ if $p^{-1}+q^{-1} = 1$ (see Hardy, Littlewood, and Polya [7], page 298 (#401)), then

$$\int_0^\infty \int_0^\infty \frac{f(x)g(y)}{x^h y^k |x-y|^{\lambda-h-k}}\,dx\,dy \leq K\left(\int_0^\infty f^p(x)\,dx\right)^{1/p}\left(\int_0^\infty g^q(x)\,dx\right)^{1/q}.$$

Here $K = K(p,q,h,k) > 0$.

[34] If $f(x)$, $g(x)$ and $h(x)$ are non-negative, and $f^*(x)$, $g^*(x)$ and $h^*(x)$ are the equi-measurable symmetrically decreasing functions (see Hardy, Littlewood, and Polya [7], page 279, (#379)), then

$$\int_{-\infty}^\infty \int_{-\infty}^\infty f(x)g(y)h(-x-y)\,dx\,dy \leq \int_{-\infty}^\infty \int_{-\infty}^\infty f^*(x)g^*(y)h^*(-x-y)\,dx\,dy.$$

[35] If $p > 1$, $p' = p/(p-1)$, $\int_0^\infty f^p(x)\,dx \leq F$ and $\int_0^\infty g^{p'}(x)\,dx \leq G$ (see Hardy, Littlewood, and Polya [7], page 226 (#316)), then (unless $f(x) = 0$ or $g(x) = 0$)

$$\int_{-\infty}^\infty \int_{-\infty}^\infty \frac{f(x)g(y)}{x+y}\,dx\,dy < \frac{\pi}{\sin(\pi/p)} F^{1/p} G^{1/p'}.$$

Other Inequalities

[36] If the function $f, g, \ldots, h$ are linearly independent functions (i.e., there do not exist constants $A, B, \ldots, C$, some not equal to zero, such that $Af+Bg+\cdots+Ch = 0$) (see Hardy, Littlewood, and Polya [7], page 134 (#182)), then

$$\begin{vmatrix} \int f^2(x)\,dx & \int f(x)g(x)\,dx & \ldots & \int f(x)h(x)\,dx \\ \vdots & \vdots & \ddots & \vdots \\ \int h(x)f(x)\,dx & \int h(x)g(x)\,dx & \ldots & \int h^2(x)\,dx \end{vmatrix} > 0.$$

Notes

[1] In this section, when no further explanation is given, functions with upper case letters are assumed to be integrals of functions with lower case letters. For example, $F(x) = \int_0^x f(x)\,dx$ and $G(x) = \int_0^x g(x)\,dx$. Also, all the integrals in this section are assumed to exist.

[2] If $f(x)$ is a real continuously differentiable function that satisfies the boundedness constraints $\int_{-\infty}^\infty x^2|f(x)|^2\,dx < \infty$ and $\int_{-\infty}^\infty |f'(x)|^2\,dx < \infty$, then for $x \geq 0$ we have

$$x|f(x)|^2 \leq 4\sqrt{\int_x^\infty t^2|f(t)|^2\,dt}\sqrt{\int_x^\infty |f'(t)|^2\,dt}.$$

This, in turn, can be used to derive the inequality

$$\int_{-\infty}^{\infty} |f(x)|^2\,dx \le 2\sqrt{\int_{-\infty}^{\infty} t^2|f(t)|^2\,dt}\sqrt{\int_{-\infty}^{\infty} |f'(t)|^2\,dt.}$$

This last inequality is known as Heisenberg's uncertainty principle in quantum mechanics.

[3] Evans *et al.* [3] contains a complete analysis of the inequality

$$\left(\int \left\{\left|f'(x)\right|^2 + (x^2-\tau)|f(x)|^2\right\}\,dx\right)^2 \\ \le K(\tau)\left(\int |f(x)|^2\,dx\right)\left(\int \left|f''(x) - (x^2-\tau)f(x)\right|^2\,dx\right).$$

[4] Pachpatte [18] derives generalizations of the inequalities

$$\int_0^{\infty} x^{-m}F_*^{2p}(x)\,dx \le \left(\frac{2p}{|m-1|}\right)^{2p}\int_0^{\infty} x^{-m}f^{2p}(x)\,dx$$
$$\int_0^{\pi} x^{-m}F_*^{2p}(x)\,dx \le \left(\frac{2p}{m-1}\right)\int_0^{\pi} x^{-m}\left|f(x) - f\left(\frac{x}{2}\right)\right|^{2p}(x)\,dx$$

with suitable constraints on f, m, and p. (Here, F_* is related to the integral of f.)

[5] The inequality $\iint\limits_G |f|^2\,dx\,dy \le \frac{1}{4\pi}\left(\int_\Gamma |f|\,d|z|\right)^2$ for functions f holomorphic in $G\cup\Gamma$ is referenced in Gamelin and Khavinson [5].

[6] Gronwalls' inequality states (see Gradshteyn and Ryzhik [6], page 1127):

> **Theorem:** Let the three piecewise continuous, nonnegative functions $\{u, v, w\}$ be defined in the interval $[0, a]$ and satisfy the inequality
>
> $$w(t) \le u(t) + \int_0^t v(\tau)w(\tau)\,d\tau,$$
>
> except at points of discontinuity of the functions. Then, except at these same points,
>
> $$w(t) \le u(t) + \int_0^t u(\tau)v(\tau)\exp\left(\int_\tau^t v(\sigma)\,d\sigma\right)\,d\tau.$$

[7] Opial [12] showed that $\int_0^h |f(x)|\, f'(x)\, dx \le \frac{1}{4} h \int_0^h (f'(x))^2\, dx$, with certain conditions on f. A comprehensive survey of Opial-type inequalities may be found in Mitrinović [11]. Yang [22] proved the generalization

> **Theorem:** If $f(s,t)$, f_s, and f_{st} are continuous functions on $[a,b] \times [c,d]$ and if $f(a,t) = f(b,t) = f_s(s,c) = f_s(s,d) = 0$ for $a \le s \le b$ and $c \le t \le d$, then
>
> $$\int_a^b \int_c^d |f(s,t)|\, |f_{st}(s,t)|\, dt\, ds \le \frac{(b-a)(d-c)}{8} \int_a^b \int_c^d |f_{st}(s,t)|^2\, dt\, ds.$$

Two other generalizations of Opial's inequality are in Pachpatte [16]. One of these generalizations is (the other is similar):

> **Theorem:** Suppose the functions p, q are positive and continuous on $\Delta = [a, X] \times [c, Y]$. Let $f = f(s,t)$, f_s, f_{st} be continuous functions on Δ with $f(a,t) = f_s(s,c) = 0$ for $a \le s \le X$ and $c \le t \le Y$. If m and n are positive integers, with $m + n > 1$, then
>
> $$\int_a^X \int_c^Y p|f|^m\, |f_{st}|^n\, dt\, ds \le K(X,Y,m,n) \int_a^X \int_c^Y q\, |f_{st}|^{m+n}\, dt\, ds \qquad (46.1)$$
>
> where $K(X,Y,m,n)$ is a finite constant that depends on the functions p and q. If $m < 0$, $n > 0$, and $m + n > 1$, then (46.1) holds with $\le$ replaced with $\ge$.

[8] Assume that $f(t)$ and $\phi(t)$ are nonnegative and measurable on $\mathbf{R}^+$, and that both a and b are in the range $(0, \infty)$. Define $\Phi(x) = \int_0^x \phi(t)\, dt$, $F_L(x) = \int_0^x f(t)\phi(t)\, dt$, $F_U(x) = \int_x^\infty f(t)\phi(t)\, dt$, and $M = (p/|c-1|)^p$. Then (see Copson [2]):

$$\int_0^b F_L^p \Phi^{-c} \phi\, dx \le M \int_0^b f^p \Phi^{p-c} \phi\, dx \qquad \text{if } p \ge 1,\ c > 1$$

$$\int_a^\infty F_L^p \Phi^{-c} \phi\, dx \ge M \int_a^\infty f^p \Phi^{p-c} \phi\, dx \qquad \text{if } 0 < p \le 1,\ c > 1 \text{ and } \Phi(x) \to 0 \text{ as } x \to \infty$$

$$\int_a^\infty F_U^p \Phi^{-c} \phi\, dx \le M \int_a^\infty f^p \Phi^{p-c} \phi\, dx \qquad \text{if } p \ge 1,\ c < 1$$

$$\int_0^b F_U^p \Phi^{-c} \phi\, dx \ge M \int_0^b f^p \Phi^{p-c} \phi\, dx \qquad \text{if } 0 < p \le 1,\ c < 1$$

$$\int_0^b F_L^p \Phi^{-1} \phi\, dx \le p^p \int_0^b f^p \Phi^{p-1} \left(\log \frac{\Phi(b)}{\Phi(x)}\right)^p \phi\, dx \qquad \text{if } p \ge 1$$

$$\int_a^\infty F_U^p \Phi^{-1} \phi\, dx \ge p^p \int_a^\infty f^p \Phi^{p-1} \left(\log \frac{\Phi(x)}{\Phi(a)}\right)^p \phi\, dx \qquad \text{if } 0 < p \le 1.$$

[9] The HELP (Hardy, Everitt, Littlewood, Polya) inequalities are of the form (see Evans and Everitt [4])

$$\left\{\int_a^b \left(p\left|f'\right|^2 + q\left|f\right|^2\right)\right\}^2 \leq K \int_a^b w\left|f\right|^2 \int_a^b w\left|w^{-1}M[f]\right|^2,$$

where p, q, and w are real-valued functions on $[a,b]$ (with $-\infty < a < b \leq \infty$), and $M[\cdot]$ denotes the second order differential expression $M[f] = -(pf')' + qf$. There are some technical conditions on p, q, and w.

References

[1] G. Brown and L. A. Shepp, "A Backward Hölder's Inequality," Problem number E 3370 in *Amer. Math. Monthly*, **98**, No. 7, August–September 1991, pages 650–652.

[2] E. T. Copson, "Some Integral Inequalities," *Proc. Roy. Soc. Edinburgh*, **75A**, No. 13, 1975/76, pages 157–164.

[3] W. D. Evans, W. N. Everitt, W. K. Hayman, and S. Ruscheweyh, "On a Class of Integral Inequalities of Hardy–Littlewood Type," *J. Analyse Math.*, **46**, 1986, pages 118–147.

[4] W. D. Evans and W. N. Everitt, "HELP Inequalities for Limit-Circle and Regular Problems," *Proc. R. Soc. London A*, 1991, **432**, pages 367–390.

[5] T. W. Gamelin and D. Khavinson, "The Isoperimetric Inequality and Rational Approximation," *Amer. Math. Monthly*, January 1989, page 22.

[6] I. S. Gradshteyn and I. M. Ryzhik, *Tables of Integrals, Series, and Products*, Academic Press, New York, 1980.

[7] H. Hardy, J. E. Littlewood, G. Polya, *Inequalities*, Cambridge Mathematical Library, Second Edition, New York, 1988.

[8] M. Izumi and S. Izumi, "On Some Inequalities for Fourier Series," *J. Anal. Math.*, **21**, 1968, pages 277-291.

[9] S. Iyanaga and Y. Kawada, *Encyclopedic Dictionary of Mathematics*, MIT Press, Cambridge, MA, 1980.

[10] A. B. Mingarelli, "A Note on Some Differential Inequalities," *Bull. Inst. Math. Acad. Sinica*, **14**, No. 3, 1986, pages 287–288.

[11] D. S. Mitrinović, *Analytic Inequalities*, Springer–Verlag, New York, 1970.

[12] Z. Opial, "Sur une inégalité," *Ann. Polon. Math.*, **8**, 1960, pages 29–32.

[13] K. Ostaszewski and J. Sochacki, "Gronwall's Inequality and the Henstock Integral," *J. Math. Anal. Appl.*, **127**, 1987, pages 370–374.

[14] B. G. Pachpatte, "On Opial-Type Integral Inequalities," *J. Math. Anal. Appl.*, **120**, No. 2, 1986, pages 547–556.

[15] B. G. Pachpatte, "On Some Variants of Hardy's Inequality," *J. Math. Anal. Appl.*, **124**, 1987, pages 495–501.

[16] B. G. Pachpatte, "On Two Independent Variable Opial-Type Integral Inequalities," *J. Math. Anal. Appl.*, **125**, 1987, pages 47–57.

[17] B. G. Pachpatte, "On Some New Integral Inequalities in Two Independent Variables," *J. Math. Anal. Appl.*, **129**, No. 2, 1988, pages 375–382.

[18] B. G. Pachpatte, "On Some Integral Inequalities Similar to Hardy's Inequality," *J. Math. Anal. Appl.*, **129**, 1988, pages 596–606.

[19] W. Squire, *Integration for Engineers and Scientists*, American Elsevier Publishing Company, New York, 1970.
[20] A. A. Tananika, "A Generalization of Wirtinger's Inequality," *Differentsial'nye Uravneniya*, **22**, No. 6, 1986, pages 1074–1076.
[21] W. Potze and H. P. Urbach, "An Inequality in L^p," *Appl. Math. Lett.*, **3**, No. 3, 1990, pages 95–96.
[22] G. S. Yang, "Inequality of Opial-type in Two Variables," *Tamkang J. Math.*, **13**, 1982, pages 255–259.

47. Integration by Parts

Applicable to A single integral.

Yields

An asymptotic expansion of the integral.

Idea

By using integration by parts an asymptotic expansion may sometimes be obtained.

Procedure

Repeatedly using the process of integration by parts (see page 161) often allows an asymptotic expansion to be obtained. The remainder term is needed to determine the error at any stage of the approximation.

Several theorems are available that can be used to state an asymptotic expansion of an integral immediately, see the Notes.

Example 1

The exponential integral is defined by

$$E_1(x) = \int_x^\infty \frac{e^{-t}}{t}\,dt. \tag{47.1}$$

We make the identification ($dv = e^{-t}dt$, $u = t^{-1}$), and then use the integration by parts formula, $\int u\,dv = uv\big| - \int v\,du$, on (47.1) to obtain

$$\begin{aligned} E_1(x) &= -\frac{e^{-t}}{t}\bigg|_{t=x}^{t=\infty} - \int_x^\infty \frac{e^{-t}}{t^2}\,dt \\ &= \frac{e^{-x}}{x} - \int_x^\infty \frac{e^{-t}}{t^2}\,dt. \end{aligned} \tag{47.2}$$

Making the identification ($dv = e^{-t}dt$, $u = t^{-2}$) in (47.2), and using integration by parts again, results in

$$E_1(x) = \frac{e^{-x}}{x} - \frac{e^{-x}}{x^2} + \int_x^\infty \frac{2e^{-t}}{t^2}\,dt.$$

Integrating by parts a total of N times results in

$$\begin{aligned} E_1(x) &= S_N(x) + R_N(x) \\ &= \left(e^{-x} \sum_{n=0}^{N-1} \frac{(-1)^n n!}{x^{n+1}} \right) + \left((-1)^N N! \int_x^\infty \frac{e^{-t}}{t^{N+1}}\, dt \right). \end{aligned} \tag{47.3}$$

Integrating by parts infinitely many times results in $e^{-x} \sum_{n=0}^{\infty} \frac{(-1)^n n!}{x^{n+1}}$. This series diverges for all values of x (since the absolute value of the ratio of successive terms in the sum is n/x, which increases as n increases), and so is not a good representation of $E_1(x)$.

However, it is not hard to bound the remainder term $R_N(x)$. We have

$$\begin{aligned} |R_N(x)| &= N! \int_x^\infty \frac{e^{-t}}{t^{N+1}}\, dt \\ &\leq N! \int_x^\infty \frac{e^{-t}}{x^{N+1}}\, dt \\ &= N! \frac{e^{-x}}{x^{N+1}} \\ &\sim \left(\sqrt{2\pi N} e^{-N} N^N \right) \frac{e^{-x}}{x^{N+1}} \\ &= \sqrt{2\pi N} \frac{e^{-x}}{x} \left(\frac{N}{ex} \right)^N \end{aligned}$$

where we have used Stirling's approximation for $N!$, which is asymptotically valid for large values of N. From this rough approximation, we conclude that, for some values of x, smaller values of N may give a smaller remainder than larger values of N.

Example 2

The complementary error function is defined by the integral $\operatorname{erfc}(x) = \frac{2}{\sqrt{\pi}} \int_x^\infty e^{-t^2}\, dt$. By repeated integration by parts, we can obtain the asymptotic expansion

$$\begin{aligned} \operatorname{erfc}(x) &\sim \frac{2e^{-x^2}}{\sqrt{\pi}} \left[\frac{1}{2x} - \frac{1}{4x^3} + \frac{3}{8x^5} - \dots \right] \\ &= \frac{2e^{-x^2}}{\sqrt{\pi}} \sum_{k=0}^{\infty} (-1)^k \frac{(1)(3)\cdots(2k-1)}{2^{k+1} x^{2k+1}}. \end{aligned}$$

Once again, this asymptotic series diverges for all values of x. However, for a fixed number of terms, the approximation becomes better as x increases. A numerical illustration of this asymptotic expansion is in Table 47.

Table 47. A numerical comparison of the complementary error function with the first term and first two terms in its asymptotic expansion.

	$\operatorname{erfc}(x)$	$\dfrac{e^{-x^2}}{x\sqrt{\pi}}$	$\dfrac{e^{-x^2}}{\sqrt{\pi}}\left[\dfrac{1}{x}-\dfrac{1}{2x^3}\right]$
$x=1$	0.15730	0.41510	0.20755
$x=2$	0.00468	0.00517	0.00452
$x=3$	0.0000221	0.0000232	0.0000219

Notes

[1] The approximation in (47.3) indicates a common trait of asymptotic sequences: For fixed x, the approximation gets worse as N increases, for fixed N, the approximation gets better as x increases.

[2] Note that, from a numerical point of view, the integral in (47.2) is more rapidly converging than the integral in (47.1).

[3] Wong [4] presents the following example where integration by parts does not lead to an asymptotic expansion. The integral $I(x) = \displaystyle\int_0^\infty \frac{dt}{(1+t)^{1/3}(x+t)}$ can be integrated by parts to obtain

$$I(x) = -\sum_{n=1}^{N-1} \frac{3^n(n-1)!}{2\cdot 5\cdots(3n-1)}\frac{1}{x^n} + \delta_N(x), \tag{47.4}$$

where $\delta_N(x) = \dfrac{3^{N-1}(N-1)!}{2\cdot 5\cdots(3N-4)}\displaystyle\int_0^\infty \frac{(1+t)^{N-4/3}}{(x+t)^N}\,dt$. Approximations arising from (47.4) are not useful, since $I(x)$ is positive but every term in (47.4) is negative.

[4] Bleistein and Handelsman [2] have several general results about integration by parts.

(A) Consider the integral $I(\lambda) = \int_a^b h(t;\lambda)f(t;\lambda)\,dt$. Assuming sufficient continuity, we have $I(\lambda) = \sum_{n=0}^N S_n(\lambda) + R_N(\lambda)$ (see Bleistein and Handelsman [2], Theorem 3.1) where

$$S_n(\lambda) = (-1)^n\left[f^{(n)}(b;\lambda)h^{(-n-1)}(b;\lambda) - f^{(n)}(a;\lambda)h^{(-n-1)}(a;\lambda)\right]$$

$$R_N(\lambda) = (-1)^{N+1}\int_a^b f^{(N+1)}(t;\lambda)h^{(-N-1)}(t;\lambda)\,dt$$

where $g^{(n)}$ denotes the n-th derivative of g if n is positive, and it denotes the $|n|$-th integral of g if n is negative. If the functions $\{\phi_n(\lambda)\}$ form an asymptotic sequence as $\lambda \to \lambda_0$, and if $\left|h^{(-n-1)}(t;\lambda)\right| \le \alpha_n(t)\phi_n(\lambda)$, where the $\{\alpha_n\}$ are continuous, then as $\lambda \to \lambda_0$ we have $I(\lambda) \sim \sum_{n=0}^{N} S_n(\lambda)$. As an example, if f is sufficiently differentiable, and if $0 \leq a < b$, then we have

$$\int_a^b t^\lambda f(t)\,dt \sim \sum_{n=0}^{N} \frac{(-1)^n}{\prod_{j=0}^{n}(\lambda + j + 1)} \left\{ f^{(n)}(b) b^{\lambda+n+1} - f^{(n)}(a) a^{\lambda+n+1} \right\}$$

as $\lambda \to \infty$.

(B) Consider the integral $I(\lambda) = \int_a^b h(\lambda t) f(t)\,dt$. Assuming

- sufficient continuity for f;
- $b - a$ is finite;
- $\lambda^{-1}\phi_{n+1}(\lambda) = o(\phi_n(\lambda))$ as $\lambda \to \infty$;
- $|h^{(-n)}(\lambda t)| \leq \alpha_n(t)\phi_n(\lambda)$, where the $\{\alpha_n\}$ are continuous,

then, as $\lambda \to \infty$ (see Bleistein and Handelsman [2], Theorem 3.2)

$$I(\lambda) \sim \sum_{n=0}^{N} \frac{(-1)^n}{\lambda^{n+1}} \left[f^{(n)}(b) h^{(-n-1)}(\lambda b) - f^{(n)}(a) h^{(-n-1)}(\lambda b) \right].$$

As an example, we have the result (if f is sufficiently smooth)

$$\int_a^b e^{i\lambda x} f(t)\,dt \sim \sum_{n=0}^{N} \frac{(-1)^n}{(i\lambda)^{n+1}} \left[f^{(n)}(b) e^{i\lambda b} - f^{(n)}(a) e^{i\lambda b} \right]$$

as $\lambda \to \infty$.

References

[1] C. M. Bender and S. A. Orszag, *Advanced Mathematical Methods for Scientists and Engineers*, McGraw–Hill, New York, 1978

[2] N. Bleistein and R. A. Handelsman, *Asymptotic Expansions of Integrals*, Dover Publications, Inc., New York, 1986, Chapter 3, pages 69–101.

[3] A. Erdélyi, *Asymptotic Expansions*, Dover Publications, Inc., New York, 1956.

[4] R. Wong, *Asymptotic Approximation of Integrals*, Academic Press, New York, 1989, pages 14–19.

48. Interval Analysis

Applicable to Ordinary integrals, or integrals containing interval expressions.

Yields

An analytical approximation with an exact bound on the error.

Idea

In interval analysis, quantities are defined by intervals with maximum and minimum values indicated by the endpoints. Definite integrals can often be approximated by an interval; intervals are better than ordinary numerical approximations since an exact bound on the error is obtained.

Procedure

We use the interval notation $[a, b]$ to indicate some number between the values of a and b. We will allow coefficients of polynomials to be intervals. For example, the interval polynomial

$$Q(x) = 1 + [2,3]x^2 + [-1,4]x^3, \tag{48.1}$$

evaluated at the point $x = y$, means that

$$\min_{\substack{2\le\eta\le 3\\ -1\le\zeta\le 4}} 1 + \eta y^2 + \zeta y^3 \le Q(y) \le \max_{\substack{2\le\eta\le 3\\ -1\le\zeta\le 4}} 1 + \eta y^2 + \zeta y^3. \tag{48.2}$$

There exists an algebra of interval polynomials. For example

$$\left(x + [2,3]x^3\right) + \left([1,2]x + [1,4]x^3\right) = [2,3]x + [3,7]x^3,$$

$$\left([1,3] + [-1,2]x\right)^2 = [1,9] + [-6,12]x + [--2,4]x^2.$$

If $P(y)$ and $Q(y)$ are interval polynomials, then at any point y we can write $P(y) \in [P_L, P_U], Q(y) \in [Q_L, Q_U]$. We say that $P(x)$ contains $Q(x)$ on some interval $[c, d]$ if $P_L \le Q_L, Q_U \le P_U$ for all $y \in [c, d]$. This is denoted by $Q(x) \subset P(x)$.

We now use capital letters to denote intervals; i.e., $F(X)$ denotes the interval $[F_L, F_H]$ where $F_L = \min_{x\in X} f(x)$ and $F_H = \max_{x\in X} f(x)$. If we define $X_i^{(n)}$ by

$$X_i^{(n)} = \left[a + \frac{i-1}{n}(x-a), a + \frac{i}{n}(x-a)\right], \qquad i = 1, 2, \dots, n, \tag{48.3}$$

then, if $f(x)$ is sufficiently smooth,

$$\int_a^x f(t)\,dt \subset \sum_i^n F\left(X_i^{(n)}\right)\left(\frac{x-a}{n}\right). \tag{48.4}$$

Define Q to the right-hand side of (48.4). The width of the interval Q, $w(Q)$, can be shown to satisfy

$$w(Q) \le \frac{K(x-a)^2}{n} \tag{48.5}$$

where K is a positive constant independent of n.

The quadrature formula in (48.4) is essentially a first order integration formula for $\int f(t)\,dt$. Higher order formulae are also available. See, for example, Corliss and Rall [4].

Example 1

This example illustrates the use of (48.4). Consider the integral $I = \int_0^1 \sin \pi x^2 \, dx$. Using (48.4) with $n = 2$, we have

$$I \subset \left(F\left(\left[0, \tfrac{1}{2}\right]\right) \tfrac{1}{2} + F\left(\left[\tfrac{1}{2}, 1\right]\right) \tfrac{1}{2} \right) \tag{48.6}$$

where $F(X) = \sin(\pi X^2)$. The expression in (48.6) can be evaluated to yield

$$\begin{aligned} I &\subset \left(F\left(\left[0, \sin^2 \tfrac{\pi}{4}\right]\right) \tfrac{1}{2} + F\left(\left[0, \sin^2 \frac{\pi}{2}\right]\right) \tfrac{1}{2} \right) \\ &= \left[0, \tfrac{1}{2}\right] \tfrac{1}{2} + [0, 1] \tfrac{1}{2} \\ &= \left[0, \tfrac{3}{4}\right]. \end{aligned}$$

Example 2

Consider the integral $I = \int_{-1}^{1} f(x) \, dx$, where $f(x) = 1 + [1, 2]x + x^2$. It is straightforward to show that

$$f(x) = \begin{cases} \left[1 + 2x + x^2, 1 + x + x^2\right] & \text{for } -1 \le x \le 0, \\ \left[1 + x + x^2, 1 + 2x + x^2\right] & \text{for } 0 \le x \le 1. \end{cases}$$

Hence, we have

$$\begin{aligned} I &= \left[\int_{-1}^{0} 1 + 2x + x^2 + \int_0^1 1 + x + x^2, \int_{-1}^{0} 1 + x + x^2 + \int_0^1 1 + 2x + x^2 \right] \\ &= \left(\tfrac{11}{6}, \tfrac{5}{2} \right). \end{aligned}$$

Notes

[1] The techniques presented in this section can be evaluated numerically. Interval arithmetic packages are available in Algol (see Guenther and Marquardt [5]), Pascal (see Wolff von Gudenberg [10]), and FORTRAN (see ACRITH [1], but see also Kahan and LeBlanc [6]).

[2] Corliss and Rall [4] describe an interval analysis program that evaluates integrals by implementing Newton–Cotes rules, Gauss rules, and Taylor series. They also consider problems in which the limits of integration are intervals.

References

[1] ACRITH *High Accuracy Subroutine Library: General Information Manual,* IBM publication # GC33-6163, Yorktown Heights, NY, 1985.

[2] O. Caprani, K. Madsen, and L. B. Rall, "Integration of Rational Functions," *SIAM J. Math. Anal.*, **12**, 1981, pages 321–341.

[3] G. Corliss and G. Krenz, "Indefinite Integration with Validation," *ACM Trans. Math. Software*, **15**, No. 4, December 1989, pages 375–393.

[4] G. F. Corliss and L. B. Rall, "Adaptive, Self-Validating Numerical Quadrature," *SIAM J. Sci. Stat. Comput.*, **8**, No. 5, 1987, pages 831–847.

[5] G. Guenther and G. Marquardt, "A Programming System for Interval Arithmetic," in K. Nickel (ed.), *Interval Mathematics 1980*, Academic Press, New York, 1980, pages 355–366.

[6] W. Kahan and E. LeBlanc, "Anomalies in the IBM ACRITH Package," *IEEE Proc. 7th Symp. on Computer Arithmetic*, 1985.

[7] R. E. Moore, *Interval Analysis*, Prentice–Hall Inc., Englewood Cliffs, NJ, 1966, Chapter 8, pages 70–80.

[8] L. B. Rall, "Integration of Rational Functions II. The Finite Case," *SIAM J. Math. Anal.*, **13**, 1982, pages 690–697.

[9] J. M. Yohe, "Software for Interval Arithmetic: A Reasonable Portable Package," *ACM Trans. Math. Software*, **5**, No. 1, March 1979, pages 50–63.

[10] J. Wolff von Gudenberg, *Floating-Point Computation in PASCAL-SC with Verified Results*, in B. Buchberger and B. F. Caviness (eds.), *EUROCAL '85*, Springer–Verlag, New York, 1985, pages 322–324.

49. Laplace's Method

Applicable to Integrals of the form $I(\lambda) = \int_a^b g(x)e^{\lambda f(x)}\,dx$, where $f(x)$ is a real-valued function.

Yields

An asymptotic approximation when $\lambda \gg 1$.

Idea

For $\lambda \to \infty$ the value of $I(\lambda)$ is dominated by the contributions at those points where $f(x)$ is a local maximum.

Procedure

Given the integral $I(\lambda)$, consider the term $e^{\lambda f(x)}$ for fixed λ. This term will have a stationary point, (i.e., a local maximum or minimum) when $de^{\lambda f(x)}/dx = 0$, or $f'(x) = 0$. The behavior of $I(\lambda)$ is dominated at the local maximums of f; points where $f'(x) = 0$ and (usually) $f''(x) < 0$.

If the stationary point x_i is an interior point (i.e., $a < x_i < b$) then $I(\lambda)$ may be approximated, in the neighborhood of this point, as

$$
\begin{aligned}
I(\lambda) &\sim \int_{x_i-\varepsilon}^{x_i+\varepsilon} e^{\lambda f(x)} g(x)\,dx \\
&\sim \int_{x_i-\varepsilon}^{x_i+\varepsilon} \exp\left[\lambda\left(f(x_i) + \frac{(x-x_i)^2 f''(x_i)}{2} + \dots\right)\right] \\
&\qquad\qquad \times \left[g(x_i) + (x-x_i)g'(x_i) + \frac{(x-x_i)^2 g''(x_i)}{2} + \dots\right] dx \\
&\sim g(x_i)e^{\lambda f(x_i)} \int_{x_i-\varepsilon}^{x_i+\varepsilon} \left(\exp\left[\lambda \frac{(x-x_i)^2 f''(x_i)}{2}\right] + \dots\right) dx \\
&\sim g(x_i)e^{\lambda f(x_i)} \int_{-\infty}^{\infty} \left(\exp\left[\lambda \frac{(x-x_i)^2 f''(x_i)}{2}\right] + \dots\right) dx \\
&\sim g(x_i)e^{\lambda f(x_i)} \left[\sqrt{\frac{2\pi}{\lambda |f''(x_i)|}}\right] + \dots
\end{aligned}
\tag{49.1}
$$

which is valid as $\lambda \to \infty$, when $f''(x_i) < 0$. If $f''(x_i) > 0$, then the point x_i is a local minimum of f, not a local maximum, and this point does not contribute to leading order. (For the case $f''(x_i) = 0$, see the Notes, below.)

For each point where f has a local maximum there will be a term in the form of (49.1). To find the asymptotic approximation to $I(\lambda)$, these terms must be summed up.

For the stationary *boundary points* (i.e., those stationary points that are on the boundary of the domain, either $x_i = a$ or $x_i = b$) there is a term in the form of (49.1), but with half the magnitude (if $f''(\,)$ is negative at that boundary point). This is because the integral in the fourth line of (49.1) becomes an integral from 0 to ∞ or $-\infty$ and not from $-\infty$ to ∞.

If either of the boundary points is not a stationary point, (i.e., $f'(a) \neq 0$ or $f'(b) \neq 0$), then these boundary points points contribute

$$
-\frac{g(a)e^{\lambda f(a)}}{\lambda f'(a)} \qquad \text{or} \qquad \frac{g(b)e^{\lambda f(b)}}{\lambda f'(b)} \tag{49.2}
$$

to the sum forming the the asymptotic approximation of $I(\lambda)$. The contributions from non-stationary boundary points will always be asymptotically smaller than the contribution from the points at which f is a local maximum. Hence, if the region of integration contains any points at which f is a local maximum, and if only the leading order behavior is desired, then the non-stationary boundary points can be ignored.

Example 1

For the integral $J(\lambda) = \int_0^{10} e^{-\lambda \cos x}\,dx$ we identify $f(x) = -\cos x$, $g(x) = 1$, $a = 0$ and $b = 10$. The stationary points are where $f'(x) = \sin x = 0$, or $x = \{0, \pi, 2\pi, 3\pi, 4\pi, \ldots\}$. We are only interested in those stationary points in the range of integration, that is $x = 0$, $x = \pi$, $x = 2\pi$, and $x = 3\pi$. Including the boundary points $x = 0$ and $x = 10$ we have four points that can potentially contribute to the leading order term in the asymptotic expansion.

$x = 0$ This is a stationary boundary point. However, since $f''(0) = 1 > 0$, this point does not contribute to leading order.

$x = \pi$ This is an interior stationary point. Since $f''(\pi) = -1 < 0$, this point contributes a term of the form in (49.1):
$$I_\pi = g(\pi)e^{\lambda f(\pi)}\sqrt{\frac{2\pi}{\lambda\,|f''(\pi)|}} = \sqrt{\frac{2\pi}{\lambda}}e^{\lambda}.$$

$x = 2\pi$ This is an interior stationary point. However, since $f''(2\pi) = 1 > 0$, this point does not contribute to leading order.

$x = 3\pi$ This is an interior stationary point that is a local maximum since $f''(3\pi) = -1 < 0$. Hence, this point contributes a term of the form in (49.1): $I_{3\pi} = g(3\pi)e^{\lambda f(3\pi)}\sqrt{\dfrac{3\pi}{\lambda\,|f''(3\pi)|}} = \sqrt{\dfrac{2\pi}{\lambda}}e^{\lambda}.$

$x = 10$ This is a non-stationary boundary point. Hence, this point contributes (from (49.2)): $I_{10} = \dfrac{g(10)e^{\lambda f(10)}}{\lambda f'(10)} = -\dfrac{e^{-\lambda \cos 10}}{\lambda \sin 10}.$

We can combine all of the leading order contributions we have to find
$$J(\lambda) \sim I_\pi + I_{3\pi} + I_{10} = \sqrt{\frac{2\pi}{\lambda}}e^{\lambda} + \sqrt{\frac{2\pi}{\lambda}}e^{\lambda} - \frac{e^{-\lambda\cos 10}}{\lambda \sin 10} = \sqrt{\frac{8\pi}{\lambda}}e^{\lambda} - \frac{e^{-\lambda\cos 10}}{\lambda \sin 10}.$$

However, since we have only kept the leading order term in the asymptotic expansion near each point, we can only keep the leading order term in the final answer (assuming no cancellation of terms has occurred). Our final result is therefore:
$$J(\lambda) \sim \sqrt{\frac{8\pi}{\lambda}}e^{\lambda} \qquad \text{as} \quad \lambda \to \infty.$$

Example 2

Consider the integral $J(\lambda) = \int_0^1 e^{\lambda x^2}\, dx$, so that $f(x) = x^2$ and $g(x) = 1$. In this case the stationary points are given by $f'(x) = 2x = 0$, or $x = 0$. The point $x = 0$ is a stationary boundary point, but it does not contribute to leading order since $f''(0) = 2 > 0$. The point $x = 1$ is a boundary point and the leading order asymptotic approximation is given by (49.2):

$$J(\lambda) \sim \frac{g(1)e^{\lambda f(1)}}{\lambda f'(1)} = \frac{e^\lambda}{2\lambda} \qquad \text{as} \quad \lambda \to \infty.$$

Example 3

Consider the integral $K(\lambda) = \int_0^1 e^{-\lambda x^2}\, dx$, so that $f(x) = -x^2$ and $g(x) = 1$. In this case the stationary points are given by $f'(x) = -2x = 0$, or $x = 0$. The point $x = 0$ is a stationary boundary point, and it contributes to leading order since $f''(0) = -2 < 0$. The leading order asymptotic approximation is given by (49.1) with a factor of $\frac{1}{2}$ (since $x = 0$ is a boundary point):

$$K(\lambda) \sim \frac{1}{2} g(1) e^{\lambda f(0)} \sqrt{\frac{2\pi}{\lambda\, |f''(0)|}} = \frac{1}{2}\sqrt{\frac{\pi}{\lambda}} \qquad \text{as} \quad \lambda \to \infty. \tag{49.3}$$

For this integral, we recognize that $K(\lambda) = \frac{1}{2}\sqrt{\frac{\pi}{\lambda}}\, \text{erf}\left(\sqrt{\lambda}\right)$, where erf is the error function. Use of the asymptotic expansion of the error function for large arguments also results in (49.3).

Example 4

An integral representation of the gamma function, for $\lambda > 0$, is

$$\Gamma(\lambda) = \int_0^\infty x^{\lambda-1} e^{-x} dx = \int_0^\infty x^{-1} e^{-x} e^{\lambda \log x} dx. \tag{49.4}$$

If λ is an integer, then $\Gamma(\lambda) = (\lambda - 1)!$ (see page 163).

From (49.4) we have $f(x) = -\log x$, but $f(x)$ has no finite stationary point about which to apply our above expansions. The change of variable $x = \lambda y$ transforms (49.4) to

$$\Gamma(\lambda) = \lambda^\lambda \int_0^\infty e^{\lambda(-y + \log y)}\, \frac{dy}{y}. \tag{49.5}$$

Now $f(y) = -y + \log y$ and $g(y) = y^{-1}$. The minimum of $f(y)$ is at $f'(y) = 0$, or $y = 1$. The value $y = 1$ is an interior stationary point of (49.5), so we have (from (49.1))

$$\Gamma(\lambda) \sim \lambda^\lambda g(1) e^{\lambda f(1)} \sqrt{\frac{2\pi}{\lambda\, |f''(1)|}} = \lambda^\lambda e^{-\lambda} \sqrt{\frac{2\pi}{\lambda}}. \tag{49.6}$$

From (49.6), and a little manipulation, we obtain the leading term in Stirling's approximation to the factorial: $\Gamma(n+1) = n! \sim \sqrt{2\pi n} n^n e^{-n}$.

Example 5

Consider the integral $I(\lambda) = \int_0^\infty e^{\lambda x - (x-1)\log x}\, dx$, for $\lambda \gg 1$. If we make the obvious identification, $f(x) = x$, then the region of maximum contribution will be around $x = \infty$. To determine this contribution, some re-scaling of the problem is required.

Looking at the whole integrand, the stationary point is given by

$$\frac{d}{dx}\left(e^{\lambda x - (x-1)\log x}\right) = 0, \quad \text{or} \quad \lambda = \frac{x-1}{x} + \log x.$$

As suggested above, the stationary point occurs at a large value of x. If x is large then, approximately, the stationary point is given by $x = e^{\lambda - 1}$. Making the change of variable $t = x/e^{\lambda-1}$, we are led to consider

$$I(\lambda) = e^{2(\lambda-1)} \int_0^\infty t e^{\left(e^{\lambda-1}\right)(t - t\log t)}\, dt = e^{2(\lambda-1)} J\left(e^{\lambda-1}\right),$$

where $J(\zeta) = \int_0^\infty t e^{\zeta(t - t\log t)}\, dt$. Since we want $\lambda \gg 1$, this corresponds to $\zeta \gg 1$. Now it is a simple matter to show that $J(\zeta) \sim e^{\zeta}\sqrt{2\pi/\zeta}$ (since the only stationary point is at $t = 1$). Hence, we obtain our final answer:

$$I(\lambda) \sim \sqrt{2\pi} e^{3(\lambda-1)/2} e^{e^{\lambda-1}} \qquad \text{as} \quad \lambda \to \infty.$$

Notes

[1] Laplace's method is an application of the method of steepest descents, see page 229. In the method of steepest descents, the function $f(x)$ can be complex valued. In the method of stationary phase (page 226), the function $f(x)$ is purely imaginary.

[2] If more terms are kept in (49.1), then we obtain the approximation

$$I(\lambda) \sim \sqrt{\frac{2\pi}{-\lambda f''}} e^{\lambda f} \left[g + \frac{1}{\lambda}\left(-\frac{g''}{2f''} + \frac{g f''''}{8\left(f''\right)^2} + \frac{g' f'''}{2\left(f''\right)^2} - \frac{5g\left(f'''\right)^2}{24\left(f''\right)^3} \right) + \dots \right]$$

where all the functions are evaluated at $x = x_i$ (see Bender and Orszag [1], page 273). If the asymptotic expansion in Example 4 were continued to higher order, then we would find

$$\Gamma(\lambda) \sim \lambda^{\lambda} e^{-\lambda} \sqrt{\frac{2\pi}{\lambda}} \left[1 + \frac{1}{12\lambda} + \frac{1}{288\lambda^2} - \dots \right].$$

This yields a better approximation to the factorial function than the one-term Stirling's approximation.

[3] Watson's lemma (page 197) applies to integrals of the form $\int_A^B e^{\lambda y} F(y)\, dy$. By an appropriate change of variable, $I(\lambda)$ can be changed to this form. For example, we can use the transformation $y = f(x)$ and then (assuming monotonicity) $A = f(a)$, $B = f(b)$, $F(y) = g(x)/f'(x)$.

[4] If the second derivative vanishes at an interior stationary point, so that the leading order behavior at a stationary points is given by

$$f(x) = f(x_i) + \frac{(x - x_i)^n}{n!} f^{(n)}(x_i)$$

for x near x_i, then the first term in the asymptotic approximation of $I(\lambda)$ becomes (assuming that n is even, and $f^{(n)}(x_i) < 0$, both of which are required for x_i to be a local maximum):

$$I(\lambda) \sim \frac{2\Gamma\left(\frac{1}{n}\right)(n!)^{1/n}}{n\left(\lambda f^{(n)}(x_i)\right)^{1/n}} g(x_i) e^{\lambda f(x_i)}. \tag{49.7}$$

[5] Multidimensional analogues of this technique are described on page 199.

[6] Skinner [3] considers uniform approximations for integrals of the form $e^{\lambda h(t)} \int_t^\infty e^{-\lambda h(x)} g(x) x^{\alpha-1}\, dx$ as $\lambda \to \infty$.

[7] Temme [4] considers uniform approximations for integrals of the form $(1/\Gamma(\lambda)) \int_\alpha^\infty t^{\lambda-1} e^{-zt} f(t)\, dt$ as $\lambda \to \infty$.

References

[1] C. M. Bender and S. A. Orszag, *Advanced Mathematical Methods for Scientists and Engineers*, McGraw–Hill, New York, 1978.

[2] N. Bleistein and R. A. Handelsman, *Asymptotic Expansions of Integrals*, Dover Publications, Inc., New York, 1986.

[3] L. A. Skinner, "Uniformly Valid Composite Expansions for Laplace Integrals," *SIAM J. Math. Anal.*, **19**, No. 4, July 1988, pages 918–925.

[4] N. M. Temme, "Incomplete Laplace Integrals: Uniform Asymptotic Expansion with Application to the Incomplete Beta Function," *SIAM J. Math. Anal.*, **18**, No. 6, November 1987, pages 1638–1663.

50. Stationary Phase

Applicable to Integrals of the form $I(\lambda) = \int_a^b g(x) e^{i\lambda f(x)}\, dx$, where $f(x)$ is a real-valued function.

Yields

An asymptotic approximation when $\lambda \gg 1$.

Idea

For $\lambda \to \infty$ the value of $I(\lambda)$ is dominated by the contributions at those points where $f(x)$ is a local minimum.

Procedure

The Riemann–Lebesgue lemma states that $\lim_{\lambda\to\infty} \int_a^b h(t)e^{i\lambda t}\,dt = 0$, provided that $\int_a^b |h(t)|\,dt$ exists. In simple terms, if the integrand is highly oscillatory, then the value of the integral is "small."

Now consider the integral

$$I(\lambda) = \int_a^b e^{i\lambda f(x)} g(x)\,dx. \tag{50.1}$$

Through an appropriate change of variables, it can be shown (in non-degenerate cases) that $I(\lambda) \to 0$ as $\lambda \to \infty$. The maximum contributions to (50.1) will come from regions where the integrand is less oscillatory. These regions are specified by the stationary points of the integrand, that is, where $f'(x) = 0$.

Let the stationary points in the interval $[a, b]$ be $\{c_i\}$. Following a derivation similar to that given for Laplace's method (see page 221), we find that the leading order contribution to $I(\lambda)$, due to the stationary point c, is (assuming that $f(c) \neq 0$, $f''(c) \neq 0$, and $g(c) \neq 0$):

$$I_c \sim g(c)\sqrt{\frac{2\pi}{\lambda|f''(c)|}} \exp\left[i\lambda f(c) - \frac{i\pi}{4}\operatorname{sgn} f''(c)\right] \tag{50.2}$$

where sgn denotes the signum (or sign) function.

The leading order asymptotic behavior of $I(\lambda)$ is then given by $\sum_i I_{c_i}$. It is difficult to obtain a better approximation than just the leading order approximation, because it requires delicate estimation of integrals. If a higher order approximation is desired, then the method of steepest descents (see page 229) should be used.

Example

The Bessel function $J_n(x)$, for integral values of n, has the integral representation (see Abramowitz and Stegun [1], 9.1.22.b)

$$\begin{aligned} J_n(\lambda) &= \frac{1}{\pi}\int_0^\pi \cos(nt - \lambda\sin t)\,dt \\ &= \frac{1}{2\pi}\sum_{\pm}\int_0^\pi e^{\pm int} e^{\mp i\lambda\sin t}\,dt. \end{aligned} \tag{50.3}$$

Each integral in the sum has the form of (50.1), with $f_\pm(t) = \mp\sin t$ and $g_\pm(t) = e^{\pm int}$. On the interval $[0, \pi]$ the only stationary point for $f_\pm$ is at $t = \pi/2$. Hence, we find $f_\pm\left(\frac{\pi}{2}\right) = \mp 1$ and $f''_\pm\left(\frac{\pi}{2}\right) = \mp 1$. Therefore, (50.2) can be evaluated to yield

$$\begin{aligned} J_n(\lambda) &\sim \frac{1}{\sqrt{2\pi\lambda}}\sum_{\pm}\exp\left(\pm i\frac{n\pi}{2}\right)\exp\left(\pm i\left[-\lambda + \frac{\pi}{4}\right]\right) \\ &= \sqrt{\frac{2}{\pi\lambda}}\cos\left(\lambda - \frac{n\pi}{2} - \frac{\pi}{4}\right) \end{aligned}$$

for $\lambda \gg 1$.

Notes

[1] Determining the asymptotic behavior of an integral by plugging into the above formulas is a dangerous approach. A simple example where this naive approach could go wrong is with the integral $I = \int_{-\infty}^{\infty} e^{i(3\lambda t^2 - 3t^3)}\,dt$. Use of the above formulas would result in a stationary point at $t = 0$, which leads to the incorrect approximation $\sqrt{\pi/3\lambda}e^{i\pi/4}$. For this integral, there are stationary points at both $t = 0$ *and* $t = \lambda$. Using the contributions from both of these stationary points results in the approximation

$$I \sim 2\sqrt{\frac{\pi}{3\lambda}}e^{i\lambda^3} \cos\left(\frac{\lambda^3}{2} - \frac{\pi}{4}\right).$$

[2] If there are no stationary points in the interval of integration, then the leading order asymptotic behavior is determined by the contribution near the limits of integration. The leading order behavior, in this case, can be determined by integration by parts.

[3] If the leading order expansion of $f(t)$, near the critical point $t = c$, is given by $f(t) = f(c) + \dfrac{f^{(n)}(c)}{n!}(t-c)^n + \dots$, then the leading order behavior of I is given by (assuming, again, that $g(c) \neq 0$):

$$I_c \sim g(c)\frac{2\Gamma(1/n)}{n}\left(\frac{n!}{\lambda\left|f^{(n)}(c)\right|}\right)^{1/n} \exp\left[i\lambda f(c) - \frac{i\pi}{2n}\operatorname{sgn} f^{(n)}(c)\right]. \quad (50.4)$$

Equation (50.2) is just this formula evaluated at $n = 2$. Note that if the stationary point is a boundary point, then the factor of 2 in (50.4) does not appear.

As an example, consider the Bessel function at large order and large argument. That is, consider the integral (see (50.3)) $J_m(m) = \pi^{-1}\int_0^\pi \cos(mt - m\sin t)\,dt$, for $m \gg 1$. Writing this as $J_m(m) = (2\pi)^{-1}\sum_\pm \int_0^\pi e^{\pm im(t-\sin t)}\,dt$, we identify $f(t) = t - \sin t$. For this integral, the only stationary point is at $c = 0$ (since $f'(c) = 0$). At this stationary point the second derivative vanishes: $f''(0) = 0$. Directly keeping terms of the next order results in

$$\begin{aligned} J_m(m) &= \operatorname{Re}\left(\frac{1}{\pi}\int_0^\pi e^{im(\sin t - t)}\,dt\right) \\ &\sim \operatorname{Re}\left(\frac{1}{\pi}\int_0^\pi e^{int^3/6}\,dt\right) \\ &= \frac{1}{\pi}2^{-2/3}3^{-1/6}\Gamma\left(\tfrac{1}{3}\right)m^{-1/3}. \end{aligned}$$

This result could also have been obtained from using (50.4). We have $g(t) = 1/\pi$, $f(t) = \sin t - t$, $f'(t) = \cos t - 1$, $f''(t) = -\sin t$, and $f'''(t) = -\cos t$. Hence, we find that $c = 0$, $n = 3$, and $|f'''(c)| = 1$. Using (50.4) and removing the factor of 2 since the stationary point is a limit of the integral we find

$$\begin{aligned} J_m(m) &\sim \operatorname{Re}\left(\frac{1}{\pi}\frac{\Gamma\left(\frac{1}{3}\right)}{3}\left(\frac{3!}{m}\right)^{1/3}\exp\left(-\frac{i\pi}{4}\right)\right) \\ &\sim \frac{1}{\pi}\frac{\Gamma\left(\frac{1}{3}\right)}{3}\left(\frac{3!}{m}\right)^{1/3}\cos\left(-\frac{\pi}{4}\right) \\ &= \frac{1}{\pi}2^{-2/3}3^{-1/6}\Gamma\left(\tfrac{1}{3}\right)m^{-1/3}. \end{aligned}$$

[4] Stationary phase is an application of the method of steepest descents, see page 229. In the method of steepest descents, the function $f(x)$ can be complex valued. In Laplace's method (page 221), the argument to the exponential is purely real.

[5] This method was first applied in Stokes [9]. A rigorous justification of the method is presented in Watson [10].

References

[1] M. Abramowitz and I. A. Stegun, *Handbook of Mathematical Functions*, National Bureau of Standards, Washington, DC, 1964.

[2] C. M. Bender and S. A. Orszag, *Advanced Mathematical Methods for Scientists and Engineers*, McGraw–Hill, New York, 1978.

[3] N. Bleistein and R. A. Handelsman, *Asymptotic Expansions of Integrals*, Dover Publications, Inc., New York, 1986, Chapter 6, pages 219–251.

[4] R. A. Handelsman and N. Bleistein, "Asymptotic Expansions of Integral Transforms with Oscillatory Kernels; A Generalization of the Method of Stationary Phase," *SIAM J. Math. Anal.*, **4**, No. 3, 1973, pages 519–535.

[5] N. Chako, "Asymptotic Expansions of Double and Multiple Integrals Occurring in Diffraction Theory," *J. Inst. Maths. Applics*, **1**, December 1965, pages 372–422.

[6] J. C. Cooke, "Stationary Phase in Two Dimensions," *IMA J. Appl. Math.*, **29**, 1982, pages 25–37.

[7] F. De Kok, "On the Method of Stationary Phase for Multiple Integrals," *SIAM J. Math. Anal.*, **2**, No. 1, February 1971, pages 76–104.

[8] J. P. McClure and R. Wong, "Two-Dimensional Stationary Phase Approximation: Stationary Point at a Corner," *SIAM J. Math. Anal.*, **22**, No. 2, March 1991, pages 500–523.

[9] G. G. Stokes, "On the Numerical Calculation of a Class of Definite Integrals and Infinite Series," *Camb. Philos. Trans.*, **9**, 1856, pages 166–187.

[10] G. N. Watson, "The Limits of Applicability of the Principle of Stationary Phase," *Proc. Camb. Philos. Soc.*, **19**, 1918, pages 49–55.

51. Steepest Descent

Applicable to Integrals of the form $\int_C e^{\lambda f(x)} g(x)\, dx$, as λ tends to infinity, where C is a contour in the complex plane and $f(x)$ and $g(x)$ may be complex.

Yields

An asymptotic approximation when $\lambda \gg 1$.

Idea

Given an integral in the form of $\int_C e^{\lambda f(z)} g(z)\, dz$ deform the contour of integration so that it is in the form of a Laplace integral (see page 221) and apply the method described there.

Procedure

Some definitions are needed before the method can be described. Given the complex analytic function $f(z)$, of the complex variable z (i.e., $z = x + iy$), let f_R and f_I denote the real and imaginary parts of that function (i.e., $f(z) = f_R(z) + i f_I(z)$). Given a point z_0, a directed curve from z_0 along which $f_R(z)$ is decreasing is called a path of descent. The path of steepest descent is the curve whose tangent is given by $-\nabla f_R$. This path is also one of the curves along which $f_I(z)$ is constant. That is, a curve of steepest descent is also a curve of constant phase.

Suppose that $f(z)$ and its first $n-1$ derivatives vanish at $z = z_0$:

$$\begin{aligned} \left.\frac{d^q f}{dz^q}\right|_{z=z_0} &= 0, \quad \text{for } q = 1, 2, \dots, n-1, \\ \left.\frac{d^n f}{dz^n}\right|_{z=z_0} &= a e^{i\alpha}, \quad \text{with } a > 0. \end{aligned} \tag{51.1}$$

That is, $f(z) = f(z_0) + f^{(n)}(z_0)(z - z_0)^n / n! + \dots$. If $z = z_0 + \varepsilon e^{i\theta}$, then the directions of constant phase (i.e., f_I is constant), from the point $z = z_0$, are given by

$$\theta_p = (2p+1)\frac{\pi}{n} - \frac{\alpha}{n} \tag{51.2}$$

for $p = 0, 1, \dots, n-1$. Furthermore, f_R decreases where $\cos(n\theta_p + \alpha)$ is positive, and increases where $\cos(n\theta_p + \alpha)$ is negative. Using $\alpha = 0$ for illustrative purposes, Figure 51.1 shows the regions of increasing and decreasing values of f_R. (A nonzero value for α would just rotate the shaded regions in Figure 51.1.)

The method of steepest descents can be succinctly described by the following steps:

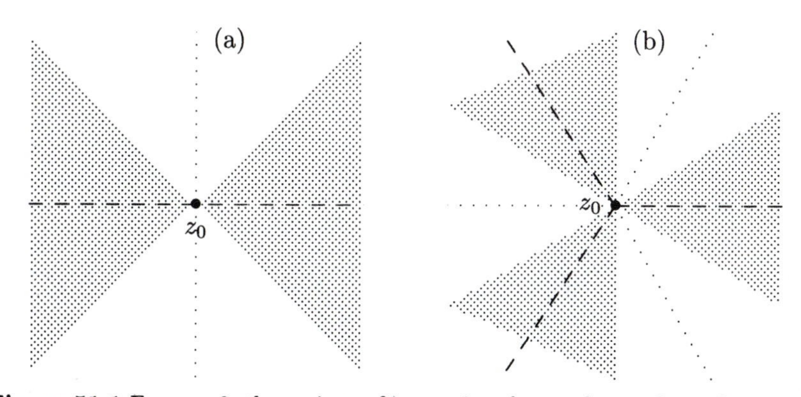

Figure 51.1 For $\alpha = 0$, the regions of increasing f_R are shown clear, the regions of decreasing f_R are shown shaded. The paths of steepest descent are shown dashed, the paths of steepest ascent are shown dotted for (a) $n = 2$ and (b) $n = 3$.

[1] Identify the possible critical points of the integrand. These are the endpoints of integration, singular points of $f(z)$ and $g(z)$ and saddle points of $f(z)$. (A saddle point is a point at which df/dz vanishes.)

[2] Determine the paths of steepest descent from each of the critical points. A picture of the complex plane with curves drawn is usually very helpful.

[3] Justify, via Cauchy's theorem, the deformation of the original contour of integration C onto one or more of the steepest descent paths that were determined that connect the endpoints of integration.

[4] The original integral is then equal to any contributions determined from the application of Cauchy's theorem and an integral along every steepest descent contour that has been used. Each of these integrals has the appropriate form so that Laplace's method may be used to determine the asymptotic value (see page 221).

Special Case

In the special case that $n = 2$, then we can formally proceed as follows. The function $f(z)$ can be expanded about the point $z = z_0$ to find $f(z) - f(z_0) = \frac{1}{2}f''(z_0)(z - z_0)^2 + \dots$. If $\alpha = \arg f''(z_0)$ and $z = z_0 + \varepsilon e^{i\theta}$, then this expansion can be written

$$\begin{aligned} f(z) - f(z_0) &= \tfrac{1}{2}\left|f''(z_0)\right| e^{i\alpha}\varepsilon^2 e^{2i\theta} + \dots \\ &= \tfrac{1}{2}\left|f''(z_0)\right| (\cos(\alpha + 2\theta) + i\sin(\alpha + 2\theta)) + \dots . \end{aligned}$$

The steepest paths are given by $\sin(\alpha + 2\theta) = 0$. The paths of steepest descent are given by $\cos(\alpha + 2\theta) < 0$. For $n = 2$, the complex plane has the appearance shown in Figure 51.1.a. If the contour can be made to pass through the saddle point, via curves of steepest descent, then the integral will have the form

$$I(\lambda) = \int_C g(z)e^{\lambda f(z)}\,dz$$
$$= \int_{\text{path into saddle}} g(z)e^{\lambda f(z)}\,dz - \int_{\text{path out of saddle}} g(z)e^{\lambda f(z)}\,dz.$$

Let A represent a point from which the contour enters the saddle, and let B represent a point that the contour approaches after leaving the saddle. Then the last integral can be written as $I(\lambda) = \int_0^A g(z)e^{\lambda f(z)}\,dz - \int_0^B g(z)e^{\lambda f(z)}\,dz$. Changing variables by $-\tau = f(z) - f(z_0)$, these integrals can be written as

$$I(\lambda) = e^{\lambda f(z_0)} \int_0^A e^{-\lambda\tau} g(z(\tau)) \frac{dz}{d\tau}\,d\tau - e^{\lambda f(z_0)} \int_0^A e^{-\lambda\tau} g(z(\tau)) \frac{dz}{d\tau}\,d\tau. \tag{51.3}$$

The Laplace type integrals in (51.3) will be dominated by the contributions near $\tau = 0$, which corresponds to $z = z_0$. Near this point we have $\tau = -(f(z) - f(z_0)) = -\frac{1}{2}f''(z_0)(z - z_0)^2 + \ldots$. Inverting this relation we find

$$(z - z_0) = \begin{cases} +\sqrt{\dfrac{2}{|f''(z_0)|}}e^{-i\alpha/2}\tau^{1/2} & \text{on } \displaystyle\int_0^A, \\ -\sqrt{\dfrac{2}{|f''(z_0)|}}e^{-i\alpha/2}\tau^{1/2} & \text{on } \displaystyle\int_0^B. \end{cases}$$

This can be used to evaluate the $dz/d\tau$ in (51.3). Substituting for this term, and extending both A and B to ∞ (to get the leading order approximation) results in

$$\begin{aligned} I(\lambda) &\sim e^{\lambda f(z_0)} \sqrt{\frac{2}{|f''(z_0)|}} e^{-i\alpha/2} g(z_0) \left(\int_0^\infty \frac{e^{-\lambda\tau}}{2\sqrt{\tau}}\,d\tau + \int_0^\infty \frac{e^{-\lambda\tau}}{2\sqrt{\tau}}\,d\tau \right) \\ &\sim g(z_0) \sqrt{\frac{2}{|f''(z_0)|}} \frac{e^{\lambda f(z_0) - i\alpha/2}}{\lambda^{1/2}}. \end{aligned} \tag{51.4}$$

This result should be used with caution because of the many assumptions in its derivation. It is usually best to work out each problem from scratch, without resorting to formulas like (51.4).

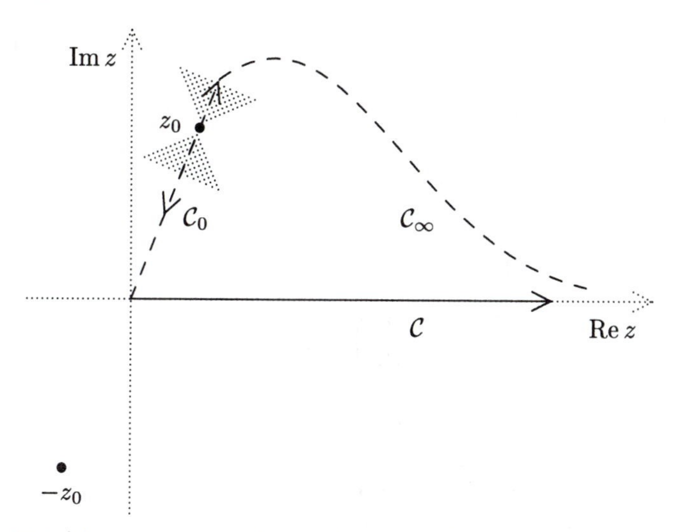

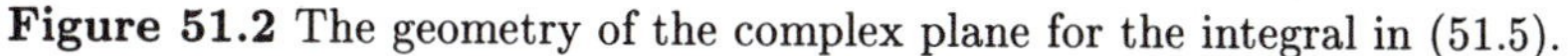
Figure 51.2 The geometry of the complex plane for the integral in (51.5).

Example 1

Consider the integral

$$I(\lambda) = \int_0^\infty e^{\lambda(z+iz-z^3)}\,dz \tag{51.5}$$

as $\lambda \to \infty$. The integration contour, $\mathcal{C}$, consists of the positive real axis. Identifying $f(z) = z + iz - z^3$, we find that the saddle points are given by $f' = 0$ or $z = \pm z_0 = \pm 2^{1/4}3^{-1/2}e^{i\pi/8}$. Near the point $z = z_0$ we can expand $f(z) - f(z_0) = \frac{1}{2}f''(z_0)(z - z_0)^2 + \dots$. Using $z = z_0 + \varepsilon e^{i\theta}$, this expansion can be written as

$$f(z) - f(z_0) = -2^{1/4}3^{1/2}\varepsilon^2 \left[\cos\left(\frac{\pi}{8} + 2\theta\right) a + i \sin\left(\frac{\pi}{8} + 2\theta\right)\right].$$

A path of constant phase has $\operatorname{Im}(f(z) - f(z_0)) = 0$, or $\sin(\pi/8 + 2\theta) = 0$. A path of descent (as opposed to a path of ascent) has $\operatorname{Re}(f(z) - f(z_0))$ decreasing, therefore $\cos(\pi/8 + 2\theta) > 0$.

Figure 51.2 shows the original contour of integration ($\mathcal{C}$), the two saddle points ($\pm z_0$), and the structure of the saddle around the point $z = z_0$. The steepest descent contours $\mathcal{C}_0$ and $\mathcal{C}_\infty$ are also shown in Figure 51.2. The immediate problem is that the given integration contour $\mathcal{C}$ is not a steepest descent contour. However, using Cauchy's theorem, we can deform the integration contour as long as we account for any singularities that we may cross. Because there are no singularities of the integrand in the region bounded by the contours, Cauchy's theorem tells us that

$$\left(\int_{\mathcal{C}} + \int_{\mathcal{C}_0} - \int_{\mathcal{C}_\infty}\right) e^{\lambda f(z)}\,dz = 0.$$

Therefore, if we can determine the integrals along the steepest decent contours then we can determine $I(\lambda)$.

Changing variables by $-\tau = f(z) - f(z_0)$, we find

$$\begin{aligned} I(\lambda) &= \int_C e^{\lambda f(z)}\,dz = \int_{C_\infty} e^{\lambda f(z)}\,dz - \int_{C_0} e^{\lambda f(z)}\,dz \\ &= e^{\lambda f(z_0)}\left(\int_{C_\infty} e^{-\lambda\tau}\,dz - \int_{C_0} e^{-\lambda\tau}\,dz\right) \\ &= e^{\lambda f(z_0)}\left(\int_{C_0} e^{-\lambda\tau}\left(\frac{dz}{d\tau}\right) d\tau - \int_{C_\infty} e^{-\lambda\tau}\left(\frac{dz}{d\tau}\right) d\tau.\right) \end{aligned} \tag{51.6}$$

At this point we have reduced the problem to calculation of Laplace type integrals. Evaluating these integrals is straightforward.

Near $z = z_0$, which is where the dominant contribution of the integrals in (51.6) comes from, we have $\tau = -\frac{1}{2}f''(z_0)(z-z_0)^2 + \ldots$. This relation can be inverted to find

$$z - z_0 = \begin{cases} +\beta\tau^{1/2} & \text{along } C_0, \\ -\beta\tau^{1/2} & \text{along } C_\infty, \end{cases}$$

where $\beta = 2^{-1/8}3^{-1/4}e^{-i\pi/16}$. Therefore, (51.6) becomes

$$\begin{aligned} I(\lambda) &\sim e^{\lambda f(z_0)}\left(\int_0^\infty e^{-\lambda\tau}\beta\frac{\tau^{-1/2}}{2}\,d\tau + \int_0^\infty e^{-\lambda\tau}\beta\frac{\tau^{-1/2}}{2}\,d\tau\right) \\ &= \beta e^{\lambda f(z_0)}\frac{\Gamma\left(\frac{1}{2}\right)}{\lambda^{1/2}} \\ &= 2^{-1/8}3^{-1/4}e^{-i\pi/16}\pi^{1/2}\frac{e^{2^{7/4}3^{-3/2}e^{3\pi i/8}\lambda}}{\lambda^{1/2}} \qquad \text{as } \lambda \to \infty. \end{aligned}$$

Example 2

Consider the integral representation of the Airy function

$$J(s) = \operatorname{Ai}(s) = \frac{2}{\pi}\int_{-\infty}^{\infty}\cos\left(\frac{t^3}{3} + st\right)dt = \frac{1}{2\pi i}\int_C \exp\left(sw - \frac{w^3}{3}\right)dw$$

as s tends to positive infinity. The contour of integration is shown in Figure 51.3 (see page 3). Since this integral is not in the form we require, we change variables by $z = w\sqrt{s}$ and $\lambda = s^{3/2}$ to obtain

$$I(\lambda) = J(s) = \frac{\lambda^{1/3}}{2\pi i}\int_C \exp\left(\lambda\left[z - \frac{z^3}{3}\right]\right)dz. \tag{51.7}$$

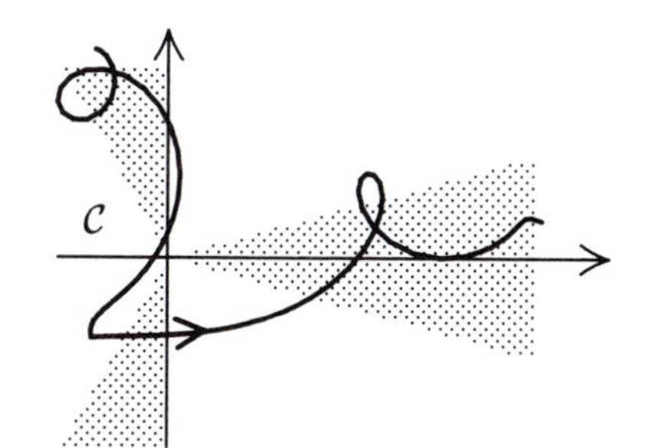

Figure 51.3 Contour of integration for (51.7).

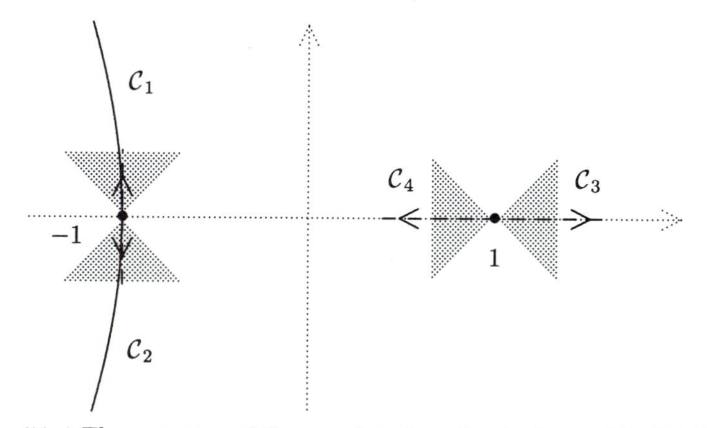

Figure 51.4 The geometry of the complex plane for the integral in (51.7).

Note that the contour remains unchanged. We can now proceed with the steps described above.

In this case, $f(z) = z - z^3/3$ and $g(z) = 1$. Since both $f(z)$ and $g(z)$ are analytic, the only critical points are the saddle points of $f(z)$ and, possibly, the endpoints of the contour of integration. The saddle points of $f(z)$ are given by $f'(z_0) = 1 - z_0^2 = 0$, or $z_0 = \pm 1$.

Expanding $f(z)$ into real and imaginary components results in

$$f(z) = z - \tfrac{1}{3}z^3 = f_R(z) + if_I(z) = x\left(1 - \tfrac{1}{3}x^2 + y^2\right) - iy\left(x^2 - \tfrac{1}{3}y^2 - 1\right).$$

The curves of steepest ascent and descent at the points $z_0 = \pm 1$, satisfy

$$f_I(z) = f_I(z = \pm 1) = f_I(x = \pm 1, y = 0) = 0.$$

That is, $y\left(x^2 - \tfrac{1}{3}y^2 - 1\right) = 0$. These curves are shown in Figure 51.4. We can now use (51.1) and (51.2) to determine exactly what the steepest descent paths are from the saddle points. Since $f''(z) = -2z$ we find that $f''(z = -1) = -2 = 2e^{i\pi}$ and $f''(z = 1) = 2 = 2e^{i0}$. Hence, $\alpha_{z=1} = \pi$ and $\alpha_{z=-1} = 0$. We conclude, therefore, that:

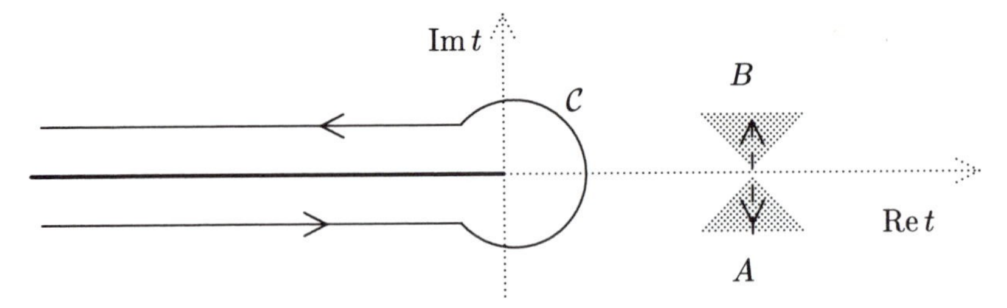

Figure 51.5 The geometry of the complex plane for the integral in (51.8).

[1] The directions of steepest descent from $z = -1$ lie along the hyperbolas $x^2 - \frac{1}{3}y^2 = 1$ (these are labeled $\mathcal{C}_1$ and $\mathcal{C}_2$ in Figure 51.4).

[2] The directions of steepest descent from $z = 1$ lie along the line $y = 0$ (these are labeled $\mathcal{C}_3$ and $\mathcal{C}_4$ in Figure 51.4).

Now that we have identified the curves of steepest descent, we would like to move our given contour to one or more of these steepest descent contours (justified, of course, by Cauchy's theorem). Since the integrand is analytic (it has no poles or branch cuts) the original contour $\mathcal{C}$ can be deformed into the new contour to obtain $\mathcal{C} = -\mathcal{C}_1 - \mathcal{C}_4 + \mathcal{C}_3$ (the plus and minus signs are needed since the contours shown in Figure 51.4 are oriented). We conclude that

$$I(\lambda) = \frac{\lambda^{1/3}}{2\pi i}\left(-\int_{\mathcal{C}_1} - \int_{\mathcal{C}_4} + \int_{\mathcal{C}_3}\right) \exp\left(\lambda\left[z - \frac{z^3}{3}\right]\right)\,dz.$$

These integrals are easily evaluated by Laplace's method. We find (see Bleistein and Handelsman [3], page 267 for details)

$$\begin{aligned} I(\lambda) &\sim -\left(\frac{\lambda^{1/6}}{4\sqrt{\pi}}e^{-2\lambda/3}\right) - \left(-\frac{\lambda^{1/6}}{4i\sqrt{\pi}}e^{2\lambda/3}\right) + \left(\frac{\lambda^{1/6}}{4i\sqrt{\pi}}e^{2\lambda/3}\right) \\ &\sim \frac{\lambda^{1/6}}{2i\sqrt{\pi}}e^{2\lambda/3} \qquad \text{as } \lambda \to \infty. \end{aligned}$$

This leads to $\operatorname{Ai}(s) \sim \dfrac{s^{-1/4}}{2\sqrt{\pi}}e^{-(2s^{2/3}/3)}$ as $s \to \infty$.

Example 3

Consider the integral for the reciprocal of the Gamma function (see Copson [5])

$$I(\lambda) = \frac{1}{\Gamma(\lambda)} = \frac{1}{2\pi i}\int_{\mathcal{C}} e^{t-\lambda t}\,dt, \tag{51.8}$$

where λ may be real or complex, the principal value of the logarithm is taken, and the contour $\mathcal{C}$ is shown in Figure 51.5. (The contour $\mathcal{C}$ starts at $-\infty$, circles the origin once in the positive sense, and returns to $-\infty$.) We are interested in (51.8) as $\lambda \to \infty$. Changing variables by $t = \lambda z$ results in the representation

$$I(\lambda) = \frac{1}{2\pi i \lambda^{\lambda-1}} \int_C e^{\lambda(z - \log z)}\, dz.$$

From this expression we have $g(z) =$ and $f(z) = z - \log z$. The only stationary point is at $f'(z_0) = 0$ or $z_0 = 0$. At the stationary point we find $f''(z_0) = 1$. A simple computation shows that the paths of steepest descent are as indicated in Figure 51.5.

The contour $\mathcal{C}$ in Figure 51.5 can be trivially moved to go through the stationary point. Hence, the asymptotic behavior of $I(\lambda)$ is determined by the two paths of steepest descent indicated in Figure 51.5. Using the substitution $z = 1 + is$, and then the expansion $\log(1+x) = x - \frac{1}{2}x^2 + \frac{1}{3}x^3 - \ldots$, we find

$$\begin{aligned}
I(\lambda) &\sim \frac{1}{2\pi i \lambda^{\lambda-1}} \left(-i \int_A^0 e^{-\lambda(\log(1+is)-is)}\, ds + i \int_0^B e^{-\lambda(\log(1+is)-is)}\, ds \right) \\
&= \frac{1}{2\pi \lambda^{\lambda-1}} \left(-\int_A^0 e^{-\lambda(\frac{1}{2}s^2+\ldots)}\, ds + \int_0^B e^{-\lambda(\frac{1}{2}s^2+\ldots)}\, ds \right) \\
&\sim \frac{1}{2\pi \lambda^{\lambda-1}} \left(-\int_{-\infty}^0 e^{-\lambda s^2/2}\, ds + \int_0^\infty e^{-\lambda s^2/2}\, ds \right) \\
&= \left(\frac{e}{\lambda}\right)^\lambda \sqrt{\frac{\lambda}{2\pi}}.
\end{aligned}$$

If the above Laplace type integrals were expanded to higher order, then we would obtain the expansion

$$I(\lambda) = \frac{1}{\Gamma(\lambda)} = \left(\frac{e}{\lambda}\right)^\lambda \sqrt{\frac{\lambda}{2\pi}} \left(1 - \frac{1}{12\lambda} + \frac{1}{288\lambda^2} + \frac{139}{51840\lambda^3} - \ldots\right). \tag{51.9}$$

Example 4

Consider the integral $I(\lambda) = \int_0^1 \log z e^{i\lambda z}\, dz$, for $\lambda \gg 1$. For this integrand, the half-plane with $\operatorname{Im} z > 0$ is a region of descent. The steepest descent contours are the lines with $\operatorname{Re} z$ equal to a constant. In this case the integration contour $\mathcal{C}$ is deformed into the 3 contours $\mathcal{C}_1$, $\mathcal{C}_2$, and $\mathcal{C}_3$ (see Figure 51.6). By Cauchy's theorem we have $I(\lambda) = \int_{\mathcal{C}} = \int_{\mathcal{C}_1} + \int_{\mathcal{C}_2} + \int_{\mathcal{C}_3}$, where the integrand is the same for each integral. In the limit of the rectangle extending vertically to infinity, we find that $I(\lambda) = \int_0^{i\infty} + \int_{1+i\infty}^1$. The first of these integrals can be evaluated exactly

$$\begin{aligned}
\int_0^{i\infty} \log z e^{i\lambda z}\, dz &= i \int_0^\infty \log(is) e^{-\lambda s}\, ds \\
&= -i \frac{\log \lambda}{\lambda} - \frac{i\gamma + \pi/2}{\lambda}
\end{aligned} \tag{51.10}$$

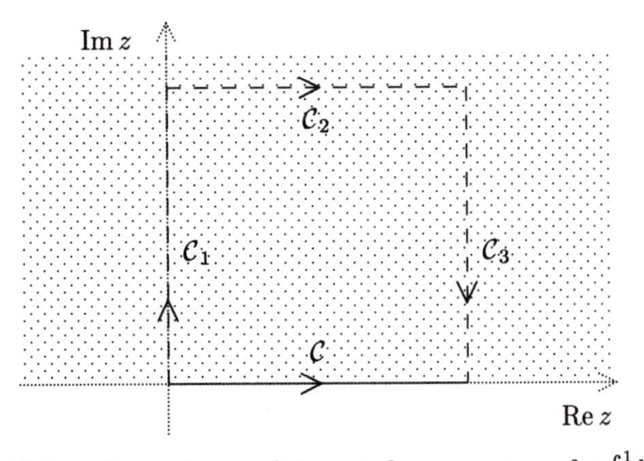

Figure 51.6 Integration contour and steepest descent contours for $\int_0^1 \log z e^{i\lambda z}\, dz$.

where γ is Euler's constant.

The second integral in the above can be evaluated by Watson's lemma. Using $\log(1+is) = -\sum_{n=1}^{\infty}(-is)^n/n$, we find

$$\begin{aligned}\int_{1+i\infty}^{1} \log z e^{i\lambda z}\, dz &= -i\int_0^{\infty} \log(1+is) e^{i\lambda(1+is)}\, ds \\ &\sim ie^{i\lambda}\sum_{n=1}^{\infty}\frac{(-i)^n(n-1)!}{\lambda^{n+1}} \qquad \text{as } \lambda\to\infty.\end{aligned} \tag{51.11}$$

Combining the results in (51.10) and (51.11) results in an asymptotic expansion of $I(\lambda)$. Full details may be found in Bender and Orszag [1], pages 281–282.

Program 51

```
s: if   mod(capN,2)=0
   then 1
   else ( if tpos then 1 else -1) $

wp[k]:= block([temp],
        if   k=0
        then w(x)
        else (temp:wp[k-1], ratsimp(diff(temp,x))))$

phip[k]:= block([temp],
          if   k=0
          then phi(x)
          else (temp:phip[k-1], ratsimp(diff(temp,x))))$

  wpz[k]:=block([temp], temp:wp[k], ev(temp,x=z0) )$

phipz[k]:=block([temp], temp:phip[k], ev(temp,x=z0) )$

alpha[n]:= (wpz[capN+n]/(capN+n)!) / (wpz[capN]/capN!) $
```

```
term(m,k,capN,kp):=((m-k-kp)*(m+1)/capN+kp) *
                    (m-kp)!/k!*alpha[m-k-kp]*D[m,m-kp]$

D[m,k]:= block([kp],
         if   m=k
         then gamma( (m+1)/capN )/(capN*m!)
         else factor( -1/(m-k)*sum( term(m,k,capN,kp), kp,0,m-k-1) ))$

hold:-s*wpz[capN]*name/capN!$

specialsum3(mp):=sum( D[2*mp,k]*phipz[k], k, 0, 2*mp)$

%e**(name*w(z0)) * sum( 2* hold**(-(2*mp+1)/capN) * specialsum3(mp),
                        mp,0,numberofterms) ;
```

Notes

[1] This is also called the *saddle point method.*

[2] Bleistein and Handelsman [3] treat the more general problem $I(\lambda) = \int_{\mathcal{C}} g(x)H(\lambda w(x))\,dx$ where $g(x)$ and $w(x)$ are analytic in the region containing the contour $\mathcal{C}$, and $\lambda \to \infty$. Here $H(z)$ is an entire transcendental function, such as $\sin z$, $\cos z$, or $\mathrm{Ai}(z)$.

[3] In many specific cases, it is often impractical to construct paths of steepest descent. In some problems it may be easier to use Perron's method (see Perron [6]), which avoids the explicit construction of steepest descent paths. See Wong [7] for details.

[4] Campbell, Fröman, and Walles [4] give explicit series for the asymptotic approximation of $\int e^{-\lambda w(z)}\phi(z)\,dz$ as $\lambda \to \infty$ for a well isolated singular point z_0. A computer program written in REDUCE is also given.

If $w^{(j)}(z_0) = 0$ for $j = 1, 2, \ldots, N-1$ and $w^{(N)}(z_0) \neq 0$ then the only inputs required for the program are $w(z)$ and $\phi(z)$ (and their derivatives), N, z_0, the number of terms desired in the expansion, and the sign of the parameter in the reparametrization of $w(z)$ near the singular point. That is, near z_0 we write $w(z) = w(z_0) - st^N$ where

$$s = \begin{cases} +1 & \text{when } N \text{ is even,} \\ +1 & \text{when } N \text{ is odd and } t > 0, \\ -1 & \text{when } N \text{ is odd and } t < 0. \end{cases}$$

The author has re-written the program in Macsyma; it is shown in Program 51. As an illustration of the Macsyma program, consider the calculation in Example 3 (with λ replace by $-\nu$). If the following program segment is placed before the code shown in Program 51:

```
capN:2$
z0:1$
numberofterms:3$
tpos:false$
w(x):=x-log(x)$
phi(x):=1$
name:nu$
```

then the following output is obtained:

```
  sqrt(2) sqrt(%pi)    sqrt(2) sqrt(%pi)    sqrt(2) sqrt(%pi)
(----------------- + ----------------- + -----------------
    sqrt(- nu)                    3/2                 5/2
                         12 (- nu)          288 (- nu)

                                    139 sqrt(2) sqrt(%pi)      nu
                                  - ---------------------) %e
                                                    7/2
                                          51840 (- nu)
```

This result must be multiplied by $\dfrac{1}{2\pi i \lambda^{\lambda-1}}$ to obtain the answer in (51.9).

References

[1] C. M. Bender and S. A. Orszag, *Advanced Mathematical Methods for Scientists and Engineers*, McGraw–Hill, New York, 1978.

[2] N. Bleistein and R. A. Handelsman, *Asymptotic Expansions of Integrals*, Dover Publications, Inc., New York, 1986.

[3] N. Bleistein and R. A. Handelsman, "A Generalization of the Method of Steepest Descent," *J. Inst. Math. Appls.*, **10**, 1972, pages 211–230.

[4] J. A. Campbell, P. E. Fröman, E. Walles, "Explicit Series Formulae for the Evaluation of Integrals by the Method of Steepest Descents," *Stud. Appl. Math.*, **77**, 1987, pages 151–172.

[5] E. T. Copson, *Asymptotic Expansions*, Cambridge University Press, New York, 1965.

[6] O. Perron, "Über die näherungsweise Berechnung von Funktionen großber Zahlen," *Sitzungsber. Bayr. Akad. Wissensch.*, Münch. Ber., 1917, pages 191–219.

[7] R. Wong, *Asymptotic Approximation of Integrals*, Academic Press, New York, 1989.

52. Approximations: Miscellaneous

Idea

This section contains some miscellaneous integral approximation techniques.

Procedure

Consider the integral $I(s) = \int_0^\infty e^{-st} f(t)g(t)\,dt$. If $f(t)$ is analytic, then it can be expanded in a Taylor series. Hence, I can be re-written as

$$\begin{aligned} I(s) &= \sum_{n=0}^{\infty} \frac{f^{(n)}(0)}{n!} \int_0^\infty e^{-st} g(t) t^n \,dt \\ &= \sum_{n=0}^{\infty} \frac{f^{(n)}(0)}{n!} (-1)^n \frac{d^n}{ds^n} \overline{g}(s) \end{aligned} \tag{52.1}$$

where $\overline{g}(s) = \int_0^\infty e^{-st} g(t)\,dt$ is the Laplace transform of $g(t)$.

Table 52. Some infinite series expansions for integrals of the form $\int_0^\infty f(t)g(t)\,dt$.

$$\int_0^\infty e^{-\alpha t} f(t)\,dt = \sum_{n=0}^\infty \frac{f^{(n)}(0)}{\alpha^{n+1}}$$

$$\int_0^\infty \sin \alpha t f(t)\,dt = \sum_{n=0}^\infty (-1)^n \frac{f^{(2n)}(0)}{\alpha^{2n+1}}$$

$$\int_0^\infty \cos \alpha t f(t)\,dt = \sum_{n=0}^\infty (-1)^n \frac{f^{(2n+1)}(0)}{\alpha^{2n+2}}$$

$$\int_0^\infty \frac{\sin \alpha t f(t)}{t}\,dt = \frac{\pi}{2} f(0) - \sum_{n=0}^\infty (-1)^n \frac{f^{(2n-1)}(0)}{(2n-1)\alpha^n}$$

Setting $s = 0$ in (52.1) allows an infinite series expansion for $I(0) = \int_0^\infty f(t)g(t)\,dt$ to be obtained. Each choice of $g(t)$ results in a different expansion, some are given in Table 52. These results are only formally correct; in practice, the resulting expressions may be asymptotically valid.

Example

Using the fourth expansion in Table 52, with $f(t) = \sin t$, we readily determine that

$$J := \int_0^\infty \frac{\sin mt \sin t}{t}\,dt = \sum_{n=0}^\infty \frac{1}{(2n+1)m^n}. \tag{52.2}$$

A table of integrals shows that $J = \frac{1}{2}\log\left(\frac{m+1}{m-1}\right)$, when $m > 1$. If this result is expanded around $m = \infty$, then the result in (52.2) is obtained.

Note

[1] This technique is from Squire [1], who credits Willis [2].

References

[1] W. Squire, *Integration for Engineers and Scientists*, American Elsevier Publishing Company, New York, 1970, pages 105–107.

[2] H. F. Willis, "A Formula for Expanding an Integral as a Series," *Phil. Mag.*, **39**, 1948, pages 455–459.

V

Numerical Methods: Concepts

53. Introduction to Numerical Methods

The last sections of this book are concerned with numerical methods for approximating definite integrals. Section V defines some of the terms used in numerical methods, describes different software libraries, and displays integrals that have appeared in the literature as test cases for numerical routines. Section VI describes a selection of many possible numerical techniques.

Note that a definite integral, say $I = \int_a^b f(x)\,dx$, can be interpreted as the solution of a differential equation. In this interpretation, $I = y(b)$ where $y' = f(x)$ and $y(a) = 0$. Hence, any numerical integration technique for differential equations can be adapted to numerically evaluate an integral. (See Zwillinger [4] for some numerical techniques for differential equations.) In fact, for an integrand that has an integrable singularity, at an unknown location in the integrand, it is probably preferable to use a differential equation technique.

One of the most comprehensive books on numerical integration techniques is Davis and Rabinowitz [1]. It contains a comprehensive bibliography, up to 1984. The book by Stroud [2] contains extensive tables of numerical methods.

The most popular computer library that the author is aware of is Quadpack (see Piessens *at al.* [3]). This library implements quadrature rules with the following desirable features:

- All nodes are within the integration interval (sometimes the nodes include the end-points).
- All weights are positive.

References

[1] P. J. Davis and P. Rabinowitz, *Methods of Numerical Integration*, Second Edition, Academic Press, Orlando, Florida, 1984.

[2] A. H. Stroud, *Approximate Calculation of Multiple Integrals*, Prentice–Hall Inc., Englewood Cliffs, NJ, 1971.

[3] R. Piessens, E. de Doncker–Kapenga, C. W. Überhuber, and D. K. Kahaner, *Quadpack*, Springer–Verlag, New York, 1983.

[4] D. Zwillinger, *Handbook of Differential Equations*, Academic Press, New York, Second Edition, 1992.

54. Numerical Definitions

Adaptive Quadrature For the integral $I = \int_0^1 f(x)\,dx$, traditional quadrature rules use estimates of the form $I \approx \widehat{I} = \sum_{i=1}^n w_i f(x_i)$ for some specific choice of the weights $\{w_i\}$ and nodes $\{x_i\}$. An adaptive quadrature technique uses an algorithm to choose the weights $\{w_i\}$ and nodes $\{x_i\}$ during the computation. Thus, an adaptive algorithm dynamically adapts the integration rule to the particular properties of the integrand.

Automatic Quadrature Routine This is a piece of software, designed to provide an approximation of specified tolerance to a given definite integral.

Closed Rule Consider an approximation to the integral $I = \int_a^b f(x)\,dx$. A quadrature rule that uses the values of $f(a)$ and $f(b)$ in the numerical approximation of I is called a closed rule or a closed formula.

Composite Rule If a numerical quadrature rule for integrating over a small parallelpiped region is repeatedly applied to find a rule applicable over a larger parallelpiped region, the new rule is called a compound or composite rule. See page 282.

Fully Symmetric A region B is said to be fully symmetric if $\mathbf{x} = (x_1, x_2, \ldots, x_d) \in B$ implies that $(\pm x_{p(1)}, \pm x_{p(2)}, \ldots, \pm x_{p(d)}) \in B$, where the set $\{p(1), p(2), \ldots, p(d)\}$ is any permutation of $\{1, 2, \ldots, d\}$.

General Purpose Integrators A numerical quadrature rule is said to be *general purpose* if the routine works for generic integrands. If a routine requires that the integrand have the form $w(x)f(x)$, for some special weight function $w(x)$, then the routine is called *special purpose.*

Nodes In the quadrature rule $\int_a^b f(x)\,dx \approx \sum_i w_i f(x_i)$, the distinct numbers $\{x_i\}$ are called the nodes (also called abscissas, or points).

Null Rule The numerical quadrature rule $\int_a^b f(x)\,dx \approx \sum_i w_i f(x_i)$ is said to be a null rule if $\sum_i w_i = 0$ and at least one of the $\{w_i\}$ values is non-zero. Furthermore, a null rule is said to have degree d if it integrates to zero all polynomials of degree less than or equal to d and fails to do so for $f(x) = x^{d+1}$.

Numerically Stable A numerical quadrature rule is said to be numerically stable if all of the weights are positive.

Open Rule Consider an approximation to the integral $I = \int_a^b f(x)\,dx$. A quadrature rule that does not use the values $f(a)$ or $f(b)$ in the numerical approximation of I is called an open rule or an open formula. This would be needed, for instance, if $a = 0$ and $f(x) = x^{-1/2}$. In this case, I has an integrable singularity, but $f(0)$ is infinite and should not be computed by a quadrature routine.

Order Given a quadrature rule of the form

$$\int_a^b f(x)\,dx = \sum_{n=1}^{N} a_n f(x_n) + E[f],$$

there exists a largest integer k such that $E[p]$ vanishes for every polynomial p of degree less than k. The number k is usually called the *order* and $k-1$ is called the *degree of precision* of the rule. We have the bound $k \leq 2N$, where equality only holds for the Gauss–Legendre rule.

Panel Rules Given an integral to evaluate $I = \int_a^b f(x)\,dx$, the interval of integration $[a, b]$ can be sub-divided into m equal length intervals, called panels. (The number m is called the *mesh ratio.*) Then one integration rule can be used on each panel, and the resulting approximation to I is called an m-panel rule. See also "composite rules," page 282.

Positive Quadrature Rule The quadrature rule $\int_0^1 f(x)\,dx \approx \sum_{i=1}^{n} w_i f(x_i)$ is called a positive quadrature rule if all of the weights are positive, $w_i > 0$.

Weights In the quadrature rule $\int_a^b f(x)\,dx \approx \sum_i w_i f(x_i)$, the numbers $\{w_i\}$ are called the weights.

55. Error Analysis

Applicable to Integration rules of the form

$$\int_a^b f(x)\,dx$$
$$\approx P[f] := \sum_{k=0}^{m_0} a_{k0} f\left(x_{k0}\right) + \sum_{k=0}^{m_1} a_{k1} f'\left(x_{k1}\right) + \cdots + \sum_{k=0}^{m_n} a_{kn} f^{(n)}\left(x_{kn}\right)$$

Yields

An estimate of the error in using $P[f]$ to approximate $\int_a^b f(x)\,dx$.

Idea

An exact formulation of the error can sometimes be obtained.

Procedure

Define the error made when using $P[f]$ to approximate $\int_a^b f(x)\,dx$ to be

$$E[f] = P[f] - \int_a^b f(x)\,dx. \tag{55.1}$$

A theorem by Peano states:

> **Theorem** (*Peano*): Suppose that $E[f] = 0$ for all polynomials f with degree no greater than n (i.e., polynomials of low degree are integrated exactly). Then, for all functions $f \in C^{n+1}[a,b]$ the error can be represented as
>
> $$E[f] = \int_a^b f^{(n+1)}(t) K(t)\,dt \tag{55.2}$$
>
> where $K(t)$ is the Peano kernel of E and is defined by
>
> $$\begin{aligned} K(t) &= \frac{1}{n!} E[g(x;t)], \\ g(x;t) &= (x-t)_+^n = \begin{cases} (x-t)^n & \text{if } x \geq t, \\ 0 & \text{if } x < t. \end{cases} \end{aligned} \tag{55.3}$$

Note that t is just a parameter in the g function and that E operates only with respect to the x variable. (When evaluating the E function in (55.3), the representation in (55.1) should be used.)

This theorem can be used to determine the error in integration rules explicitly. For many integration rules, the Peano kernel has a constant sign on $[a,b]$. Using the mean value theorem (see page 83) on (55.2) then allows the error to be written as

$$E[f] = f^{(n+1)}(\xi) \int_a^b K(t)\,dt, \tag{55.4}$$

for some $\xi \in [a, b]$. Hence, by determining $\int_a^b K(t)\,dt$, a simple, exact representation of the error is obtained.

There is an easy way to determine the integral of $K(t)$. Since (55.4) is presumed to be valid for all $f \in C^{n+1}[a, b]$ we can use $f(x) = x^{n+1}$ in (55.4) to obtain $E[x^{n+1}] = (n + 1)! \int_a^b K(t)\,dt$ so that

$$E[f] = \frac{E[x^{n+1}]}{(n+1)!} f^{(n+1)}(\xi) \tag{55.5}$$

for some $\xi \in [a, b]$.

Example

Consider using the Peano theorem to find the error in Simpson's rule. In Simpson's rule we have the approximation

$$\int_{-1}^{1} f(x)\,dx \approx \tfrac{1}{3}f(-1) + \tfrac{4}{3}f(0) + \tfrac{1}{3}f(1)$$

and hence,

$$E[f] = \left[\tfrac{1}{3}f(-1) + \tfrac{4}{3}f(0) + \tfrac{1}{3}f(1)\right] - \int_{-1}^{1} f(x)\,dx.$$

For this integration rule, all polynomials of degree less than or equal to three are integrated exactly. Therefore, we can apply the theorem with $n = 3$. The Peano kernel becomes

$$\begin{aligned} K(t) &= \tfrac{1}{6}E[g(x;t)] \\ &= \tfrac{1}{6}E[(x-t)_+^3] \\ &= \tfrac{1}{6}\left\{\left[\tfrac{1}{3}(-1-t)_+^3 + \tfrac{4}{3}(0-t)_+^3 + \tfrac{1}{3}(1-t)_+^3\right] - \int_{-1}^{1}(x-t)_+^3\,dx\right\}. \end{aligned}$$

When t is in the range $[-1, 1]$ this expression can be simplified to yield

$$\begin{aligned} K(t) &= \tfrac{1}{6}\left\{\left[\tfrac{1}{3}\cdot 0^3 + \tfrac{4}{3}(-t)_+^3 + \tfrac{1}{3}(1-t)^3\right] - \int_{t}^{1}(x-t)^3\,dx\right\} \\ &= \tfrac{1}{6}\left\{\left[\tfrac{4}{3}(-t)_+^3 + \tfrac{1}{3}(1-t)^3\right] - \tfrac{1}{4}(1-t)^4\right\}. \end{aligned}$$

The term $(-t)_+^3$ depends on whether or not $t \geq 0$. Specifically,

$$(-t)_+^3 = \begin{cases} 0 & \text{if } t \geq 0 \\ -t^3 & \text{if } t < 0\,. \end{cases}$$

Using this in the formula for $K(t)$ we finally arrive at

$$K(t) = \begin{cases} \frac{1}{72}(1-t)^3(1+3t) & \text{if } \quad 0 \leq t \leq 1 \\ K(-t) & \text{if } -1 \leq t \leq 0. \end{cases}$$

With Peano's kernel explicitly determined, we observe that it has a constant sign (that is, positive) on the interval $[-1, 1]$. Hence, by carrying out the integration in (55.4) we find

$$\begin{aligned} E[f] &= f^{(4)}(\xi) \int_{-1}^{1} K(t)\, dt \\ &= \tfrac{1}{90} f^{(4)}(\xi) \end{aligned}$$

for some ξ in the range $[-1, 1]$. We could also have obtained the factor of $\frac{1}{90}$, using (55.5):

$$\begin{aligned} \frac{E[x^4]}{4!} &= \frac{1}{24}\left(\left[\tfrac{1}{3}(-1)^4 + \tfrac{4}{3}0^4 + \tfrac{1}{3}1^4\right] - \int_{-1}^{1} x^4\, dx\right) \\ &= \tfrac{1}{24}\left(\tfrac{1}{3} + \tfrac{1}{3} - \tfrac{2}{5}\right) \\ &= \tfrac{1}{90}. \end{aligned}$$

Notes

[1] The usual way of estimating the error using a quadrature routine is to use two different rules, say A and B, and then estimate the error by $|A - B|$, or some scaling of this. In the case of adaptive quadrature (see page 277), the subdivision process provides additional information that can be used. See, for example, Espelid and Sorevik [3].

[2] Interval analysis is a technique in which an interval which contains the numerical value of an integral is obtained, see page 218.

[3] For most automatic quadrature routines, an absolute error ε_a and a relative error ε_r are input. For the integral $I = \int_a^b f(x)\, dx$, the routine will compute a sequence of values $\{R_{n_k}, E_{n_k}\}$. Here, R_{n_k} is an estimate of I using n_k values of the integrand, and E_{n_k} is the associated error estimate. The routine will terminate (and return a value), when the error criteria

$$|R_{n_k} - I| \leq E_{n_k} \leq \max(\varepsilon_a, \varepsilon_r |R_{n_k}|)$$

is achieved.

[4] Piessens *at al.* [10], Section 2.2.4.1, has a summary of the asymptotic expansion of integration errors. Piessens *at al.* [10] (page 40) also has the (pessimistic) error bound:

> Let $\left|f^{(k+1)}(x)\right| \leq M_{k+1}$ for $a \leq x \leq b$. Then the absolute error of the positive quadrature rule of precision $d > k$ satisfies:

$$|E| = \left| \int_a^b f(x)\, dx - \sum_i w_i f(x_i) \right| \leq \frac{2e^k 3^{k+1}}{k+1} \frac{(b-a)^{k+2}}{d^k} \frac{M_{k+1}}{d-k}.$$

[5] Some integration rules have errors that can be represented as sums of terms with each term involving derivatives of the integrand at an endpoint of integration. For example, the composite trapezoidal rule has this property (see page 342).

Given an integrand, suppose that a transformation can be found such that the derivatives of the new integrand vanish at the endpoints. Then, using this transformation with a rule of the above type should result in small errors. See Davis and Rabinowitz [2] for details.

[6] Some quadrature rules have the error decreasing exponentially with the number of nodes. For example, the "tanh rule" of Schwartz is given by

$$\int_{-1}^{1} f(x)\,dx = h \sum_{k=-N}^{N} \frac{f\left(\tanh \frac{1}{2}hk\right)}{2\cosh^2 \frac{1}{2}hk} + O\left(e^{-c\sqrt{N}}\right)$$

for some $c > 0$, where $h = h(N)$ is given by a certain formula (and $h \sim cN^{-1/2}$). See Kahaner *et al.* [6].

References

[1] J. Berntsen and T. O. Espelid, "Error Estimation in Automatic Quadrature Routines," *ACM Trans. Math. Software*, **17**, No. 2, 1991, pages 233–252.

[2] P. J. Davis and P. Rabinowitz, *Methods of Numerical Integration*, Second Edition, Academic Press, Orlando, Florida, 1984, Chapter 4, pages 142–144 and 271–343.

[3] T. O. Espelid and T. Sorevik, "A Discussion of a New Error Estimate for Adaptive Quadrature," *BIT*, **29**, 1989, pages 283–294.

[4] K.-J. Förster and K. Petras, "Error Estimates in Gaussian Quadrature for Functions of Bounded Variation," *SIAM J. Math. Anal.*, **28**, No. 3, June 1991, pages 880–889.

[5] W. Gautschi, E. Tychopoulos, and R. S. Varga, "A Note on the Contour Integral Representation of the Remainder Term of a Gauss–Chebyshev Quadrature Rule," *SIAM J. Numer. Anal.*, **27**, No. 1, 1990, pages 219–224.

[6] D. Kahaner, C. Moler, and S. Nash, *Numerical Methods and Software*, Prentice–Hall Inc., Englewood Cliffs, NJ, 1989.

[7] C. Schneider, "Error Bounds for the Numerical Evaluation of Integrals with Weights," in *Numerical Integration III*, Birkhäuser, Basel, 1988, pages 226–236.

[8] J. Stoer and R. Bulirsch, *Introduction to Numerical Analysis*, translated by R. Bartels, W. Gautschi, and C. Witzgall, Springer–Verlag, New York, 1976, pages 123–127.

[9] A. H. Stroud, *Approximate Calculation of Multiple Integrals*, Prentice–Hall Inc., Englewood Cliffs, NJ, 1971, Chapter 5, pages 137–192.

[10] R. Piessens, E. de Doncker–Kapenga, C. W. Überhuber, and D. K. Kahaner, *Quadpack*, Springer–Verlag, New York, 1983.

56. Romberg Integration / Richardson Extrapolation

Applicable to Numerical techniques for integrals.

Yields

A procedure for increasing the accuracy.

Procedure

Suppose that a grid with a characteristic spacing of h is used to approximate the integral $I = \int_a^b f(x)\,dx$ numerically. Then the numerical approximation to the integral, $I(h)$, can be represented by

$$I(h) = I + R_m h^m + O(h^{m+1}), \tag{56.1}$$

where m is the order of the method, and the other terms represent the error (see page 245). For example, the trapezoidal rule has the parameters $m = 3$ and $R_3 = \frac{1}{12} f''(\zeta)$, with $a < \zeta < b$.

If the numerical integration scheme is kept the same, but the characteristic width of the grid is changed from h to k, then

$$I(k) = I + R_m k^m + O(k^{m+1}). \tag{56.2}$$

Equations (56.1) and (56.2) can be combined to yield the approximation

$$I(h,k) := \frac{k^m I(h) - h^m I(k)}{k^m - h^m} = I + O(kh^m, hk^m). \tag{56.3}$$

Note that $I(h,k)$ is one more order accurate than either $I(h)$ or $I(k)$. This process, known as Richardson extrapolation, may be repeated as often as is desired. If we write $k = rh$, then (56.3) may be written in the alternative form $I(h,k) := I(h) - \left[\frac{I(h) - I(rh)}{1 - r^m}\right] + \ldots$. In this formulation it is clear that if r is close to one, there may be roundoff problems.

When Richardson extrapolation is used with the composite trapezoidal rule, and the number of nodes (nearly) doubles with each iteration (i.e., $r \approx \frac{1}{2}$), then we obtain the Romberg integration technique. For the integral $I = \int_a^b f(x)\,dx$, the m-th approximation is given by using $2^m + 1$ nodes:

$$I \approx I_{0,m} = \frac{b-a}{2^m}\left[\frac{1}{2} f_0 + f_1 + \ldots + f_{2^m - 1} + \frac{1}{2} f_{2^m}\right],$$

Table 56.1. Results of using Romberg integration on (56.5).

nodes	$I_{0,m}$	$I_{1,m}$	$I_{2,m}$	$I_{3,m}$	$I_{4,m}$	$I_{5,m}$
3	0.905330					
5	0.964924	0.984789				
9	0.987195	0.994619	0.997895			
17	0.995372	0.998097	0.999257	0.999710		
33	0.998338	0.999327	0.999737	0.999897	0.999960	
65	0.999406	0.999762	0.999907	0.999964	0.999986	0.999994

where $f_i = f(x_i)$ and $x_i = a + \frac{i}{2^m}(b-a)$. The error in this approximation is given by (see page 342)

$$\begin{aligned} \text{error} = &\sum_{i=1}^{j} \frac{B_{2i}h^{2i}}{(2i)!}\left[g^{(2i-1)}(b) - g^{(2i-1)}(a)\right] \\ &+ (b-a)h^{2j-2}\frac{B_{2j+2}}{(2j+2)!}g^{(2j+2)}(\eta) \end{aligned} \tag{56.4}$$

where the $\{B_i\}$ are the Bernoulli numbers and $a < \eta < b$.

Note that only even powers of h appear in the error formula (56.4). Because of this, each extrapolation step increases the accuracy by two orders. The extrapolated values of the integral are then given by

$$I_{i,m} = \frac{I_{i-1,m-1} - \left(\frac{1}{2}\right)^{2i} I_{i-1,m-1}}{1 - \left(\frac{1}{2}\right)^2} \qquad \text{for } i = 1, 2, \ldots.$$

It should be noted that the values $\{I_{1,m}\}$ are identical to the values obtained from using the composite Simpson's rule.

Example

Given the integral

$$I = \frac{3}{2}\int_0^1 \sqrt{x}\,dx = 1 \tag{56.5}$$

we might choose to approximate the value by using Romberg integration. Table 56.1 has the result of using Romberg integration on this integral.

Note that the initial data has (at best) about 11 bits of precision (that is, $-\log_2 0.000594$), yet the fully extrapolated result has more than 17 bits of precision (that is, $-\log_2 0.000006$).

Table 56.2. Results of using Richardson extrapolation on (56.5).

nodes	$R_m^{(0)}$	$R_m^{(1)}$	$R_m^{(2)}$	$R_m^{(3)}$	$R_m^{(4)}$	$R_m^{(5)}$
3	0.90533					
5	0.96492	1.02452				
7	0.98052	0.99611	0.98664			
9	0.98720	0.99387	0.99312	0.99355		
11	0.99076	0.99433	0.99449	0.99440	0.99436	
13	0.99293	0.99510	0.99536	0.99521	0.99515	0.99512
15	0.99437	0.99580	0.99603	0.99590	0.99585	0.99582
17	0.99537	0.99638	0.99657	0.99646	0.99642	0.99640

Notes

[1] This method also works for non-uniform grids if every interval is subdivided.

[2] While Romberg is intrinsically a recursive computation, it is possible to write explicit formulas for the successive approximations. See Davis and Rabinowitz [3] for details.

[3] A benefit of doubling the number of nodes at each iteration of Romberg integration is that the integrand values at the old nodes can be re-used. Press *et al.* [6] present a modification of Romberg integration where the number of nodes must be tripled before the values at the old nodes can be re-used. (This is because the integration rule they use for $\int_{x_1}^{x_N} f(x)\,dx$ depends on the values $f_{m+1/2}$, where m is an integer.)

[4] A liability of doubling the number of nodes at each iteration of Romberg integration is that the number of nodes increases exponentially. Hence, some modifications of Romberg integration use a different sequence of node numbers. The sequences of node numbers $\{1, 2, 3, 6, 9, 18, 27, 54, \ldots\}$, $\{1, 2, 3, 4, 6, 8, 12, \ldots\}$, and $\{1, 2, 3, 4, 5, \ldots\}$ are discussed in Davis and Rabinowitz [3] and Fairweather and Keast [4]. Richardson extrapolation was used on problem (56.5) with the sequence of nodes $\{3, 5, 7, 9, \ldots\}$; the numerical results are in Table 56.2.

[5] In some cases, the order of the method, and hence the value of m in (56.1), will be unknown. Richardson extrapolation method may still be used by first estimating the value of m numerically, or the Aitken Δ^2 transformation may be used. The Aitken Δ^2 transformation uses three successive terms of the form $S_n = S_\infty + \alpha h^n$ to estimate S_∞ via

$$S_\infty = \frac{S_{n+1}S_{n-1} - S_n^2}{S_{n+1} + S_{n-1} - 2S_n}. \tag{56.6}$$

This transformation may be repeatedly applied; see Bender and Orszag [1] for details. If $2n+1$ terms are used, not just three, then this transformation is known as the Shanks transformation.

Table 56.3 has the results of applying the Aitken transformation to the integral in (56.5), when the trapezoidal rule has been used with n nodes (with $n = 3, 5, 7, \ldots$). The superscripts refer to the number of times the

Table 56.3. Results of using the Aitken transformation on (56.5).

nodes	$S_n^{(0)}$	$S_n^{(1)}$	$S_n^{(2)}$	$S_n^{(3)}$
3	0.90533			
5	0.96492	0.98604		
7	0.98052	0.99220	0.99690	
9	0.98720	0.99486	0.99795	0.99919
11	0.99076	0.99629	0.99851	
13	0.99293	0.99716		
15	0.99437			

Aitken transformation has been applied; the initial quadrature result has a superscript of zero.

[6] Bulirsch and Stoer [2] have a modification of this method. They compute trapezoidal sums for several values of h, fit a rational function of h^2 to this data (a Padé approximant), and then extrapolate to $h = 0$. On a practical level, nonlinear transformations such as those using Padé approximants are much more useful that Romberg integration. See also Press *et al.* [6].

[7] Note that the term R_m appearing in both (56.1) and (56.2) might conceivably not represent the same quantity. However, when an explicit error formula is available, such as the one in (56.4), then it is seen that the error terms are completely specified at every order.

[8] Richardson extrapolation is sometimes called *deferred approach to the limit.* In more detail, Richardson extrapolation proceeds as follows. If I is any function of h which can be represented as

$$I = I(h) + \sum_{i=k}^{m} a_i h^i + c_m(h) h^{m+1}$$

and the $\{c_i\}$ are known, then we may define $b_i = a_i(r^i - r^k)/(1 - r^k)$ and $d_m(h) = (c_m(rh)r^{m+1} - c_m(h)r^k)/(1 - r^k)$. This results in

$$I = \frac{I(rh) - r^k I(h)}{1 - r^k} + \sum_{i=k+1}^{m} b_i h^i + d_m(h) h^{m+1}.$$

References

[1] C. M. Bender and S. A. Orszag, *Advanced Mathematical Methods for Scientists and Engineers*, McGraw–Hill, New York, 1978, page 369.

[2] R. Bulirsch and J. Stoer, “Handbook Series Numerical Integration: Numerical Quadrature by Extrapolation,” *Numer. Math.*, **8**, 1966, pages 93–104.

[3] P. J. Davis and P. Rabinowitz, *Methods of Numerical Integration*, Second Edition, Academic Press, Orlando, Florida, 1984, pages 45–47 and 436–446.

[4] G. Fairweather and P. Keast, “An Investigation of Romberg Quadrature,” *ACM Trans. Math. Software*, **4**, 1978, pages 316–322.

[5] D. C. Joyce, "Survey of Extrapolation Processes in Numerical Analysis," *SIAM Review*, **13**, 1971, pages 435–490.

[6] W. H. Press, B. P. Flannery, S. Teukolsky, and W. T. Vetterling, *Numerical Recipes*, Cambridge University Press, New York, 1986, pages 123–127.

[7] A. Sidi, "The Numerical Evaluation of Very Oscillatory Infinite Integrals by Extrapolation," *Math. of Comp.*, **38**, 1982, pages 517–529.

57. Software Libraries: Introduction

When approximating the value of an integral numerically, it is best to use software prepared by experts whenever possible. Many commercial computer libraries and isolated computer routines for evaluating integrals are readily available.

No endorsement of the software referenced in this book is intended, nor does it necessarily imply that unnamed integrators are not worth trying. However, if you are using a integrator on anything bigger than a pocket calculator, you should consider using one of the cited packages. In the literature there now appears to be commercially available "software" for integration with no error control, a user-specified step size, and no warning messages. We strongly advise against using such programs, even on a personal computer. For all but trivial integrals, such programs cannot be sufficiently reliable for accurate computational results.

A taxonomy and description of mathematical and statistical routines in use at the National Institute of Standards and Technology (NIST) has been constructed, see Boisvert *et al.* [1] and [2]. The NIST guide, called GAMS (for *Guide to Available Mathematical Software*), was assembled to help users of the NIST computer facilities. The GAMS taxonomy for integration routines is reproduced on page 258. The integration routines listed in GAMS are reproduced on page 260.

There are many computer routines that are not described in the NIST guide, some of which are described in the Notes.

Notes

[1] Appendix 2 of Davis and Rabinowitz [3] contains a collection of FORTRAN integration routines.

[2] Appendix 3 of Davis and Rabinowitz [3] contains a bibliography of papers describing ALGOL, FORTRAN, and PL/1 integration routines.

[3] In the books by Press *et al.* (see, for example, [10]), there are a collection of C, FORTRAN, and PASCAL integration routines.

[4] Quadpack, a library of FORTRAN routines for one-dimensional integration by Piessens *at al.* [9], has the following system for naming its routines:

- The first letter is always "Q";
- The second letter is "N" for a non-adaptive integrator or "A" for an adaptive integrator;
- The third letter is "G" is for a general (user-defined) integrand or "W" for one of several weight functions;
- The fourth letter is
  "S" if routine handles singularities well,
  "P" if the location of points of special difficulty (such as singularities) can be input by the user,
  "I" if the region of integration is infinite,
  "O" if the weight function is oscillatory (i.e., is a sine or cosine function),
  "F" for a Fourier integral,
  "C" for a Cauchy principal-value integral,
  "E" for an extended parameter list (allows more user control).

Figures Figure 57.1 and Figure 57.2 show decision trees from Piessens *at al.* [9] that lead the user to the correct Quadpack routines.

[5] Quadlib [11] is a PC package derived from Quadpack. It contains the high-level double precision Quadpack subroutines along with an extensive collection of sample programs, test programs, and skeleton files.

[6] Many scientific software routines, including those for integration, may be obtained for free (via electronic mail) from a variety of computer networks. For example, all of the Quadpack [9] routines are available. To receive instructions on how to obtain this software, send the mail message "`send index`" to one of the following five addresses:

netlib@research.att.com
netlib@ornl.gov
netlib@nac.no
netlib@draci.cs.uow.edu.au
uunet!research!netlib

See the article by Dongarra and Grosse [5] for details.

[7] Many of the symbolic manipulation languages (see page 117) can also evaluate integrals numerically.

[8] Note that there are times when it is not wise to use an automatic quadrature routine, see Lyness [8].

References

[1] R. F. Boisvert, S. E. Howe, D. K. Kahaner, and J. L. Springmann, *Guide to Available Mathematical Software*, NISTIR 90-4237, Center for Computing and Applied Mathematics, National Institute of Standards and Technology, Gaithersburg, MD, March 1990.

[2] R. F. Boisvert, S. E. Howe, and D. K. Kahaner, "GAMS: A Framework for the Management of Scientific Software," *ACM Trans. Math. Software*, **11**, No. 4, December 1985, pages 313–355.

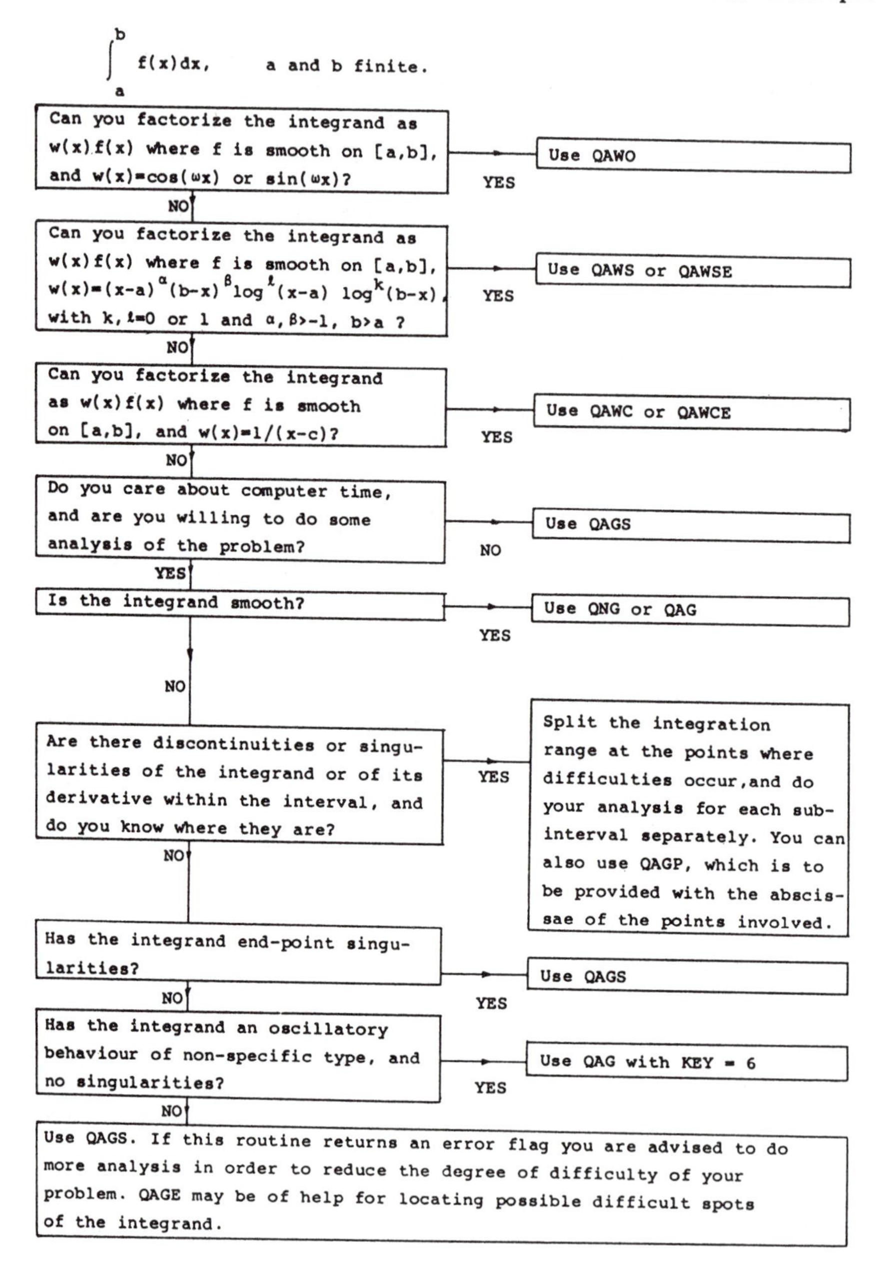

Figure 57.1. Decision tree for finite-range integration from Piessens *at al.* [9].

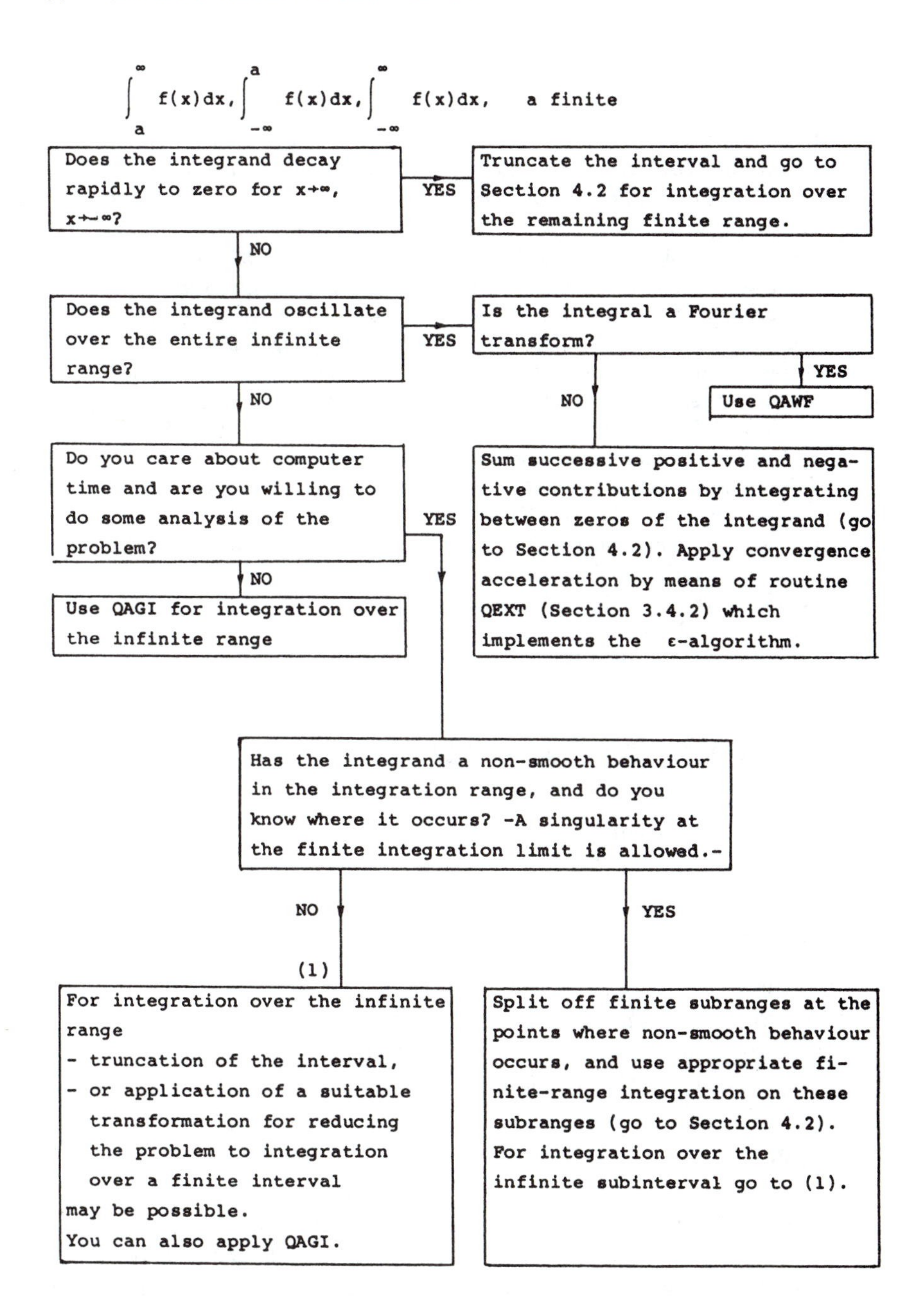

Figure 57.2. Decision tree for infinite-range integration from Piessens *at al.* [9].

[3] P. J. Davis and P. Rabinowitz, *Methods of Numerical Integration*, Second Edition, Academic Press, Orlando, Florida, 1984, Appendices 2 and 3, pages 480–517.

[4] L. M. Delves and B. G. S. Doman "The Design of a Generic Quadrature Library in ADA," *Numerical Integration*, P. Keast and G. Fairweather (eds.), D. Reidel Publishing Co., Boston, 1987, pages 307–319.

[5] J. J. Dongarra and E. Grosse, "Distribution of Mathematical Software Via Electronic Mail," *Comm. of the ACM*, **30**, No. 5, May 1987, pages 403–407.

[6] D. Kahaner, "Sources of Information on Quadrature Software," in W. R. Cowell (ed.), *Sources and Development of Mathematical Software*, Prentice–Hall Inc., Englewood Cliffs, NJ, 1984, Chapter 7, pages 134–164.

[7] D. K. Kahaner, "Development of Useful Quadrature Software, with Particular Emphasis on Microcomputers," in P. Keast and G. Fairweather (eds.), *Numerical Integration: Recent Developments, Software and Applications*, D. Reidel Publishing Co., Boston, 1987, pages 343–369.

[8] J. N. Lyness, "When Not to Use an Automatic Quadrature Routine," *SIAM Review*, **25**, No. 1, January 1983, pages 63–87.

[9] R. Piessens, E. de Doncker–Kapenga, C. W. Überhuber, and D. K. Kahaner, *Quadpack*, Springer–Verlag, New York, 1983.

[10] W. H. Press, B. P. Flannery, S. Teukolsky, and W. T. Vetterling, *Numerical Recipes*, Cambridge University Press, New York, 1986.

[11] *Quadlib*, McGraw–Hill Book Company, New York.

58. Software Libraries: Taxonomy

Applicable to Definite integrals.

Yields

The software category code appropriate for a specific numerical integration problem.

Procedure

A taxonomy for integration software has been developed as part of the GAMS project at the National Institute of Standards and Technology [1]. This section contains the taxonomy of integration routines from GAMS [2].

Example

Suppose that a multidimensional integration must be performed numerically. If the integrand is known analytically, and the integration region is a hyper-rectangle, then software in the category code **H2b1** is required. Most modern software comes with the appropriate category codes, which makes it easy to determine if some software is appropriate for a specific problem. This category code can be used when searching electronic databases, such as MathSci [3] or Netlib [4].

Taxonomy

H2.	Quadrature (numerical evaluation of definite integrals)
H2a.	One dimensional integrals
H2a1.	Finite interval (general integrand)
H2a1a.	Integrand available via user-defined procedure
H2a1a1.	Automatic (user need only specify desired accuracy)
H2a1a2.	Nonautomatic
H2a1b.	Integrand available only on a grid
H2a1b1.	Automatic (user need only specify desired accuracy)
H2a1b2.	Nonautomatic
H2a2.	Finite interval (specific or special type integrands including weight functions, oscillating and singular integrands, principal values integrals, splines, etc.)
H2a2a.	Integrand available via user-defined procedure
H2a2a1.	Automatic (user need only specify desired accuracy)
H2a2a2.	Nonautomatic
H2a2b.	Integrand available only on a grid
H2a2b1.	Automatic (user need only specify desired accuracy)
H2a2b2.	Nonautomatic
H2a3.	Semi-infinite interval (including e^{-x} weight function)
H2a3a.	Integrand available via user-defined procedure
H2a3a1.	Automatic (user need only specify desired accuracy)
H2a3a2.	Nonautomatic
H2a4.	Infinite interval (including e^{-x^2} weight function)
H2a4a.	Integrand available via user-defined procedure
H2a4a1.	Automatic (user need only specify desired accuracy)
H2a4a2.	Nonautomatic
H2b.	Multidimensional integrals
H2b1.	One or more hyper-rectangular regions (includes iterated integrals)
H2b1a.	Integrand available via user-defined procedure
H2b1a1.	Automatic (user need only specify desired accuracy)
H2b1a2.	Nonautomatic
H2b1b.	Integrand available only on a grid
H2b1b1.	Automatic (user need only specify desired accuracy)
H2b1b2.	Nonautomatic
H2b2.	n dimensional quadrature on a nonrectangular region
H2b2a.	Integrand available via user-defined procedure
H2b2a1.	Automatic (user need only specify desired accuracy)
H2b2a2.	Nonautomatic
H2b2b.	Integrand available only on a grid
H2b2b1.	Automatic (user need only specify desired accuracy)
H2b2b2.	Nonautomatic
H2c.	Service routines (compute weight and nodes for quadrature formulas)

Note

[1] GAMS [2] does not list software for all of the above categories.

References

[1] R. F. Boisvert, S. E. Howe, and D. K. Kahaner, "GAMS: A Framework for the Management of Scientific Software," *ACM Trans. Math. Software*, **11**, No. 4, December 1985, pages 313–355.

[2] R. F. Boisvert, S. E. Howe, D. K. Kahaner, and J. L. Springmann, *Guide to Available Mathematical Software*, NISTIR 90-4237, Center for Computing and Applied Mathematics, National Institute of Standards and Technology, Gaithersburg, MD, March 1990.

[3] *MathSci Disc*, CD-ROM product, Amer. Math. Soc., Providence, Rhode Island, 1992.

[4] J. J. Dongarra and E. Grosse, "Distribution of Mathematical Software Via Electronic Mail," *Comm. of the ACM*, **30**, No. 5, May 1987, pages 403–407.

59. Software Libraries: Excerpts from GAMS

Applicable to Definite integrals.

Yields

Software appropriate for a specific integration problem.

Procedure

A taxonomy and description of mathematical and statistical routines in use at the National Institute of Standards and Technology (NIST) has been constructed, see Boisvert *et al.* [1]. The NIST guide, called GAMS [2], was assembled to help users of the NIST computer facilities. Since NIST has many software packages, many people outside of NIST have found the guide useful despite its limited coverage. This section contains the GAMS listing of integration routines, by software category code (see page 258).

The following computer libraries (mostly in FORTRAN) are referred to†. Their inclusion does not constitute an endorsement. Nor does it necessarily imply that unnamed packages are not worth trying.

- CMLIB [3]
- Collected Algorithms of the ACM
- IMSL [4]
- NAG [5]
- NMS [6]
- PORT [7]
- Scientific Desk [8]
- SCRUNCH [9]

† Identification of commercial products does not imply recommendation or endorsement by NIST.

Excerpts from GAMS

H2a1a1:	**Automatic 1-D finite interval quadrature (user need only specify required accuracy), integrand available via user-defined procedure**

Collected Algorithms of the ACM

A614 INTHP: a Fortran subroutine for automatic numerical integration in H_p. The functions may have singularities at one or both endpoints of an interval. Each of finite, semi-infinite, and infinite intervals are admitted. (See K. Sikorski, F. Stenger, and J. Schwing, ACM TOMS 10 (1984) pp. 152–160.)

CMLIB Library (Q1DA Sublibrary)

Q1DA Automatic integration of a user-defined function of one variable. Special features include randomization and singularity weakening.

Q1DAX Flexible subroutine for the automatic integration of a user-defined function of one variable. Special features include randomization, singularity weakening, restarting, specification of an initial mesh (optional), and output of smallest and largest integrand values.

Q1DB Automatic integration of a user-defined function of one variable. Integrand must be a Fortran FUNCTION but user may select name. Special features include randomization and singularity weakening. Intermediate in usage difficulty between Q1DA and Q1DAX.

CMLIB Library (QUADPKS Sublibrary)

QAG Automatic adaptive integrator, will handle many non-smooth integrands using Gauss–Kronrod formulas.

QAGE Automatic adaptive integrator, can handle most non-smooth functions, also provides more information than QAG.

QAGS Automatic adaptive integrator, will handle most non-smooth integrands including those with endpoint singularities, uses extrapolation.

QAGSE Automatic adaptive integrator, can handle integrands with endpoint singularities, provides more information than QAGS.

QNG Automatic non-adaptive integrator for smooth functions, using Gauss–Kronrod–Patterson formulas.

IMSL Subprogram Library

DCADRE Numerical integration of a function using cautious adaptive Romberg extrapolation.

IMSL MATH/LIBRARY Subprogram Library

QDAG Integrate a function using a globally adaptive scheme based on Gauss–Kronrod rules.

QDAGS Integrate a function (which may have endpoint singularities).

QDNG Integrate a smooth function using a nonadaptive rule.

IMSL STAT/LIBRARY Subprogram Library

QDAGS Integrate a function (which may have endpoint singularities).

JCAM Software Library

DEFINT Uses double exponential transformation of Mori to compute definite integral automatically to user specified accuracy.

NAG Subprogram Library

D01AHF Computes a definite integral over a finite range to a specified relative accuracy using a method described by Patterson.

D01AJF Is a general-purpose integrator which calculates an approximation to the integral of a function F(x) over a finite interval (A,B).

D01ARF Computes definite and indefinite integrals over a finite range to a specified relative or absolute accuracy, using a method described by Patterson.

D01BDF Calculates an approximation to the integral of a function over a finite interval (A,B). It is non-adaptive and as such is recommended for the integration of smooth functions. These exclude integrands with singularities, derivative singularities or high peaks on (A,B), or which oscillate too strongly on (A,B).

NMS Subprogram Library

Q1DA Automatic integration of a user-defined function of one variable. Special features include randomization and singularity weakening.

PORT Subprogram Library

ODEQ Finds the integral of a set of functions over the same interval by using the differential equation solver ODES1. For smooth functions.

QUAD Finds the integral of a general user defined EXTERNAL function by an adaptive technique to given absolute accuracy.

RQUAD Finds the integral of a general user defined EXTERNAL function by an adaptive technique. Combined absolute and relative error control.

Scientific Desk PC Subprogram Library

H2A1 Automatically evaluates the definite integral of a user defined function of one variable.

H2A1U Automatically evaluates the definite integral of a user defined function of one variable.

SCRUNCH Subprogram Library

SIMP Calculates an estimate of the definite integral of a user supplied function by adaptive quadrature. In BASIC.

H2a1a2:	**Nonautomatic 1-D finite interval quadrature, integrand available via user-defined procedure**

CMLIB Library (QUADPKS Sublibrary)

QK15 Evaluates integral of given function on an interval with a 15 point Gauss–Kronrod formula and returns error estimate.

QK21 Evaluates integral of given function on an interval with a 21 point Gauss–Kronrod formula and returns error estimate.

QK31 Evaluates integral of given function on an interval with a 31 point Gauss–Kronrod formula and returns error estimate.

QK41 Evaluates integral of given function on an interval with a 41 point Gauss–Kronrod formula and returns error estimate.

QK51 Evaluates integral of given function on an interval with a 51 point Gauss–Kronrod formula and returns error estimate.

QK61 Evaluates integral of given function on an interval with a 61 point Gauss–Kronrod formula and returns error estimate.

NAG Subprogram Library

D01BAF Computes an estimate of the definite integral of a function of known analytical form, using a Gaussian quadrature formula with a specified number of abscissae. Formulae are provided for a finite interval (Gauss–Legendre), a semi-infinite interval (Gauss–Laguerre, Gauss–Rational), and an infinite interval (Gauss–Hermite).

NMS Subprogram Library

QK15 Evaluates integral of given function on an interval with a 15 point Gauss–Kronrod formula and returns error estimate.

H2a1b2: Nonautomatic 1-D finite interval quadrature, integrand available only on a grid

NAG Subprogram Library

D01GAF Integrates a function which is specified numerically at four or more points, over the whole of its specified range, using third-order finite-difference formulae with error estimates, according to a method due to Gill and Miller.

NMS Subprogram Library

PCHQA Integrates piecewise cubic from A to B given N-arrays X,F,D. Usually used in conjunction with PCHEZ to form cubic, but can be used independently, especially if the abscissae are equally spaced.

PORT Subprogram Library

CSPQU Finds the integral of a function defined by pairs (x,y) of input points. The x's can be unequally spaced. Uses spline interpolation.

Scientific Desk PC Subprogram Library

H2A1T Computes the integral of the array f between x(i) and x(j), given n points in the plane (x(k),f(k)), k=1,...,n.

H2a2a1:	**Automatic 1-D finite interval quadrature (user need only specify required accuracy) (special integrand including weight functions, oscillating and singular integrands, principal value integrals, splines, etc.), integrand available via user-defined procedure**

CMLIB Library (BSPLINE Sublibrary)

BFQAD Integrates function times derivative of B-spline from X1 to X2. The B-spline is in B representation.

PFQAD Computes integral on (X1,X2) of product of function and the ID-th derivative of B-spline which is in piecewise polynomial representation.

CMLIB Library (QUADPKS Sublibrary)

QAGP Automatic adaptive integrator, allows user to specify location of singularities or difficulties of integrand, uses extrapolation.

QAGPE Automatic adaptive integrator for function with user specified endpoint singularities, provides more information that QAGP.

QAWC Cauchy principal value integrator, using adaptive Clenshaw–Curtis method (real Hilbert transform).

QAWCE Cauchy principal value integrator, provides more information than QAWC (real Hilbert transform).

QAWO Automatic adaptive integrator for integrands with oscillatory sine or cosine factor.

QAWOE Automatic integrator for integrands with explicit oscillatory sine or cosine factor, provides more information than QAWO.

QAWS Automatic integrator for functions with explicit algebraic and/or logarithmic endpoint singularities.

QAWSE Automatic integrator for integrands with explicit algebraic and/or logarithmic endpoint singularities, provides more information than QAWS.

QMOMO Computes integral of k-th degree Chebyshev polynomial times one of a selection of functions with various singularities.

IMSL MATH/LIBRARY Subprogram Library

QDAGP Integrate a function with singularity points given.

QDAWC Integrate a function F(x)/(x-c) in the Cauchy principal value sense.

QDAWO Integrate a function containing a sine or a cosine.

QDAWS Integrate a function with algebraic-logarithmic singularities.

NAG Subprogram Library

D01AKF Is an adaptive integrator, especially suited to oscillating, non-singular integrands, which calculates an approximation to the integral of a function F(x) over a finite interval (A,B).

D01ALF Is a general purpose integrator which calculates an approximation to the integral of a function F(x) over a finite interval (A,B), where the integrand may have local singular behavior at a finite number of points within the integration interval.

D01ANF Calculates an approximation to the cosine or the sine transform of a function G over (A,B), i.e., the integral of G(x)cos(ωx) or G(x)sin(ωx) over (A,B) (for a user-specified value of ω).

D01APF Is an adaptive integrator which calculates an approximation to the integral of a function G(x)W(x) over (A,B) where the weight function W has end-point singularities of algebraic-logarithmic type (see input parameter KEY).

D01AQF Calculates an approximation to the Hilbert transform of a function G(x) over (A,B), i.e., the integral of G(x)/(x-c) over (A,B), for user-specified values of A,B,C.

PORT Subprogram Library

BQUAD Adaptively integrates functions which have discontinuities in their derivatives. User can specify these points.

H2a2a2:	**Nonautomatic 1-D finite interval quadrature (special integrand including weight functions, oscillating and singular integrands, principal value integrals, splines, etc.), integrand available via user-defined procedure**

CMLIB Library (QUADPKS Sublibrary)

QC25C Uses 25 point Clenshaw–Curtis formula to estimate integral of F(x)W(x) where W(x)=1/(x-c).

QC25F Clenshaw–Curtis integration rule for function with cos or sin factor, also uses Gauss–Kronrod formula.

QC25S Estimates integral of function with algebraic-logarithmic singularities using 25 point Clenshaw–Curtis formula and gives error estimate.

QK15W Evaluates integral of given function times arbitrary weight function on interval with 15 point Gauss–Kronrod formula and gives error estimate.

H2a2b1:	**Automatic 1-D finite interval quadrature (user need only specify required accuracy) (special integrand including weight functions, oscillating and singular integrands, principal value integrals, splines, etc.), integrand available only on a grid**

CMLIB Library (BSPLINE Sublibrary)

BSQAD Computes the integral of a B-spline from X1 to X2. The B-spline must be in B representation.

PPQAD Computes the integral of a B-spline from X1 to X2. The B-spline must be in piecewise polynomial representation.

IMSL Subprogram Library

DCSQDU Cubic spline quadrature.

IMSL MATH/LIBRARY Subprogram Library

BSITG Evaluate the integral of a spline, given its B-spline representation.

NAG Subprogram Library

E02AJF Determines the coefficients in the Chebyshev series representation of the indefinite integral of a polynomial given in Chebyshev series form.

E02BDF Computes the definite integral of a cubic spline from its B-spline representation.

PORT Subprogram Library

BSPLI Obtains the integrals of basis splines, from the left-most mesh point to a specified set of points.

SPLNI Integrates a function described previously by an expansion in terms of B-splines. Several integrations can be performed in one call.

Scientific Desk PC Subprogram Library

E3HIN Evaluates the definite integral of a piecewise cubic Hermite function over an arbitrary interval.

E3INT Evaluates the definite integral of a piecewise cubic Hermite function over an interval whose endpoints are data points.

H2a3a1: Automatic 1-D semi-infinite interval quadrature (user need only specify required accuracy) (including e^{-x}) weight function), integrand available via user-defined procedure

Collected Algorithms of the ACM

A614 INTHP: a Fortran subroutine for automatic numerical integration in H_p. The functions may have singularities at one or both end-points of an interval. Each of finite, semi-infinite, and infinite intervals are admitted. (See K. Sikorski, F. Stenger, and J. Schwing, ACM TOMS 10 (1984) pp. 152–160.)

A639 OSCINT: a Fortran subprogram for the automatic integration of some infinitely oscillating tails. That is, the evaluation of the integral from a to infinity of h(x)j(x), where h(x) is ultimately positive, and j(x) is either a circular function (e.g., cosine) or a first-kind Bessel function of fractional order. (See J. Lyness and G. Hines, ACM TOMS 12 (1986) pp. 24–25.)

CMLIB Library (QUADPKS Sublibrary)

QAGI Automatic adaptive integrator for semi-infinite or infinite intervals. Uses nonlinear transformation and extrapolation.

QAGIE Automatic integrator for semi-infinite or infinite intervals and general integrands, provides more information than QAGI.

QAWF Automatic integrator for Fourier integrals on (a,∞) with factors sin(ωx), cos(ωx) by integrating between zeros.

QAWFE Automatic integrator for Fourier integrals, with sin(ωx) factor on (a,∞), provides more information than QAWF.

IMSL MATH/LIBRARY Subprogram Library

QDAGI Integrate a function over an infinite or semi-infinite interval.

QDAWF Compute a Fourier integral.

JCAM Software Library

DEHINT Uses double exponential transformation of Mori to compute semi-infinite range integral automatically to user specified accuracy.

NAG Subprogram Library

D01AMF Calculates an approximation to the integral of a function F(x) over an infinite or semi-infinite interval (A,B).

NMS Subprogram Library

QAGI Automatic adaptive integrator for semi-infinite or infinite intervals. Uses nonlinear transformation and extrapolation.

H2a3a2:	**Nonautomatic 1-D semi-infinite interval quadrature) (including e^{-x} weight function), integrand available via user-defined procedure**

CMLIB Library (QUADPKS Sublibrary)

QK15I Evaluates integral of given function on semi-infinite or infinite interval with a transformed 15 point Gauss–Kronrod formula and gives error estimate.

NAG Subprogram Library

D01BAF Computes an estimate of the definite integral of a function of known analytical form, using a Gaussian quadrature formula with a specified number of abscissae. Formulae are provided for a finite interval (Gauss–Legendre), a semi-infinite interval (Gauss–Laguerre, Gauss–rational), and an infinite interval (Gauss-Hermite).

H2a4a1:	**Automatic 1-D infinite interval quadrature (user need only specify required accuracy) (including e^{-x^2}) weight function), integrand available via user-defined procedure**

Collected Algorithms of the ACM

A614 INTHP: a Fortran subroutine for automatic numerical integration in H_p. The functions may have singularities at one or both end-points of an interval. Each of finite, semi-infinite, and infinite intervals are admitted. (See K. Sikorski, F. Stenger, and J. Schwing, ACM TOMS 10 (1984) pp. 152–160.)

CMLIB Library (QUADPKS Sublibrary)

QAGI Automatic adaptive integrator for semi-infinite or infinite intervals. Uses nonlinear transformation and extrapolation.

QAGIE Automatic integrator for semi-infinite or infinite intervals and general integrands, provides more information than QAGI.

NAG Subprogram Library

D01AMF Calculates an approximation to the integral of a function F(x) over an infinite or semi-infinite interval (A,B).

NMS Subprogram Library

QAGI Automatic adaptive integrator for semi-infinite or infinite intervals. Uses nonlinear transformation and extrapolation.

H2a4a2: Nonautomatic 1-D infinite interval quadrature (including e^{-x^2}) weight function), integrand available via user-defined procedure

CMLIB Library (QUADPKS Sublibrary)

QK15I Evaluates integral of given function on semi-infinite or infinite interval with a transformed 15 point Gauss–Kronrod formula and gives error estimate.

NAG Subprogram Library

D01BAF Computes an estimate of the definite integral of a function of known analytical form, using a Gaussian quadrature formula with a specified number of abscissae. Formulae are provided for a finite interval (Gauss–Legendre), a semi-infinite interval (Gauss–Laguerre, Gauss–rational), and an infinite interval (Gauss–Hermite).

H2b1a1: Automatic n-D quadrature (user need only specify required accuracy) on one or more hyper-rectangular regions, integrand available via user-defined procedure

CMLIB Library (ADAPT Sublibrary)

ADAPT Computes the definite integral of a user specified function over a hyper-rectangular region in 2 through 20 dimensions. User specifies tolerance. A restarting feature is useful for continuing a computation without wasting previous function values.

IMSL Subprogram Library

DBLIN Numerical integration of a function of two variables.

DMLIN Numerical integration of a function of several variables over a hyper-rectangle (Gaussian method).

IMSL MATH/LIBRARY Subprogram Library

QAND Integrate a function on a hyper-rectangle.

TWODQ Compute a two-dimensional iterated integral using internal calls to a one-dimensional automatic integrator.

NAG Subprogram Library

D01DAF Attempts to evaluate a double integral to a specified absolute accuracy by repeated applications of the method described by Patterson.

D01EAF Computes approximations to the integrals of a vector of similar functions, each defined over the same multi-dimensional hyper-rectangular region. The routine uses an adaptive subdivision strategy, and also computes absolute error estimates.

D01FCF Attempts to evaluate a multidimensional integral (up to 15 dimensions), with constant and finite limits, to a specified relative accuracy, using an adaptive subdivision strategy.

D01GBF Returns an approximation to the integral of a function over a hyper-rectangular region, using a Monte–Carlo method. An approximate relative error estimate is also returned. This routine is suitable for low accuracy work.

H2b1a2: Nonautomatic n-D quadrature on one or more hyper-rectangular regions, integrand available via user-defined procedure

NAG Subprogram Library

D01FBF Computes an estimate of a multidimensional integral (from 1 to 20 dimensions), given the analytic form of the integrand and suitable Gaussian weights and abscissae.

D01FDF Calculates an approximation to a definite integral in up to 30 dimensions, using the method of Sag and Szekeres. The region of integration is an n-sphere, or by built-in transformation via the unit n-cube, any product region.

D01GCF Calculates an approximation to a definite integral in up to 20 dimensions, using the Korobov–Conroy number theoretic method.

H2b1b2: Nonautomatic n-D quadrature on one or more hyper-rectangular regions, integrand available only on a grid

IMSL Subprogram Library

DBCQDU Bicubic spline quadrature.

IMSL MATH/LIBRARY Subprogram Library

BS2IG Evaluate the integral of a tensor-product spline on a rectangular domain, given its tensor-product B-spline representation.

BS3IG Evaluate the integral of a tensor-product spline in three dimensions over a three-dimensional rectangle, given its tensor-product B-spline representation.

H2b2a1:	**Automatic n-D quadrature on a nonrectangular region (user need only specify required accuracy), integrand available via user-defined procedure**

Collected Algorithms of the ACM

A584 CUBTRI: a Fortran subroutine for adaptive cubature over a triangle. (See D. P. Laurie, ACM TOMS 8 (1982) pp. 210–218.)

A612 TRIEX: a Fortran subroutine for integration over a triangle. Uses an adaptive subdivisional strategy with global acceptance criteria and incorporates the epsilon algorithm to speed convergence. (see E. de Doncker and I. Robinson, ACM TOMS 10 (1984) pp. 17–22.)

CMLIB Library (TWODQ Sublibrary)

TWODQ Automatic (adaptive) integration of a user specified function f(x,y) on one or more triangles to a prescribed relative or absolute accuracy. Two different quadrature formulas are available within TWODQ. This enables a user to integrate functions with boundary singularities.

NAG Subprogram Library

D01JAF Attempts to evaluate an integral over an n-dimensional sphere (n=2, 3, or 4), to a user specified absolute or relative accuracy, by means of a modified Sag–Szekeres method. The routine can handle singularities on the surface or at the center of the sphere, and returns an error estimate.

Scientific Desk PC Subprogram Library

H2B2A Computes the two-dimensional integral of a function f over a region consisting of n triangles.

H2b2a2:	**Nonautomatic n-D quadrature on a nonrectangular region, the integrand available via user-defined procedure**

JCAM Software Library

DTRIA Computes an approximation to the double integral of f(u,v) over a triangle in the uv-plane by using an n^2 point, generalized Gauss–Legendre product rule of polynomial degree precision 2n-2. From "Computation of Double Integrals over a Triangle," by F. G. Lether, Algorithm 007, J. Comp. Appl. Math. 2(1976), pp. 219–224.

NAG Subprogram Library

D01PAF Returns a sequence of approximations to the integral of a function over a multi-dimensional simplex, together with an error estimate for the last approximation.

H2c:	**Service routines for quadrature (compute weight and nodes for quadrature formulas)**

Collected Algorithms of the ACM

A647 Fortran subprograms for the generation of sequences of quasirandom vectors with low discrepancy. Such sequences may be used to reduce error bounds for multidimensional integration and global optimization. (See B. L. Fox, ACM TOMS 12 (1986) pp. 362–376.)

A655 IQPACK: Fortran routines for the stable evaluation of the weights and nodes of interpolatory and Gaussian quadratures with prescribed simple or multiple knots. (See S. Elhay and J. Kautsky, ACM TOMS 13 (1987) pp. 399–415.)

A659 A Fortran implementation of Sobol's quasirandom sequence generator for multivariate quadrature and optimization. (See P. Bratley and B. L. Fox, ACM TOMS 14 (1988) pp. 88–100.)

IMSL MATH/LIBRARY Subprogram Library

FQRUL Compute a Fejer quadrature rule with various classical weight functions.

GQRCF Compute a Gauss, Gauss–Radeau or Gauss–Lobatto quadrature rule given the recurrence coefficients for the monic polynomials orthogonal with respect to the weight function.

GQRUL Compute a Gauss, Gauss–Radeau or Gauss–Lobatto quadrature rule with various classical weight functions.

RECCF Compute recurrence coefficients for various monic polynomials.

RECQR Compute recurrence coefficients for monic polynomials given a quadrature rule.

NAG Subprogram Library

D01BBF Returns the weights and abscissae appropriate to a Gaussian quadrature formula with a specified number of abscissae. The formulae provided are Gauss-Legendre, Gauss–rational, Gauss–Laguerre and Gauss–Hermite.

D01BCF Returns the weights (normal or adjusted) and abscissae for a Gaussian integration rule with a specified number of abscissae. Six different types of Gauss rule are allowed.

PORT Subprogram Library

GAUSQ Finds the abscissae and weights for Gauss quadrature on the interval (a,b) for a general weight function with known moments.

GQ0IN Finds the abscissae and weights for Gauss–Laguerre quadrature on the interval $(0,+\infty)$.

GQM11 Finds the abscissae and weights for Gauss–Legendre quadrature on the interval $(-1,1)$.

Notes

[1] In the excerpts section, ACM TOMS stands for *ACM Trans. Math. Software.*

[2] Software is not listed for all the taxonomy classes that have been established.

[3] The author thanks Dr. Ronald Boisvert of NIST for making part of GAMS available electronically.

References

[1] R. F. Boisvert, S. E. Howe, and D. K. Kahaner, "GAMS: A Framework for the Management of Scientific Software," *ACM Trans. Math. Software*, **11**, No. 4, December 1985, pages 313–355.

[2] R. F. Boisvert, S. E. Howe, D. K. Kahaner, and J. L. Springmann, *Guide to Available Mathematical Software*, NISTIR 90-4237, Center for Computing and Applied Mathematics, National Institute of Standards and Technology, Gaithersburg, MD, March 1990.

[3] CMLIB, this is a collection of code from many sources that NIST has combined into a single library. The relevant sublibraries are

(A) CDRIV and SDRIV, see D. Kahaner, C. Moler, and S. Nash, *Numerical Methods and Software*, Prentice–Hall Inc., Englewood Cliffs, NJ, 1989.

(B) DEPAC: Code developed by L. Shampine and H. A. Watts.

(C) FISHPAK: Code developed by P. N. Swartztrauber and R. A. Sweet.

(D) VHS3: Code developed by R. A. Sweet.

[4] IMSL Inc., 2500 Park West Tower One, 2500 City West Blvd., Houston, TX 77042.

[5] NAG, Numerical Algorithms Group, Inc., 1400 Opus Place, Suite 200, Downers Grove, IL, 60515.

[6] NMS, this is an internal name at NIST. The code is from D. Kahaner, C. Moler, and S. Nash, *Numerical Methods and Software*, Prentice–Hall Inc., Englewood Cliffs, NJ, 1989.

[7] PORT, see P. Fox, *et al.*, *The PORT Mathematical Subroutine Library Manual*, Bell Laboratories, Murray Hill, NJ, 1977.

[8] Scientific Desk is distributed by M. McClain, NIST, Bldg 225 Room A151, Gaithersburg, MD 20899.

[9] SCRUNCH, these are old, unsupported codes in BASIC. The codes are translations of Fortran algorithms from G. Forsythe, M. Malcom, and C. Moler, *Computer Methods for Mathematical Computations*, Prentice–Hall Inc., Englewood Cliffs, NJ, 1977.

60. Testing Quadrature Rules

Applicable to Numerical approximations to integrals.

Idea

Many integrals have been used as examples to test quadrature rules.

Procedure

As new quadrature rules are developed, they are compared to existing quadrature rules in terms of accuracy and efficiency. Many authors have introduced example integrals to indicate the performance of their algorithms and implementations. We tabulate some of those integrals.

- Lyness [4] uses the test integral $I(\lambda) = \int_1^2 \frac{0.1}{(1-\lambda)^2 + 0.01}\, dx$.

- Piessens *et al.* [5] uses the test integrals (the numbers correspond to their original numbering, numbers 4–6 represent previous integrals with different parameters):

1) $$\int_0^1 x^\alpha \log\left(\frac{1}{x}\right) dx = \frac{1}{(1+\alpha)^2}$$

2) $$\int_0^1 \frac{4^{-\alpha}}{\left(x - \frac{\pi}{4}\right)^2 + 16^{-\alpha}}\, dx = \tan^{-1}\left((4-\pi)4^{\alpha-1}\right) + \tan^{-1}\left(\pi 4^{\alpha-1}\right)$$

3) $$\int_0^\pi \cos\left(2^\alpha \sin x\right) dx = \pi J_0\left(2^\alpha\right)$$

7) $$\int_0^1 \left|x - \tfrac{1}{3}\right|^\alpha dx = \frac{\left(\frac{2}{3}\right)^{\alpha+1} + \left(\frac{1}{3}\right)^{\alpha+1}}{1+\alpha}$$

8) $$\int_0^1 \left|x - \frac{\pi}{4}\right|^\alpha dx = \frac{\left(1 - \frac{\pi}{4}\right)^{\alpha+1} + \left(\frac{\pi}{4}\right)^{\alpha+1}}{1+\alpha}$$

9) $$\int_{-1}^1 \frac{1}{\sqrt{1-x^2}} \frac{1}{1+x+2^{-\alpha}}\, dx = \frac{\pi}{\sqrt{(1+2^{-\alpha})^2 - 1}}$$

10) $$\int_0^{\pi/2} \sin^{\alpha-1}(x)\, dx = \frac{2^{\alpha-2}\Gamma^2\left(\frac{\alpha}{2}\right)}{\Gamma(\alpha)}$$

11) $$\int_0^1 \log^{\alpha-1}\left(\frac{1}{x}\right) dx = \Gamma(\alpha)$$

12) $$\int_0^1 e^{20(x-1)} \sin\left(2^\alpha x\right) dx = \frac{20\sin\left(2^\alpha\right) - 2^\alpha \cos\left(2^\alpha\right) + 2^\alpha e^{-20}}{400 + 4^\alpha}$$

13) $$\int_0^1 \frac{\cos\left(2^\alpha x\right)}{\sqrt{x(1-x)}}\, dx = \pi \cos\left(2^{\alpha-1}\right) J_0\left(2^{\alpha-1}\right)$$

14) $$\int_0^\infty \frac{e^{-2^{-\alpha}x}}{\sqrt{x}}\, dx = \frac{\sqrt{\pi}}{\left(1+4^{-\alpha}\right)^{-1/4}} \cos\left(\frac{\tan^{-1}\left(2^\alpha\right)}{2}\right)$$

15) $$\int_0^\infty x^2 e^{-2^{-\alpha}x}\, dx = 2^{3\alpha+1}$$

16) $$\int_0^\infty \frac{x^{\alpha-1}}{(1+10x)^2}\, dx = \frac{(1-\alpha)\pi}{10^\alpha \sin(\pi\alpha)}$$

17) $$\int_0^5 \frac{2^{-\alpha}}{(x-2)\left((x-1)^2+4^{-\alpha}\right)}\,dx = \frac{2^{-\alpha}\log\left(\frac{3}{2}\right) - 2^{-\alpha-1}\log\frac{16+4^{-\alpha}}{1+4^{-\alpha}} - \tan^{-1}\left(2^{\alpha+2}\right) - \tan^{-1}\left(2^{\alpha}\right)}{1+4^{-\alpha}}$$

- Berntsen *et al.* [1] uses the test integrals:

1) $\int_0^1 |x-\lambda|^{\alpha_1}\,dx$ — *feature:* singularity
2) $\int_0^1 f_2(x)\,dx$ — *feature:* discontinuous
3) $\int_0^1 e^{-\alpha_3|x-\lambda|}\,dx$ — *feature:* C_0 function
4) $\displaystyle\int_1^2 \frac{10^{\alpha_4}}{(x-\lambda)^2+10^{2\alpha_4}}\,dx$ — *feature:* one peak
5) $\displaystyle\int_1^2 \sum_{i=1}^4 \frac{10^{\alpha_5}}{(x-\lambda_i)^2+10^{2\alpha_5}}\,dx$ — *feature:* four peaks
6) $\int_0^1 2B(x-\lambda)\cos(B(x-\lambda)^2)\,dx$ — *feature:* nonlinear oscillation

where $B = \dfrac{10^{\alpha_6}}{\max(\lambda^2,(1-\lambda)^2)}$ and $f_2(x) = \begin{cases} 0 & \text{if } x \le \lambda \\ \exp(\alpha_2 x) & \text{otherwise} \end{cases}$.

- Hunter and Smith [3] use the principal-value integrals:

1) $$\fint_0^{\pi/2} \frac{\cos(\cos t)}{k^2-\sin^2 t}\,dt$$
2) $$\fint_0^{\infty} \frac{e^{-t^2}}{t^2-\lambda^2}\,dt$$

where $0 < k < 1$ and $\lambda > 0$.

- Corliss and Rall [2] have a collection of test problems that exercise their interval analysis integration package:

1) $$\int_{[0,0.1]}^{[3.1,3.2]} \sin x\,dx$$
2) $\int_0^1 (B\sin(Bx) - A\sin(Ax))\,dx$
3) $$\int_{0.6}^{0.7} \frac{dx}{1-x}$$
4) $\int_0^4 \sqrt{x}\,dx$
5) $\int_0^1 f(x)\,dx$
6) $\int_{0.3}^1 1\,dx$
7) $\int_0^1 (x-2)\,dx$
8) $$\int_0^1 \frac{dx}{1+x^4}$$

where $A = [0.0.1]$ and $B = [3.1, 3.2]$, and $f(x) = \begin{cases} 0, & x < 0.3 \\ 1, & x \geq 0.3 \end{cases}$.

References

[1] J. Berntsen, T. O. Espelid, and T. Sørevik, "On the Subdivision Strategy in Adaptive Quadrature Algorithms," *J. Comput. Appl. Math.*, **35**, 1991, pages 119–132.

[2] G. F. Corliss and L. B. Rall, "Adaptive, Self-Validating Numerical Quadrature," *SIAM J. Sci. Stat. Comput.*, **8**, No. 5, 1987, pages 831–847.

[3] D. B. Hunter and H. V. Smith, "The Evaluation of Cauchy Principal Value Integrals Involving Unknown Poles," *BIT*, **29**, No. 3, 1989, pages 512–517.

[4] J. N. Lyness, "When Not to Use an Automatic Quadrature Routine," *SIAM Review*, **25**, No. 1, January 1983, pages 63–87.

[5] R. Piessens, E. de Doncker–Kapenga, C. W. Überhuber, and D. K. Kahaner, *Quadpack*, Springer–Verlag, New York, 1983, pages 83–84.

[6] I. Robinson, "A Comparison of Numerical Integration Programs," *J. Comput. Appl. Math.*, **2**, 1979, pages 207–223.

61. Truncating an Infinite Interval

Applicable to Integrals that have an infinite limit of integration.

Yields

An approximating integral, with a bound on the error.

Idea

By truncating an infinite integral, a numerical routine may have an easier computation.

Procedure

An infinite integral can always be truncated to a finite interval. Estimating the error made in the truncation process establishes the usefulness of the truncation.

Example

Consider the integral $I = \int_0^\infty \frac{x}{1+x} e^{-x^2}\, dx$. If we truncate the upper limit of integration to be, say, α, then we have

$$I \approx J_\alpha = \int_0^\alpha \frac{x}{1+x} e^{-x^2}\, dx.$$

In this case we can estimate the error made in the truncation process:

$$\begin{aligned} I - J_\alpha &= \int_\alpha^\infty \frac{x}{1+x} e^{-x^2}\, dx \\ &< \int_\alpha^\infty x e^{-x^2}\, dx \\ &= \left(-\tfrac{1}{2} e^{-x^2} \right) \Big|_\alpha^\infty \\ &= \tfrac{1}{2} e^{-\alpha^2} \end{aligned} \tag{61.1}$$

where we have used the approximation $\dfrac{x}{1+x} < x$ for $x > 0$.

If we were to approximate I by numerically approximating J_α, then we would need $\alpha \geq 4.3$ to insure that $I - J_\alpha \leq 10^{-8}$.

Reference

[1] P. J. Davis and P. Rabinowitz, *Methods of Numerical Integration*, Second Edition, Academic Press, Orlando, Florida, 1984, page 205.

VI

Numerical Methods: Techniques

62. Adaptive Quadrature

Applicable to Integrals in any number of dimensions.

Yields

A numerical quadrature scheme.

Idea

If a numerical quadrature scheme does not result in a sufficiently accurate numerical approximation, then sampling the integrand at more nodes should increase the accuracy. However, the additional sampling only needs to be performed in problem areas (where the error estimates are large).

Procedure

Adaptive quadrature is an automatic procedure for increasing the accuracy of a numerical approximation to an integral by increasing the number of samples of the integrand. Additional samples only need to be taken where the quadrature scheme is having numerical difficulties. Hence, the overall scheme is given by the following steps:

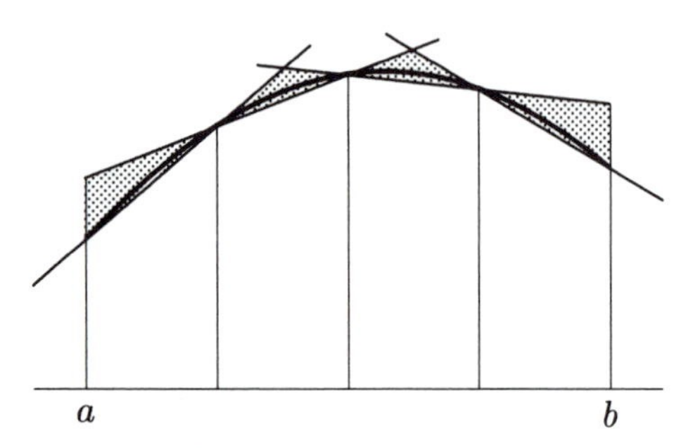

Figure 62. The geometry of an adaptive integration computation. The function $f(x)$ lies in the shaded triangles; the area of these triangles provide error estimates.

[1] Start with a given integral to be integrated over a given interval.
[2] Use a quadrature rule to approximate the integral over the entire interval; this is the global approximation. Estimate the error in this approximation; this is the global error. Place the interval, and the estimated error on that interval, onto a list.
[3] If the global error is not small enough, then:
  (A) Choose an interval from the list of intervals (presumably, choose the one with the largest estimated error).
  (B) Subdivide the chosen interval.
  (C) Approximate the integral over each of the new sub-intervals, and estimate the error in these approximations.
  (D) Update the list of intervals and the estimated error on each sub-interval.
  (E) Update the global approximation to the integral and the estimate of the global error. Go to step [3].
[4] Done.

Figure 62 shows an early stage in an adaptive integration computation. The adaptive strategy is to subdivide the largest shaded triangle. For the figure shown, the left-most region is the next to be sub-divided.

Rice [12] determines that there are between 1 and 10 million adaptive algorithms that are potentially interesting and significantly different from one another. This number arises from the following:

- There are at least six different processor components. That is, the integration rules chosen for each sub-interval can be the same, or the sub-intervals can have different rules of the same order, or many other possibilities.
- For each choice of how the rules are to be chosen, there are at least five possibilities for each choice of rule. For example, using only three nodes per interval, there are the following methods: Simpson's, 3-point

Gauss, 3-point Tschebyscheff–Gauss, 3-point Tschebyscheff, and open Newton–Cotes.

- There are at least three ways in which to determine a bound on the error.
- There are many types of integrands that the routines could identify and handle specially. For instance, power-law singularities or discontinuous integrands might be identified. Note, though, that sometimes singularities at the endpoints of the region of integration can be ignored and the computation will still converge (see, for example, Myerson [10]).
- There are at least six different data structures for maintaining the list of intervals used at a given stage in an adaptive algorithm, including ordered lists, stacks, queues, and "boxes." For each data structure, there are several significant variations.

Rice [12] arrives at the pessimistic conclusion that there are potentially 1 million research papers to be written, each with a novel algorithm that, for some test cases, is superior to the other 999,999 algorithms.

Example

Consider numerically evaluating the integral $I = \int_0^1 4x^3\,dx = 1$ using adaptive quadrature. We must choose an integration rule and a measure of the error. Using the trapezoidal rule in the form

$$\int_a^b f(x)\,dx = \frac{b-a}{2}\,(f(a)+f(b)) + E$$

we find the error estimate $E = -\dfrac{(b-a)^3}{12} f^{(2)}(\zeta)$ for some $\zeta \in [a,b]$. In this example $f^{(2)}(x) = 24x$ so we can bound the error by $|E| \le 2(b-a)^3 b$. An adaptive numerical computation can then proceed as follows:

[1] We start with one interval equal to the entire range of integration:

interval	$\{[0,1]\}$
integral estimates	$\{2\}$
error estimates	$\{2\}$
$I \approx 2$	total error ≤ 2.

[2] We choose to subdivide the interval with the largest estimated error. Subdividing $[0,1]$ we find:

intervals	$\{[0,\frac{1}{2}],[\frac{1}{2},1]\}$
integral estimates	$\{0.125, 1.125\}$
error estimates	$\{0.125, 0.250\}$
$I \approx 1.25$	total error ≤ 0.375.

[3] Now the largest error is in the interval $[\frac{1}{2},1]$, so we subdivide it to find:

intervals	$\{[0,\frac{1}{2}],[\frac{1}{2},\frac{3}{4}],[\frac{3}{4},1]\}$
integral estimates	$\{0.125, 0.2734, 0.7109\}$
error estimates	$\{0.125, 0.0234, 0.0313\}$
$I \approx 1.1093$	total error ≤ 0.1797.

[4] Now the largest error is in the interval $[0, \frac{1}{2}]$, so we subdivide it to find:

intervals	$\{[0, \frac{1}{4}], [\frac{1}{4}, \frac{1}{2}], [\frac{1}{2}, \frac{3}{4}], [\frac{3}{4}, 1]\}$
integral estimates	$\{0.0078, 0.0703, 0.2734, 0.7109\}$
error estimates	$\{0.0078, 0.0156, 0.0234, 0.0313\}$
$I \approx 1.0624$	total error ≤ 0.0781.

[5] Now the largest error is in the interval $[\frac{3}{4}, 1]$, so we subdivide it to find:

intervals	$\{[0, \frac{1}{4}], [\frac{1}{4}, \frac{1}{2}], [\frac{1}{2}, \frac{3}{4}], [\frac{3}{4}, \frac{7}{8}], [\frac{7}{8}, 1]\}$
integral estimates	$\{0.0078, 0.0703, 0.2734, 0.2729, 0.4175\}$
error estimates	$\{0.0078, 0.0156, 0.0234, 0.0034, 0.0039\}$
$I \approx 1.0419$	total error ≤ 0.0541.

At this point we can conclude that I lies in the range $[0.99, 1.10]$.

Notes

[1] The subdivision procedure used in most adaptive quadrature codes is a simple bisection of the chosen interval. Berntsen *et al.* [2] present an algorithm in which a subdivision strategy results in three non-equally sized sub-intervals.

[2] When an adaptive algorithm is used, the nodes at which the integrand is evaluated cannot be determined beforehand. Therefore, adaptive techniques are inappropriate for tabulated integrands.

[3] Most of the routines in Quadpack (see Piessens *at al.* [11]) are based on adaptive algorithms.

[4] Corliss and Rall [4], who combine interval analysis methods (see page 218) with adaptive integration, estimate that the increase in CPU time for modest accuracy requests is about a factor of 3–5, and for stringent accuracy requests the factor is about 3–15. However, with a stringent accuracy request, the width of the final interval is only a few units in the last place (ULP).

References

[1] J. Berntsen, "Practical Error Estimation in Adaptive Multidimensional Quadrature Routines," *J. Comput. Appl. Math.*, **25**, No. 3, 1989, pages 327–340.

[2] J. Berntsen, T. O. Espelid, and T. Sørevik, "On the Subdivision Strategy in Adaptive Quadrature Algorithms," *J. Comput. Appl. Math.*, **35**, 1991, pages 119–132.

[3] J. Berntsen, T. O. Espelid, and A. Genz, "An Adaptive Algorithm for the Approximate Calculation of Multiple Integrals," *ACM Trans. Math. Software*, **17**, No. 4, December 1991, pages 437–451.

[4] G. F. Corliss and L. B. Rall, "Adaptive Self-Validating Numerical Quadrature," *SIAM J. Sci. Stat. Comput.*, **8**, No. 5, 1987, pages 831–847.

[5] M. C. Eiermann, "Automatic, Guaranteed Integration of Analytic Functions," *BIT*, **29**, 1989, pages 270–282.

[6] P. J. Davis and P. Rabinowitz, *Methods of Numerical Integration*, Second Edition, Academic Press, Orlando, Florida, 1984, 418–434.

[7] T. O. Espelid and T. Sorevik, "A Discussion of a New Error Estimate for Adaptive Quadrature," *BIT*, **29**, No. 2, 1989, pages 283–294.

[8] D. K. Kahaner and O. W. Rechard, "TWODQD an Adaptive Routine for Two-Dimensional Integration," *J. Comput. Appl. Math.*, **17** No. 1–2, 1987, pages 215–234.
[9] W. M. McKeeman, "Algorithm 145, Adaptive Numerical Integration by Simpson's Rule," *Comm. ACM*, **5**, No. 12, December 1962, page 604.
[10] G. Myerson, "On Ignoring the Singularity," *SIAM J. Numer. Anal.*, **28**, No. 6, December 1991, pages 1803–1807.
[11] R. Piessens, E. de Doncker–Kapenga, C. W. Überhuber, and D. K. Kahaner, *Quadpack*, Springer–Verlag, New York, 1983.
[12] J. R. Rice, "A Metalgorithm for Adaptive Quadrature," *J. ACM*, **22**, No. 1, January 1973, pages 61–82.
[13] H. D. Shapiro, "Increasing Robustness in Global Adaptive Quadrature through Interval Selection Heuristics," *ACM Trans. Math. Software*, **10**, No. 2, 1984, pages 117–139.

63. Clenshaw–Curtis Rules

Applicable to One-dimensional definite integrals.

Yields

A numerical quadrature scheme.

Idea

A function can be approximated by a (finite) linear combination of basis functions. If the integrated value of the basis functions is known, then an approximation of the integral is obtained.

Procedure

Let $\{\phi_n(x)\}$ represent a set of functions for which the integrals $s_n := \int_a^b \phi_n(x)\,dx$ are known. To approximate the value of $I = \int_a^b f(x)\,dx$, we approximate the integrand by

$$f(x) \approx a_1\phi_1(x) + a_2\phi_2(x) + \ldots + a_N\phi_N(x).$$

Knowing the $\{a_k\}$ values, we find that $I \approx a_1s_1 + a_2s_2 + \ldots + a_Ns_N$.

For the Clenshaw–Curtis rules, we take $\{a = -1,\, b = 1\}$ and use the Tschebyscheff polynomials $\phi(x) = T_n(x) = \cos\left(n\cos^{-1}x\right)$. The first few such polynomials are $T_1(x) = 1$, $T_2(x) = x$, and $T_3(x) = 2x^2 - 1$. In this case we can write the $\{a_k\}$ analytically as

$$a_k = \frac{2}{\pi}\int_0^{\pi} f(\cos\theta)\cos(k\theta)\,d\theta. \tag{63.1}$$

To determine the value of a_k numerically, as defined by (63.1), a technique may be used that is exact for trigonometric polynomials (see page 322). (That is, if f is a polynomial, then the exact value of a_k will be returned.)

Example

Consider the numerical evaluation of the integral $I = 4/\pi \int_0^1 \sqrt{1-x^2}\,dx$. (The exact value is $I = 1$.) For the Tschebyscheff polynomials we can compute

$$\begin{aligned} s_1 &= \int_0^1 T_1(x)\,dx = \int_0^1 1\,dx = 1, \\ s_2 &= \int_0^1 T_2(x)\,dx = \int_0^1 x\,dx = \frac{1}{2}, \\ s_2 &= \int_0^1 T_3(x)\,dx = \int_0^1 (2x^2-1)\,dx = -\frac{1}{3}, \\ s_3 &= \int_0^1 T_4(x)\,dx = \int_0^1 (4x^3-3x)\,dx = -\frac{1}{2}, \\ &\vdots \end{aligned}$$

For our specific integrand, we can evaluate the integral in (63.1) to obtain

$$\begin{aligned} a_k &= \frac{2}{\pi}\int_0^\pi \left(\frac{4}{\pi}\sqrt{1-\cos^2\theta}\right)\cos(k\theta)\,d\theta \\ &= \frac{8}{\pi^2}\int_0^\pi \sin\theta\cos(k\theta)\,d\theta \\ &= \begin{cases} 0 & \text{if } k \text{ is odd,} \\ \dfrac{16}{\pi^2(1-k^2)} & \text{if } k \text{ is even.} \end{cases} \end{aligned}$$

Therefore, we can approximate I by the series

$$I \approx 1 \cdot 0 + \frac{1}{2}\cdot\frac{16}{\pi^2} - \frac{1}{3}\cdot 0 + \frac{1}{2}\cdot\frac{16}{3\pi^2} + \cdots.$$

The partial sum of this series after 2 terms is $I \approx \frac{8}{\pi^2} \approx 0.81$. After 4 terms we obtain $I \approx \frac{32}{3\pi^2} \approx 1.08$.

Note

[1] This method can be modified to account for integrals with weight functions; see Piessens *at al.* [4].

References

[1] C. W. Clenshaw and A. R. Curtis, "A Method for Numerical Integration on an Automatic Computer," *Numer. Math.*, **2**, 1960, pages 197–205.

[2] P. J. Davis and P. Rabinowitz, *Methods of Numerical Integration*, Second Edition, Academic Press, Orlando, Florida, 1984, pages 86–87, 193–196, and 446–449.

[3] R. B. Kearfott, "A Sinc Approximation for the Indefinite Integral," *Math. of Comp.*, **41**, 1983, pages 559–572.

[4] R. Piessens, E. de Doncker–Kapenga, C. W. Überhuber, and D. K. Kahaner, *Quadpack*, Springer–Verlag, New York, 1983, Section 2.2.3, pages 28–39.

64. Compound Rules

Applicable to Definite integrals in any number of dimensions.

Yields

A numerical quadrature scheme.

Idea

If a numerical quadrature scheme is known for integration over a small parallelpiped region, then the rule may be repeatedly applied to find a rule applicable over a larger parallelpiped region.

Procedure

Suppose we seek to approximate an integral numerically over a large region. We may subdivide the region into smaller regions (often called "panels"), and then apply a quadrature scheme in each of the smaller regions. The resulting quadrature rule, in the larger region, is called a compound rule.

Example

Simpson's rule, in its most elementary form, approximates a one dimensional integral using three nodes via

$$\int_a^b f(x)\,dx \simeq \frac{b-a}{6}\left[f(a) + 4f\left(\frac{a+b}{2}\right) + f(b)\right]. \tag{64.1}$$

To apply Simpson's rule on a large interval, say from c to d, the larger region can be subdivided into smaller regions and Simpson's rule applied in each of the smaller regions.

If the interval $[c,d]$ is subdivided into two equally spaced intervals, $[c,e]$ and $[e,d]$ (with $e=(d+c)/2$), then

$$\begin{aligned}
I &= \int_c^d f(x)\,dx \\
&= \int_c^e f(x)\,dx + \int_e^d f(x)\,dx, \\
&\simeq \frac{e-c}{6}\left[f(c) + 4f\left(\frac{c+e}{2}\right) + f(e)\right] + \frac{d-e}{6}\left[f(e) + 4f\left(\frac{e+d}{2}\right) + f(d)\right], \\
&= \frac{d-c}{12}\left[f(c) + 4f\left(c+\frac{d-c}{4}\right) + 2f\left(c+2\frac{d-c}{4}\right) + 4f\left(c+3\frac{d-c}{4}\right) + f(d)\right], \\
&= \frac{h}{3}\left[f_0 + 4f_1 + 2f_2 + 4f_3 + f_4\right],
\end{aligned} \tag{64.2}$$

where $f_n = f(c+nh)$ and $h=(e-c)/2=(d-c)/4$.

When manipulating integration rules, as above, it is often easier to just use subscripted variables. For example, the above derivation could be written as

$$\begin{aligned} I &= \int_c^d f(x)\,dx \\ &= \int_c^e f(x)\,dx + \int_e^d f(x)\,dx, \\ &= \frac{h}{3}\left(f_0 + 4f_1 + f_2\right) + \frac{h}{3}\left(f_2 + 4f_3 + f_4\right) \\ &= \frac{h}{3}\left[f_0 + 4f_1 + 2f_2 + 4f_3 + f_4\right]. \end{aligned}$$

If the original interval $[c, d]$ had been divided into three equally sized intervals, instead of two, then the compound rule obtained would have been

$$\begin{aligned} I &= \int_c^d f(x)\,dx \\ &\simeq \frac{k}{3}\left(f_0 + 4f_1 + f_2\right) + \frac{k}{3}\left(f_2 + 4f_3 + f_4\right) + \frac{k}{3}\left(f_4 + 4f_5 + f_5\right) \\ &= \frac{k}{3}\left[f_0 + 4f_1 + 2f_2 + 4f_3 + 2f_4 + 4f_5 + f_6\right] \end{aligned} \tag{64.3}$$

where $f_m = f(c+mk)$ and $k = (d-c)/6$. The extended form of Simpson's rule, that is, the rule applied to n equal-width intervals, is:

$$\begin{aligned} \int_c^d f(x)\,dx \simeq \frac{k}{3}\Big[& f_0 + 4\left(f_1 + f_3 + \cdots + f_{2n-1}\right) \\ & + 2\left(f_2 + f_4 + \cdots + f_{2n-2}\right) + f_{2n}\Big] \end{aligned} \tag{64.4}$$

where $f_m = f(c+mk)$ and $k = (d-c)/(2n)$.

Notes

[1] A compound rule is sometimes called a *composite rule.*

[2] Compound rules also exist for simplices. See, for example, Lyness [3] or De Donker [2].

References

[1] P. J. Davis and P. Rabinowitz, *Methods of Numerical Integration*, Second Edition, Academic Press, Orlando, Florida, 1984, pages 379–384.

[2] E. De Doncker, "New Euler–Maclaurin Expansions and Their Application to Quadrature Over the s-Dimensional Simplex," *Math. of Comp.*, **33**, 1978, pages 1003–1018.

[3] J. M. Lyness, "Quadrature over a Simplex: Part 1. A Representation for the Integrand Function," and "Quadrature over a Simplex: Part 2. A Representation for the Error Functional," *SIAM J. Numer. Anal.*, **15**, 1978, pages 122–133 and 870–887.

[4] J. N. Lyness and A. C. Genz, "On Simplex Trapezoidal Rule Families," *SIAM J. Numer. Anal.*, **17**, 1980, pages 126–147.

[5] A. Ralston, "A Family of Quadrature Formulas which Achieve High Accuracy in Composite Rules," *J. ACM*, **6**, 1959, pages 384–394.

65. Cubic Splines

Applicable to One-dimensional definite integrals.

Yields

A numerical quadrature scheme.

Idea

Given data values of a function (not necessarily equally spaced), a cubic spline can be fit to those values, and then the integral of the cubic spline can be determined.

Procedure

Let $[a, b]$ be a finite interval and assume that we have the points $\{x_i\}$ with $a \leq x_1 < x_2 < \cdots < x_{n+1} \leq b$ and $n \geq 2$. Given the data values $\{f(x_i)\}$, let $S(x)$ be the natural cubic spline which interpolates those data values. That is,

- $S(x)$ is a cubic polynomial in every interval $[x_i, x_{i+1}]$ (say this polynomial is $S_i(x)$).
- $S(x)$ matches the data values at the nodes $S_i(x_i) = f(x_i)$ for $i = 1, \dots, n$, and $S_n(x_{n+1}) = f(x_{n+1})$.
- At the nodes, S and its first and second derivatives are continuous: $S_{i-1}(x_i) = S_i(x_i)$, $S'_{i-1}(x_i) = S'_i(x_i)$, and $S''_{i-1}(x_i) = S''_i(x_i)$ for $i = 2, \dots, n$.
- There is no curvature as the ends of the spline: $S''(a) = S''(b) = 0$ (this is what makes the spline "natural").

(In this definition we have implicitly assumed that $a = x_1$ and $b = x_{n+1}$, if this is not the case, then the range for i changes.) See Figure 65.

$$\begin{array}{ccccccc} & S_1 & S_2 & \cdots & S_{n-1} & S_n & \\ x_1 & x_2 & x_3 & & x_{n-1} & x_n & x_{n+1} \end{array}$$

Figure 65. Location of the nodes $\{x_i\}$ and the cubic polynomials $\{S_i\}$.

We can now approximate the integral of f by the integral of S and introduce an error E in doing so:

$$\int_a^b f(x)\,dx = \int_a^b S(x)\,dx + E. \tag{65.1}$$

Example

Consider approximating the integral $I = \int_0^1 \sin \pi x\,dx$. We choose to use the equally-spaced points $\{x_i\} = \{0, \frac{1}{4}, \frac{1}{2}, \frac{3}{4}, 1\}$. Hence, we have the data values $\{(0,0), (\frac{1}{4}, \frac{\sqrt{2}}{2}), (\frac{1}{2}, 1), (\frac{3}{4}, \frac{\sqrt{2}}{2}), (1,0)\}$.

We choose to represent the cubic on the interval $[x_i, x_{i+1}]$ in the form $S_i(x) = a_i(x-x_i)^3 + b_i(x-x_i)^2 + c_i(x-x_i) + d_i$. Using this notation the first few equations for the unknowns $\{a_i, b_i, c_i, d_i \mid i = 1, \ldots, 4\}$ are

$$\begin{array}{lcl}
S_1(0) = f(0) = 0 & \Longrightarrow & d_1 = 0 \\
S_1''(0) = 0 & \Longrightarrow & b_1 = 0 \\
S_1\left(\frac{1}{4}\right) = f\left(\frac{1}{4}\right) = \frac{\sqrt{2}}{2} & \Longrightarrow & \frac{1}{64}a_1 + \frac{1}{16}b_1 + \frac{1}{4}c_1 + d_1 = \frac{\sqrt{2}}{2} \\
S_1\left(\frac{1}{4}\right) = S_2\left(\frac{1}{4}\right) & \Longrightarrow & \frac{1}{64}a_1 + \frac{1}{16}b_1 + \frac{1}{4}c_1 + d_1 = d_2 \\
S_1'\left(\frac{1}{4}\right) = S_2'\left(\frac{1}{4}\right) & \Longrightarrow & \frac{3}{16}a_1 + \frac{1}{2}b_1 + c_1 = c_2 \\
S_1''\left(\frac{1}{4}\right) = S_2''\left(\frac{1}{4}\right) & \Longrightarrow & \frac{3}{2}a_1 + b_1 = 2b_2 \\
S_2\left(\frac{1}{2}\right) = S_3\left(\frac{1}{2}\right) & \Longrightarrow & \ldots \\
\ldots. & \Longrightarrow & \ldots.
\end{array}$$

Completing this list of equations, and solving the resulting linear system, results in the approximation

$$\begin{array}{ll}
\text{interval } \left[0, \frac{1}{4}\right]: & -4.8960(x-0)^3 + 3.1340(x-0) \\
\text{interval } \left[\frac{1}{4}, \frac{1}{2}\right]: & -2.0288(x-\frac{1}{4})^3 - 3.6720(x-\frac{1}{4})^2 + 2.2164(x-\frac{1}{4}) + 0.7071 \\
\text{interval } \left[\frac{1}{2}, \frac{3}{4}\right]: & \;\;2.0288(x-\frac{1}{2})^3 - 5.1936(x-\frac{1}{2})^2 + 1 \\
\text{interval } \left[\frac{3}{4}, 1\right]: & \;\;4.8960(x-\frac{3}{4})^3 - 3.6720(x-\frac{3}{4})^2 - 2.2164(x-\frac{3}{4}) + 0.7071.
\end{array} \tag{65.2}$$

Now we can determine the approximation to the integral:

$$\int_{x_1}^{x_{n+1}} f(x)\,dx \approx \frac{h^4}{4}\sum_{i=1}^{n} a_i + \frac{h^3}{3}\sum_{i=1}^{n} b_i + \frac{h^2}{2}\sum_{i=1}^{n} c_i + h\sum_{i=1}^{n} d_i. \tag{65.3}$$

Evaluating (65.3), when the coefficients are given in (65.2), results in the approximation $I \approx 0.6362$. (Note that the exact value is $I = \frac{2}{\pi} \approx 0.6366$.)

Notes

[1] Among all functions S that are of class $C^2[a,b]$ which interpolate the data values, it is the cubic spline approximation that minimizes the "total curvature": $C = \int_a^b |S''(x)|^2 \, dx$.

[2] Cubic splines may be numerically computed by the software in Press *et al.* [5].

[3] Use of cubic splines does not result in a conventional quadrature rule. If the numerical approximation is written in the form $\int_a^b f(x)\,dx \approx \sum_i w_i f(x_i)$, then the weights $\{w_i\}$ depend on all of the function values $\{f(x_i)\}$.

References

[1] G. H. Behforooz and N. Papamichael, "End Conditions for Interpolatory Cubic Splines with Unequally Spaced Knots," *J. Comput. Appl. Math.*, **6**, 1980, pages 59–65.

[2] P. J. Davis and P. Rabinowitz, *Methods of Numerical Integration*, Second Edition, Academic Press, Orlando, Florida, 1984, 62–70.

[3] P. Dierckx, "Algorithm 003. An Algorithm for Smoothing, Differentiation and Integration of Experimental Data Using Spline Functions," *J. Comput. Appl. Math.*, **1**, 1975, pages 165–184.

[4] D. Kahaner, C. Moler, and S. Nash, *Numerical Methods and Software*, Prentice–Hall Inc., Englewood Cliffs, NJ, 1989.

[5] W. H. Press, B. P. Flannery, S. Teukolsky, and W. T. Vetterling, *Numerical Recipes*, Cambridge University Press, New York, 1986.

[6] P. Rabinowitz, "Numerical Integration Based on Approximating Splines," *J. Comput. Appl. Math.*, **33**,No. 1, 1990, pages 73–83.

[7] M. N. El Tarazi and S. Sallam, "On Quartic Splines with Application to Quadratures," *Computing*, **38**, 1987, pages 355–361.

66. Using Derivative Information

Applicable to Definite integrals.

Yields

A numerical quadrature scheme.

Idea

A quadrature rule can use the value of the integrand at the nodes, and it can also use the value of the derivative of the integrand at the nodes.

Procedure

A quadrature rule in the form

$$\int_a^b f(x)\,dx \simeq \sum_{j=1}^{N} w_j f(x_j) + \sum_{k=1}^{M} v_j f'(x_k)$$

can often be determined by making the rule exact for the polynomials $\{1, x, x^2, \ldots\}$.

Example

Consider an integration rule of the form

$$\int_a^b f(x)\,dx \approx \alpha f(0) + \beta f(1) + \gamma f'(0) + \delta f'(1).$$

Making this rule be exact for $f(x) = \{1, x, x^2, x^3\}$, we obtain the following set of simultaneous equations:

$$\begin{array}{lcl} f(x) = 1 & \Longrightarrow & 1 = \alpha + \beta \\ f(x) = x & \Longrightarrow & \frac{1}{2} = \beta + \gamma + \delta \\ f(x) = x^2 & \Longrightarrow & \frac{1}{3} = \beta + 2\delta \\ f(x) = x^3 & \Longrightarrow & \frac{1}{4} = \beta + 3\delta. \end{array}$$

These equations can be solved to obtain the approximation:

$$\int_0^1 f(x)\,dx \simeq \tfrac{1}{2}\left[f(0) + f(1)\right] + \tfrac{1}{12}\left[f'(0) - f'(1)\right].$$

This rule is known as the corrected trapezoidal rule.

Notes

[1] Some of the common quadrature rules can be improved by including derivative terms. The corrected trapezoidal rule is given in the example. The corrected midterm rule is

$$\int_0^1 f(x)\,dx \simeq f(\tfrac{1}{2}) + \tfrac{1}{24}\left[f'(1) - f'(0)\right].$$

[2] When either the corrected midterm rule or the corrected trapezoidal rule is compounded (see page 282), the derivative at the nodes in common cancel, so that only the derivatives at the end points remain. For example, the compounded corrected trapezoidal rule is

$$\int_a^b f(x)\,dx = \frac{h}{2}\left[f_0 + 2f_1 + 2f_2 + \ldots + 2f_{n-1} + f_n\right] + \frac{h^2}{12}\left[f'(a) - f'(b)\right] + E$$

where $f_i = f(a + ih)$ and $h = (b - a)/n$. It can be shown that the error is bounded by $|E| \leq \dfrac{1}{720} h^4 (b - a) \max_{a \leq x \leq b} \left|f^{(4)}(x)\right|$ (Davis and Rabinowitz [2], page 132).

[3] The trapezoidal rule, corrected by using both f' and f'' terms, takes the form

$$\int_a^b f(x)\,dx = \frac{h}{2}\left[f(a) + f(b)\right] + \frac{h^2}{10}\left[f'(a) - f'(b)\right] + \frac{h^3}{120}\left[f''(a) + f''(b)\right] + E$$

where $E = -\dfrac{h^7}{100800} f^{(6)}(\zeta)$ and $a < \zeta < b$ (Davis and Rabinowitz [2], page 133).

References

[1] R. A. Cicenia, "Numerical Integration Formulae Involving Derivatives," *J. Inst. Math. Appls.*, **24**, 1979, 347–352.

[2] P. J. Davis and P. Rabinowitz, *Methods of Numerical Integration*, Second Edition, Academic Press, Orlando, Florida, 1984, pages 132–134.

[3] J. D. Lambert and A. R. Mitchell, "The Use of Higher Derivatives in Quadrature Formulae," *Comput. J.*, **5**, 1962–1963, pages 322–327.

67. Gaussian Quadrature

Applicable to One-dimensional definite integrals.

Yields

Integration rules on a finite interval using non-uniformly spaced nodes.

Idea

A general expression for approximating an integral is proposed. The unknown constants in this expression are determined by making the quadrature rule exact for polynomials of low degree.

Procedure

Given the integral $I = \int_a^b g(x)\,dx$ we first map it to the interval $[-1, 1]$; this will minimize the algebra later. We have

$$I = \int_a^b g(x)\,dx = \frac{b-a}{2}\int_{-1}^{1} g\left(\frac{(b-a)t + (b+a)}{2}\right)\,dt = \int_{-1}^{1} f(t)\,dt,$$

so that we can now focus on $I = \int_{-1}^{1} f(t)\,dt$. Now we search for an approximate integration rule of the form

$$I \approx \widehat{I} = \sum_{i=1}^{n} \alpha_i f(t_i). \tag{67.1}$$

Since this integration rule has $2n$ unknown constants (the weights $\{\alpha_i\}$ and the nodes $\{t_i\}$), we can hope to choose these constants so that the integration rule is exact for all polynomials of degree less than or equal to $2n-1$. It turns out this is always possible.

For example, with $n = 2$ we have

$$\int_{-1}^{1} f(t)\,dt \approx \alpha_1 f(t_1) + \alpha_2 f(t_2).$$

If this formula is to be exact for polynomials of degree less than or equal to 3, then $\{\alpha_1, \alpha_2\}$ and $\{t_1, t_2\}$ must satisfy the simultaneous algebraic equations:

Table 67. Values used in Gaussian quadrature formulas.

Number of terms	Nodes $\{t_i\}$	Weights $\{\alpha_i\}$	Valid for polynomials up to degree
2	−0.57735027	1	3
	0.57735027	1	
3	−0.77459667	0.55555855	5
	0	0.88888889	
	0.77459667	0.55555555	
4	−0.86113631	0.34785485	7
	−0.33998104	0.65214515	
	0.33998104	0.65214515	
	0.86113631	0.34785485	

$$\begin{aligned}
f(t) &= 1 & &\Longrightarrow & \int_{-1}^{1} 1\,dt &= 2 = \alpha_1 + \alpha_2 \\
f(t) &= t & &\Longrightarrow & \int_{-1}^{1} t\,dt &= 0 = \alpha_1 t_1 + \alpha_2 t_2 \\
f(t) &= t^2 & &\Longrightarrow & \int_{-1}^{1} t^2\,dt &= \frac{2}{3} = \alpha_1 t_1^2 + \alpha_2 t_2^2 \\
f(t) &= t^3 & &\Longrightarrow & \int_{-1}^{1} t^3\,dt &= 0 = \alpha_1 t_1^3 + \alpha_2 t_2^3.
\end{aligned}$$

These equations have the unique solution $\{\alpha_1 = \alpha_2 = 1,\ t_1 = -t_2 = 1/\sqrt{3} \approx 0.5773\}$. Hence, we have the approximation

$$\int_{-1}^{1} f(t)\,dt \approx f(-0.5773) + f(0.5773).$$

To obtain Gaussian quadrature formulas for larger values of n, we must find the solutions to a large set of nonlinear algebraic equations. The results of such a calculation are shown in Table 67.

Example

Consider the integral $J = \dfrac{1}{2\sin 1}\displaystyle\int_{-1}^{1} \cos x\,dx = 1$. We represent the numerical approximation obtained by using Gaussian quadrature with n nodes by J_n. Using 2, 3, and 4 nodes we obtain the approximations

$$\begin{aligned}
J &\approx J_2 = 0.9957702972 \\
J &\approx J_3 = 1.0000365711 \\
J &\approx J_4 = 0.9999998148.
\end{aligned}$$

Notes

[1] The values of the $\{t_i\}$ in (67.1) turn out to be roots of the Legendre polynomial $P_n(x)$. These polynomials are defined by the recurrence relation

$$(n+1)P_{n+1}(x) = (2n+1)xP_n(x) - nP_{n-1}(x)$$

with the initial conditions: $P_0(x) = 1$ and $P_1(x) = x$. Then, for example, we can calculate $P_2(x) = \frac{3}{2}x^2 - \frac{1}{2}$. The roots of $P_2(x)$ are at $x = \pm 1/\sqrt{3} \approx \pm 0.5773$. The next Legendre polynomial is $P_3(x) = \frac{5}{2}x^3 - \frac{3}{2}x$; its roots are at $x = 0$ and $x = \pm\sqrt{\frac{3}{5}} \approx \pm 0.7746$.

[2] The values of the weights $\{\alpha_i\}$ in (67.1) are also functions of Legendre polynomials. If x_i is the i-th root of $P_n(x)$, then the corresponding weight, α_i, is given by $\alpha_i = \dfrac{2}{\left(1-x_i^2\right)\left(P_n'(x_i)\right)^2}$. For example, when $n = 2$ we find $\alpha_i = \dfrac{2}{\left(1-x_i^2\right)(3x_i)^2}$. When $x_i = \pm 1/\sqrt{3}$ this results in $\alpha_i = 1$.

[3] See the section on generalized Gaussian quadrature (page 291) for the analogous technique applied to integrals of the form $\int_a^b w(x)f(x)\,dx$, when $w(x)$ is a positive weighting function.

[4] Newton–Cotes rules (see page 319) are also interpolatory, but the nodes are chosen to be equidistant from one another.

[5] Several modifications of the Gaussian principle have been developed, in which some of the nodes or weights, or both, are specified in advance. The *Radeau formulas* use one of the endpoints, the *Lobatto formulas* use both of the endpoints.

The simplest Radeau formula has the form $\int_0^1 f(x)\,dx \approx w_1 f(x_1) + w_2 f(0)$, for some unknown $\{x_1, w_1, w_2\}$. Making this quadrature rule exact for $f(x) = x^k$ (for $k = 0, 1, 2$) results in the quadrature rule $\int_0^1 f(x)\,dx \approx \frac{3}{4} f(\frac{2}{3}) + \frac{1}{4} f(0)$.

[6] The Tschebyscheff weight function $w(x) = (1-x^2)^{-1/2}$ is the only weight function (up to a linear transformation) for which all the weights in a n-point Gauss quadrature formula are equal. See Peherstorfer [5] and the section on Tschebyscheff rules (page 331).

References

[1] M. Abramowitz and I. A. Stegun, *Handbook of Mathematical Functions*, National Bureau of Standards, Washington, DC, 1964, Table 25.4, page 919.

[2] W. Cheney and D. Kincaid, *Numerical Mathematics and Computing*, Second Edition, Brooks/Cole Pub. Co., Monterey, CA, pages 193–197.

[3] P. J. Davis and P. Rabinowitz, *Methods of Numerical Integration*, Second Edition, Academic Press, Orlando, Florida, 1984, Section 2.7, pages 95–132.

[4] W. Gautschi, E. Tychopoulos, and R. S. Varga, "A Note on the Contour Integral Representation of the Remainder Term for a Gauss–Chebyshev Quadrature Rule," *SIAM J. Numer. Anal.*, **27**, No. 1, 1990, pages 219–224.

[5] F. Peherstorfer, "Gauss–Tschebyscheff Quadrature Formulas," *Numer. Math.*, **58**, 1990, pages 273–286.

68. Gaussian Quadrature: Generalized

Applicable to Integrals that have a positive weighting function.

Yields

Integration rules using non-uniformly spaced nodes.

Idea

A general expression for approximating an integral is suggested. The unknown constants in this expression are determined by making the quadrature rule exact for polynomials of low degree.

Procedure

Consider the integral

$$I[f] = \int_a^b w(x) f(x)\, dx$$

where $w(x)$ is a specified positive function. To numerically approximate $I[f]$ for different functions $f(x)$, we choose to use a quadrature rule of the form

$$I[f] \approx \sum_{i=1}^{n} w_i f(x_i) \tag{68.1}$$

where the weights $\{w_i\}$ and the nodes $\{x_i\}$ are to be determined. These values are determined by requiring that (68.1) be exact when $f(x)$ is a polynomial of low degree.

In the section on Gaussian quadrature (page 289) the same formulation was used when $w(x) = 1$ and a and b were finite. In that section it was shown that the nodes $\{x_i\}$ and the weights $\{w_i\}$ are related to the roots of Legendre polynomials. In the more general case considered here, it can be shown that the weights $\{w_i\}$ satisfy a polynomial specified by the weight function $w(x)$.

In particular, if the inner product of two functions f and g is defined by $(f, g) := \int_a^b w(x) f(x) g(x)\, dx$, and j is any positive integer, then there exist j polynomials $\{p_k \mid k = 1, 2, \ldots, j\}$ with the k-th polynomial being of degree k, such that $(p_{k_1}, p_{k_2}) = 0$ when $k_1 \neq k_2$. Such a set of polynomials can be constructed via

$$\begin{aligned} p_0(x) &= 1, \\ p_{i+1}(x) &= (x - \delta_{i+1}) p_i(x) - \gamma_{i+1}^2 p_{i-1}(x) \quad \text{for } i \geq 0, \end{aligned} \tag{68.2}$$

where $p_{-1}(x) = 0$ and

$$\delta_{i+1} = \frac{(xp_i, p_i)}{(p_i, p_i)} \quad \text{for } i \geq 0,$$
$$\gamma_{i+1}^2 = \begin{cases} 0 & \text{for } i = 0, \\ \dfrac{(p_i, p_i)}{(p_{i-1}, p_{i-1})} & \text{for } i \geq 1. \end{cases} \tag{68.3}$$

We can now state what the values of $\{w_i\}$ and $\{x_i\}$ are. The $\{x_i \mid i = 1, 2, \ldots, n\}$ are the roots of the polynomial $p_n(x)$ and the $\{w_i \mid i = 1, 2, \ldots, n\}$ are the unique solution of the (nonsingular) linear system of n simultaneous equations

$$\sum_{i=1}^{n} p_k(x_i) w_i = \begin{cases} (p_0, p_0) & \text{for } k = 0, \\ 0 & \text{for } k = 1, 2, \ldots, n-1. \end{cases}$$

The nodes and weights appearing in (68.1) may also be found by finding the eigenstructure of a specific matrix. If the tri-diagonal matrix J_n is defined by

$$J_n = \begin{pmatrix} \delta_1 & \gamma_2 & & & 0 \\ \gamma_2 & \delta_2 & \ddots & & \\ & \ddots & \ddots & \ddots & \\ & & \ddots & \delta_{n-1} & \gamma_n \\ 0 & & & \gamma_n & \delta_n \end{pmatrix}$$

then the eigenvalues of J_n will be the nodes in the quadrature formula, $\{x_i\}$. Corresponding to each of these eigenvalues is an associated eigenvector, $\mathbf{v}_i = (v_i^{(1)}, v_i^{(2)}, \ldots, v_i^{(n)})^{\mathrm{T}}$ (so that $J_n \mathbf{v}_i = x_i \mathbf{v}_i$). The weights in the quadrature formula are then given by $w_i = \left(v_i^{(1)}\right)^2$. See Stoer and Bulirsch [3] for details.

The roots, $\{x_i \mid i = 1, 2, \ldots, n\}$, turn out to be real, simple, and to lie in the interval (a, b). The weights satisfy $w_i \geq 0$ and the relation

$$\int_a^b w(x) p(x)\, dx = \sum_{i=1}^{n} w_i p(x_i)$$

holds for all polynomials $p(x)$ of degree $2n - 1$ or less. For some specific weight functions, $w(x)$, and intervals $[a, b]$, the polynomials $\{p_k \mid k = 1, 2, \ldots, n\}$ turn out to be classical polynomials. For example, we find

interval	$w(x)$	Orthogonal polynomials
$[-1,1]$	$(1-x^2)^{-1/2}$	$T_k(x)$, Tschebyscheff polynomials
$[0,\infty]$	e^{-x}	$L_k(x)$, Laguerre polynomials
$[-\infty,\infty]$	e^{-x^2}	$H_k(x)$, Hermite polynomials

In Table 68 may be found a more complete list of orthogonal polynomials. That table also describes the corresponding nodes and weights of each quadrature rule.

The error term in such an approximation is given by the following theorem.

Theorem: If $f \in C^{2n}[a,b]$ then

$$E[f] = \int_a^b w(x)f(x)\,dx - \sum_{i=1}^{n} w_i f(x_i) = \frac{f^{(2n)}(\xi)}{(2n)!}(p_n, p_n)$$

for some $\xi \in (a,b)$.

Table 68. Some specific generalized Gaussian quadrature rules.

68.1 Gauss' formula

Approximation: $\int_{-1}^{1} f(x)\,dx = \sum_{i=1}^{n} w_i f(x_i) + R_n$

Nodes: x_i is the i-th root of $P_n(x)$ (Legendre polynomial)

Weights: $w_i = \dfrac{2}{(1-x_i^2)\left(P_n'(x_i)\right)^2}$

Error term: $R_n = \dfrac{2^{2n+1}(n!)^4}{(2n+1)[(2n)!]^3} f^{(2n)}(\xi)$ with $-1 < \xi < 1$

Reference: Abramowitz and Stegun [1], 25.4.29

68.2 Radeau's integration formula

Approximation: $\int_{-1}^{1} f(x)\,dx = \dfrac{2}{n^2} f(-1) + \sum_{i=1}^{n-1} w_i f(x_i) + R_n$

Nodes: x_i is the i-th root of $\dfrac{P_{n-1}(x) + P_n(x)}{x+1}$

Weights: $w_i = \dfrac{1}{(1-x_i)\left(P_{n-1}'(x_i)\right)^2}$

Error term: $R_n = \dfrac{2^{2n-1} n[(n-1)!]^4}{[(2n-1)!]^3} f^{(2n-1)}(\xi)$ with $-1 < \xi < 1$

Reference: Abramowitz and Stegun [1], 25.4.31

68.3 Lobatto's integration formula

Approximation: $\displaystyle\int_{-1}^{1} f(x)\,dx = \frac{2}{n(n-1)}\left[f(1)+f(-1)\right] + \sum_{i=2}^{n-1} w_i f(x_i) + R_n$

Nodes: x_i is the $(i-1)$st root of $P'_{n-1}(x)$

Weights: $w_i = \dfrac{2}{n(n-1)\left(P_{n-1}(x_i)\right)^2}$

Error term: $R_n = -\dfrac{n(n-1)^3 2^{2n-1}[(n-2)!]^4}{(2n-1)[(2n-2)!]^3} f^{(2n-2)}(\xi)$ with $-1 < \xi < 1$

Reference: Abramowitz and Stegun [1], 25.4.32

68.4 Weight function x^k

Approximation: $\displaystyle\int_{-1}^{1} x^k f(x)\,dx = \sum_{i=1}^{n} w_i f(x_i) + R_n$

Polynomials: $q_n(x) = \sqrt{k+2n+1}\,P_n^{(k,0)}(1-2x)$

Nodes: x_i is the i-th root of $q_n(x)$

Weights: $w_i = \left(\displaystyle\sum_{j=0}^{n-1} q_j^2(x_i)\right)^{-1}$

Error term: $R_n = \dfrac{f^{(2n)}(\xi)}{(k+2n+1)(2n)!}\left[\dfrac{n!(k+n)!}{(k+2n)!}\right]^2$ with $0 < \xi < 1$

Reference: Abramowitz and Stegun [1], 25.4.33

68.5 Weight function $\sqrt{1-x}$

Approximation: $\displaystyle\int_{0}^{1} f(x)\sqrt{1-x}\,dx = \sum_{i=1}^{n} w_i f(x_i) + R_n$

Nodes: $x_i = 1-\xi_i^2$ where ξ_i is the i-th positive root of $P_{2n+1}(x)$

Weights: $w_i = 2\xi_i^2 w_i^{(2n+1)}$ where $\{w_i^{(2n+1)}\}$ are the Gaussian weights of order $2n+1$

Error term: $R_n = \dfrac{2^{4n+3}[(2n+1)!]^4}{(2n)!(4n+3)[(4n+2)!]^2} f^{(2n)}(\xi)$ with $0 < \xi < 1$

Reference: Abramowitz and Stegun [1], 25.4.34

68.6 Weight function $\dfrac{1}{\sqrt{1-x}}$

Approximation: $\displaystyle\int_{0}^{1} \frac{f(x)}{\sqrt{1-x}}\,dx = \sum_{i=1}^{n} w_i f(x_i) + R_n$

Nodes: $x_i = 1-\xi_i^2$ where ξ_i is the i-th positive root of $P_{2n}(x)$

Weights: $w_i = 2w_i^{(2n)}$ where $\{w_i^{(2n)}\}$ are the Gaussian weights of order $2n$

Error term: $R_n = \dfrac{2^{4n+1}}{4n+1}\dfrac{[(2n)!]^3}{[(4n)!]^2} f^{(2n)}(\xi)$ with $0 < \xi < 1$

Reference: Abramowitz and Stegun [1], 25.4.36

68.7 Weight function $\frac{1}{\sqrt{1-x^2}}$

Approximation: $\displaystyle\int_{-1}^{1} \frac{f(x)}{\sqrt{1-x^2}}\,dx = \sum_{i=1}^{n} w_i f(x_i) + R_n$

Nodes: $x_i = \cos\frac{(2i-1)\pi}{2n}$

Weights: $w_i = \frac{\pi}{n}$

Error term: $R_n = \frac{\pi}{(2n)!2^{2n-1}} f^{(2n)}(\xi)$ with $-1 < \xi < 1$

Reference: Abramowitz and Stegun [1], 25.4.38

68.8 Weight function $\sqrt{1-x^2}$

Approximation: $\displaystyle\int_{-1}^{1} f(x)\sqrt{1-x^2}\,dx = \sum_{i=1}^{n} w_i f(x_i) + R_n$

Nodes: $x_i = \cos\frac{(i+1)\pi}{n+1}$

Weights: $w_i = \frac{\pi}{n+1}\sin^2\frac{(i+1)\pi}{n+1}$

Error term: $R_n = \frac{\pi}{(2n)!2^{2n+1}} f^{(2n)}(\xi)$ with $-1 < \xi < 1$

Reference: Abramowitz and Stegun [1], 25.4.40

68.9 Weight function $\frac{2}{1-x}$

Approximation: $\displaystyle\int_{0}^{1} f(x)\sqrt{\frac{x}{1-x}}\,dx = \sum_{i=1}^{n} w_i f(x_i) + R_n$

Nodes: $x_i = \cos^2\left(\frac{2i-1}{2n+1}\frac{\pi}{2}\right)$

Weights: $w_i = \frac{2\pi}{2n+1}x_i$

Error term: $R_n = \frac{\pi}{(2n)!2^{4n+1}} f^{(2n)}(\xi)$ with $0 < \xi < 1$

Reference: Abramowitz and Stegun [1], 25.4.42

68.10 Weight function e^{-x}

Approximation: $\displaystyle\int_{0}^{\infty} e^{-x} f(x)\,dx = \sum_{i=1}^{n} w_i f(x_i) + R_n$

Nodes: x_i is the i-th root of $L_n(x)$ (Laguerre polynomial)

Weights: $w_i = \frac{(n!)^2 x_i}{(n+1)^2 L_{n+1}^2(x_i)}$

Error term: $R_n = \frac{(n!)^2}{(2n)!} f^{(2n)}(\xi)$ with $0 < \xi < \infty$

Reference: Abramowitz and Stegun [1], 25.4.45

68.11 Weight function e^{-x^2}

Approximation: $\int_{-\infty}^{\infty} e^{-x^2} f(x)\,dx = \sum_{i=1}^{n} w_i f(x_i) + R_n$

Nodes: x_i is the i-th root of $H_n(x)$ (Hermite polynomial)

Weights: $w_i = \dfrac{2^{n-1} n! \sqrt{\pi}}{n^2 H_{n-1}^2(x_i)}$

Error term: $R_n = \dfrac{n!\sqrt{\pi}}{2^n (2n)!} f^{(2n)}(\xi)$ with $-\infty < \xi < \infty$

Reference: Abramowitz and Stegun [1], 25.4.46

Notes

[1] The technical conditions required on $w(x)$ for the orthogonal polynomials to exist are (see Stoer and Bulirsch [3]):

- $w(x) \geq 0$ is measurable on the (finite or infinite) interval $[a, b]$;
- All the moments $\int_a^b x^k w(x)\,dx$ for $k = 0, 1, \ldots$ exist and are finite;
- If $s(x)$ is a polynomial which is nonnegative on the interval $[a, b]$, then $\int_a^b w(x)s(x)\,dx = 0$ implies that $s(x)$ is identically zero.

[2] Press and Teukolsky [2] discuss the numerical development of Gaussian quadrature rules when the weight function desired is not one of those in Table 68.

[3] Depending on which polynomials are used to obtain the quadrature rule, the rules obtained by this technique are called Gauss–Hermite rules, Gauss–Jacobi rules, Gauss–Laguerre rules, Gauss–Legendre rules, etc.

[4] Frequently, Gaussian integration rules of successively higher order are tried when approximating an integral. Unfortunately, it is most often the case that information about the function cannot be re-used when using a higher order rule; that is, the roots of the polynomials at each order do not overlap. See the section on Kronrod rules (page 298) for a solution to this problem.

References

[1] M. Abramowitz and I. A. Stegun, *Handbook of Mathematical Functions*, National Bureau of Standards, Washington, DC, 1964.

[2] W. H. Press and S. A. Teukolsky, "Orthogonal Polynomials and Gaussian Quadrature with Nonclassical Weight Functions," *Comp. in Physics*, Jul/Aug 1990, pages 423–426.

[3] J. Stoer and R. Bulirsch, *Introduction to Numerical Analysis*, translated by R. Bartels, W. Gautschi, and C. Witzgall, Springer–Verlag, New York, 1976, pages 142–151.

[4] A. H. Stroud and D. Secrest, *Gaussian Quadrature Formulas*, Prentice–Hall Inc., Englewood Cliffs, NJ, 1966.

69. Gaussian Quadrature: Kronrod's Extension

Applicable to One-dimensional definite integrals.

Yields

A sequence of integration rules starting with a Gaussian rule.

Idea

When an integral is to be evaluated by different Gaussian integration rules, none of the values of the integrand, except possibly for the $x = 0$ value, can be re-used. It is possible to devise interpolatory rules that re-use all of the nodes in a Gaussian rule.

Procedure

Consider the integral $I = \int_a^b f(x)\,dx$. Suppose the Gaussian n-point rule, G_n, is used to approximate I numerically. Later, it may be of interest to approximate I numerically using m nodes (with $m > n$). If the Gaussian m-point rule, G_m, is used, then the only values of f that were obtained using G_n that can be re-used is, possibly, the value $x = 0$. (The value $x = 0$ can be re-used only if both n and m are odd). If f is an expensive function to compute, then it would be useful to re-use many values from the G_n computation.

It is possible to start with the nodes from the rule G_n, $\{x_i \mid i = 1, 2, \dots, n\}$, and add new nodes $\{y_i \mid i = 1, 2, \dots, n+1\}$ so that all polynomials of degree $3n + 1$ are integrated exactly, if n is even (degree $3n + 2$, if n is odd). The Kronrod rule, which uses the nodes $\{x_i, y_j\}$ to evaluate polynomials of maximal degree exactly, will be called K_{2n+1}. Note that the weights corresponding to the nodes $\{x_i\}$ in K_{2n+1} will not be the same as the weights corresponding to the nodes $\{x_i\}$ in G_n.

As an example, Table 69 shows the rules G_7 and K_{15}.

Notes

[1] Piessens *et al.* [11] contains the numerical values of the nodes and weights for the following rules: $\{G_7, K_{15}\}$, $\{G_{10}, K_{21}\}$, $\{G_{15}, K_{31}\}$, $\{G_{20}, K_{41}\}$, $\{G_{25}, K_{51}\}$, and $\{G_{30}, K_{61}\}$.

[2] The technique in this section may be continued; after the n-point Gaussian rule (G_n) is used, and the $2n+1$-point Kronrod rule (K_{2n+1}) is used, additional nodes may be added to interpolate higher order polynomials. Nodes and weights for the sequence of rules $\{G_3, K_7, P_{15}, P_{31}, \dots, P_{255}\}$ are given in Patterson [8]. (Here, the rule P_k is exact for polynomials of degree $(3k+1)/2$.)

Similarly, the sequence of rules $\{G_{10}, K_{21}, P_{43}, P_{87}\}$ is given in Piessens *et al.* [11].

[3] Kronrod extensions also exist for Gaussian rules with the weight function $(1 - x^2)^\mu$ (for $-\frac{1}{2} \leq \mu \leq \frac{3}{2}$).

Table 69. Values used in 7-point Gaussian and 15-point Kronrod quadrature formulas. The formulas are symmetric, only the positive nodes are shown. (That is, if $w_i f(t_i)$ appears, then so does $w_i f(-t_i)$.)

Name	Nodes	Weights
7-point Gaussian	0.94910	0.12958
	0.74153	0.27970
	0.40584	0.38183
	0	0.41795
15-point Kronrod	0.94910	0.02293
	0.94910	0.06309
	0.86486	0.10479
	0.74153	0.14065
	0.58608	0.16900
	0.40584	0.19036
	0.20778	0.20443
	0	0.20948

[4] Instead of using a higher order rule, the same rule can be re-applied with a smaller interval size. For even more accuracy, an extrapolation technique can be used, see page 249.

[5] Favati *et al.* [3] derive a set of symmetric, closed, interpolatory quadrature formulas on the interval $[-1, 1]$ with positive weights and increasing precision. These formulas re-use previously computed functional values. They obtain a tree of quadrature rules having 74 elements, 27 leaves, and a maximum height of 14. That is, the sequence of height 14 is a collection of quadrature rules that use $(2, 3, 5, 7, 9, 13, 19, 27, 41, 57, 85, 117, 181, 249)$ nodes, and each rule re-uses all the nodes from the previous rule.

Favati *et al.* [3] performed extensive numerical tests using their new rules in place of the Gauss–Kronrod rules in the routines QAG and QAGS, in the computer library Quadpack. For one-dimensional and two-dimensional integrals, the resulting programs appear to be faster, more reliable, and to require fewer function evaluations.

[6] Rabinowitz [12] considers the numerical evaluation of integrals of the form $\int_{-1}^{1} \frac{\omega(x) f(x)}{x - \lambda}\, dx$ with $\omega(x) = (1 - x^2)^{\mu - 1/2}$ for $0 \leq \mu \leq 2$.

References

[1] P. J. Davis and P. Rabinowitz, *Methods of Numerical Integration*, Second Edition, Academic Press, Orlando, Florida, 1984, pages 106–109.

[2] F. Caliò and W. Gautschi, "On Computing Gauss–Kronrod Quadrature Formula," *Math. of Comp.*, **47**, No. 176, 1986, pages 639–650, S57–S63.

[3] P. Favati, G. Lotti, and F. Romani, "Interpolatory Integration Formulas for Optimal Composition," and "Algorithm 691: Improving QUADPACK Automatic Integration Routines," *ACM Trans. Math. Software*, **17**, No. 2, June 1991, pages 207–217 and 218–232.

[4] W. Gautschi and S. E. Notaris, "An Algebraic Study of Gauss–Kronrod Quadrature Formulae for Jacobi Weight Functions," *Math. of Comp.*, **51**, No. 183, 1988, pages 231–248.

[5] A. S. Kronrod, *Nodes and Weights of Quadrature Formulas*, Consultants Bureau, NY, 1965.

[6] G. Monegato, "Stieltjes Polynomials and Related Quadrature Rules," *SIAM Review*, **24**, 1982, pages 137–158.

[7] S. E. Notaris, "Gauss-Kronrod Quadrature Formulae for Weight Functions of Bernstein–Szego Type. II," *J. Comput. Appl. Math.*, **29**, No. 2, 1990, pages 161–169.

[8] T. N. L. Patterson, "The Optimal Addition of Points to Quadrature Formulae," *Math. of Comp.*, **22**, 1968, pages 847–856.

[9] F. Peherstorfer, "Weight Functions Admitting Repeated Positive Kronrod Quadrature," *BIT*, **30**, No. 1, 1990, pages 145–151.

[10] F. Peherstorfer, "On Stieltjes Polynomials and Gauss–Kronrod quadrature," *Math. of Comp.*, **55**, No. 192, 1990, pages 649–664.

[11] R. Piessens, E. de Doncker–Kapenga, C. W. Überhuber, and D. K. Kahaner, *Quadpack*, Springer–Verlag, New York, 1983.

[12] P. Rabinowitz, "A Stable Gauss–Kronrod Algorithm for Cauchy Principal-Value Integrals." *Comput. Math. Appl. Part B*, **12**, No. 5–6, 1986, pages 1249–1254.

70. Lattice Rules

Applicable to One-dimensional and multidimensional integrals on the unit cube.

Yields

A numerical approximation scheme.

Idea

A lattice rule uses all the nodes on a lattice that lie within and on the boundary of the unit cube.

Procedure

A lattice rule is a numerical scheme for approximating the value of an integral over the multidimensional unit cube. The integral is assumed to have the form $I = \int_{C^s} f(x)\,dx$, where C^s is the closed s-dimensional unit cube

$$C^s = \{(x_1, \dots, x_s) \mid 0 \le x_i \le 1, \quad i = 1, 2, \dots, s\}.$$

Every lattice rule can be written in the form

$$I \approx \frac{1}{n_1 n_2 \cdots n_m} \sum_{j_m=1}^{n_m} \cdots \sum_{j_1=1}^{n_1} \overline{f}\left(\frac{j_1 \mathbf{z}_1}{n_1} + \dots \frac{j_m \mathbf{z}_m}{n_m}\right) \tag{70.1}$$

where $\overline{f}$ is the periodic extension of f (see next paragraph), $\{\mathbf{z}_1, \mathbf{z}_2, \ldots, \mathbf{z}_m\}$ are vectors with integer components, and $\{n_1, n_2, \ldots, n_m\}$ are given integers called *invariants*. (It is generally assumed that n_{i+1} divides n_i for $i = 1, 2, \ldots, m-1$.)

Let $\{x\}$ denote the fractional part of x (e.g., $\{3.4\} = 0.4$), and let $\{\mathbf{x}\} = (\{x_1, x_2, \ldots, x_s\}) = (\{x_1\}, \{x_2\}, \ldots, \{x_s\})$. Then the periodic extension of f is defined by

$$\overline{f}(\mathbf{x}) := f(\{\mathbf{x}\}), \qquad \text{when} \quad \{x_j\} \neq 0, \quad \text{for all} \quad j = 1, 2, \ldots, s.$$

Hence, $\overline{f}$ coincides with f in the interior of the unit cube. At the points on the boundary of the unit cube, $\overline{f}$ is generally not continuous. At these points, $\overline{f}$ is defined by

$$\begin{aligned}\overline{f}(x_1, x_2, \ldots, x_s) \\ &= s^{-s} \lim_{\varepsilon \to 0+} \sum_{r_1} \sum_{r_2} \cdots \sum_{r_s} \overline{f}\,(x_1 + \varepsilon r_1, x_2 + \varepsilon r_2, \ldots, x_s + \varepsilon r_s)\end{aligned}$$

with each r_i taking only the values ± 1. If f is continuous on C^s, then this limit exists and is a symmetrical average of the values of f at corresponding points on opposite faces of the boundary. For example, in the one-dimensional case we find

$$\overline{f}(0) = \overline{f}(1) = \tfrac{1}{2}(f(0) + f(1)). \tag{70.2}$$

Sloan and Lyness [8] consider quadrature rules for the s-dimensional hypercube of the form

$$I \approx \frac{1}{n_1 n_2} \sum_{j_2=1}^{n_2} \sum_{j_1=1}^{n_1} \overline{f}\left(\frac{j_1 \mathbf{z}_1}{n_1} + \frac{j_2 \mathbf{z}_2}{n_2}\right)$$

which cannot be expressed in an analogous form with a single sum. These rules are called rank-2 lattice rules.

If m summations are required to represent the rule, as shown in (70.1), then the rule has rank m. The number of nodes used by such a rule is $\prod_{i=1}^{m} n_i$, and the rule may be expressed in a canonical form with m independent summations. Under this classification, an N-node number-theoretic rule (see page 312) is a rank $m = 1$ rule with $\{n_i\} = \{N, 1, 1, \cdots, 1\}$, and the product trapezoidal rule (see page 323) using N^s nodes is a rank $m = s$ rule with $\{n_i\} = \{N, N, \cdots, N\}$.

Example

In the case of one dimension, (70.1) becomes (where the single subscript has been suppressed)

$$I \approx J = \frac{1}{n}\sum_{j=1}^{n} \overline{f}\left(\frac{j\mathbf{z}}{n}\right).$$

Combining this with (70.2) and using $\mathbf{z} = 1$, we are led to the trapezoidal rule in the more familiar form

$$J = \frac{1}{N}\sum_{j=1}^{N} \overline{f}\left(\frac{j}{N}\right) = \frac{1}{N}\left(\frac{1}{2}f(0) + \sum_{j=1}^{N-1} f\left(\frac{j}{N}\right) + \frac{1}{2}f(1)\right).$$

The only N-point one-dimensional lattice rule is the trapezoidal rule.

Notes

[1] The s-dimensional product-trapezoidal rule is defined by

$$\begin{aligned} I &\approx \frac{1}{n^s}\sum_{j_s=1}^{n} \cdots \sum_{j_2=1}^{n}\sum_{j_1=1}^{n} \overline{f}\left(\frac{(j_1, j_2, \dots, j_s)}{n}\right) \\ &= \frac{1}{n^s}\sum_{j_s=1}^{n} \cdots \sum_{j_2=1}^{n}\sum_{j_1=1}^{n} \overline{f}\left(\frac{j_1\mathbf{e}_1}{n} + \frac{j_2\mathbf{e}_2}{n} + \cdots + \frac{j_s\mathbf{e}_s}{n}\right) \end{aligned} \tag{70.3}$$

where $\mathbf{e}_j$ is a unit vector with a one in the x_j direction, and zeros elsewhere.

[2] For the one-dimensional integration of a periodic function, the trapezoidal rule is an efficient choice. However, for s-dimensional integration of a periodic function over a hypercube, the s-dimensional product trapezoidal rule is not generally cost effective. Other lattice rules can be more effective.

[3] Lattice rules often have many representations. For example, the two-dimensional product of the 3-point and 4-point trapezoidal rules may be written in the two ways

$$\frac{1}{12}\sum_{j_1=1}^{4}\sum_{j_2=1}^{3} \overline{f}\left(j_1\frac{(1,0)}{4} + j_2\frac{(0,1)}{3}\right) = \frac{1}{12}\sum_{j=1}^{12} \overline{f}\left(j\frac{(3,4)}{12}\right) \tag{70.4}$$

Figure 70.a shows this lattice rule. As another example, all of the following represent the same rule:

$$\begin{aligned} \frac{1}{25}\sum_{j_1=1}^{5}\sum_{j_2=1}^{5} \overline{f}\left(j_1\frac{(1,2)}{5} + j_2\frac{(3,1)}{5}\right) &= \frac{1}{5}\sum_{j=1}^{5} \overline{f}\left(j\frac{(1,2)}{5}\right) \\ &= \frac{1}{5}\sum_{j=1}^{5} \overline{f}\left(j\frac{(2,4)}{5}\right) \\ &= \frac{1}{10}\sum_{j=1}^{10} \overline{f}\left(j\frac{(2,4)}{10}\right). \end{aligned} \tag{70.5}$$

Figure 70.b shows this lattice rule.

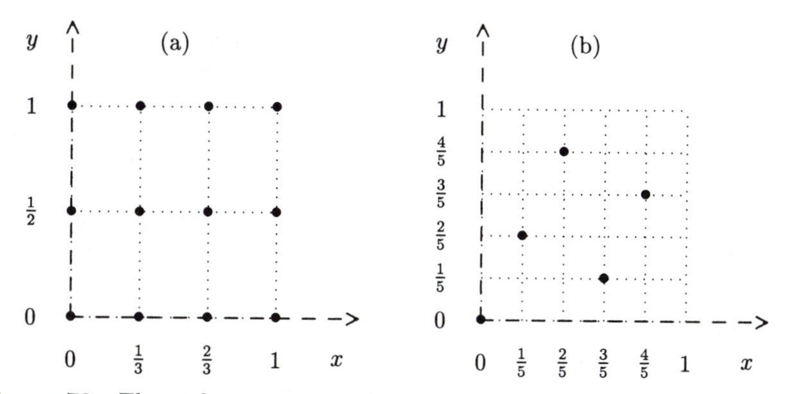

Figure 70. The nodes appearing in some lattice rules: (a) the lattice rule in (70.4); (b) the lattice rule in (70.5).

[4] Sloan and Walsh [10] consider lattice rules of rank 2, i.e., rules of the form

$$I \approx \frac{1}{n^2 r} \sum_{j_2=1}^{n} \sum_{j_1=1}^{nr} \overline{f}\left(\frac{j_1 \mathbf{z}_1}{nr} + \frac{j_2 \mathbf{z}_2}{n}\right),$$

where $n > 1$ and $r \geq 1$. According to a criterion introduced for number-theoretic rules (see page 312) they find the "best rules" of the above form. Lyness and Sørevik [4] use a different measure to determine optimal lattices. Both papers contain numerical results and identify useful rules.

[5] Lyness and Sørevik [3] investigate the number of distinct s-dimensional lattice rules that employ precisely N nodes, $\nu_s(N)$. They give many results for $\nu_s(N)$ including:

(A) $\nu_s(MN) \leq \nu_s(M)\nu_s(N)$,

(B) Equality holds in (A) if and only if M and N are relatively prime.

(C) If the prime factorization of L is $L = \prod_j p_j^{t_j}$, then $\nu_s(L) = \prod_j \nu_s\left(p_j^{t_j}\right)$ where $\nu_s\left(p^t\right) = \prod_{i=1}^{s-1}\left(\frac{p^{i+t}-1}{p^i - 1}\right)$ for $s \geq 1$ and $t \geq 0$.

[6] For lattice rules of the form $\frac{1}{N}\sum_{j=0}^{N-1} f\left(\left\{\frac{j\mathbf{p}}{N}\right\}\right)$, Haber [1] calculates several "good" choices for $\mathbf{p}$ of the form $\mathbf{p} = (1, b, b^2, \ldots, b^{s-1})$ for various values of N and s. Here, goodness is defined by how well the integral rule performs on the test function $\prod_{i=1}^{s} f(x_i)$, with $f(x_i) = 1 + \pi^2(x^2 - x + \frac{1}{6})$.

[7] Zinterhof [12] presents a fast method for generating lattices that are nearly as good as the optimal lattices. The lattices so generated are tested on the integrands

$$H_1(x_1, \ldots, x_s) = \prod_{i=1}^{s} \left(1 - 2\log\left(2\sin \pi x_i\right)\right)$$

$$H_2(x_1, \ldots, x_s) = \prod_{i=1}^{s} \left(1 - \frac{\pi^2}{6} + \frac{\pi^2}{2}\left(1 - 2\left\{x_i\right\}\right)^2\right)$$

in C^s. These integrands are chosen because their Fourier transforms are $\widehat{H}_1(\mathbf{m}) = R(\mathbf{m})^{-1}$ and $\widehat{H}_2(\mathbf{m}) = R(\mathbf{m})^{-2}$, where $\mathbf{m} = (m_1, \ldots, m_s)$ and $R(\mathbf{m}) = \prod_{i=1}^{s} \max(1, |m_i|)$.

[8] Sloan and Lyness [9] give some of the properties of the projections of rules into lower dimensions.

[9] Worlet [11] introduces some new families of integration lattices. These have a better order of convergence that previously known constructions.

[10] The results in Niederreiter [5] suggest that in the search for efficient lattice rules, one should concentrate on lattice rules with large first invariant.

References

[1] S. Haber, "Parameters for Integrating Periodic Functions of Several Variables," *Math. of Comp.*, **41**, 1983, pages 115–129.

[2] J. N. Lyness, "An Introduction to Lattice Rules and Their Generator Matrices," *IMA J. Num. Analysis*, **9**, No. 3, 1989, pages 405–419.

[3] J. N. Lyness and T. Sørevik, "The Number of Lattice Rules," *BIT*, **29**, No. 3, 1989, pages 527–534.

[4] J. N. Lyness and T. Sørevik, "A Search Program for Finding Optimal Integration Lattices," *Computing*, **47**, 1991, pages 103–120.

[5] H. Niederreiter, "The Existence of Efficient Lattice Rules for Multidimensional Numerical Integration," *Math. of Comp.*, **58**, No. 197, January 1992, pages 305–314.

[6] I. H. Sloan, "Lattice Methods for Multiple Integration," *J. Comput. Appl. Math.*, **12–13**, 1985, pages 131–143.

[7] I. H. Sloan and P. J. Kachoyan, "Lattice Methods for Multiple Integration: Theory, Error Analysis and Examples," *SIAM J. Numer. Anal.*, **24**, No. 1, 1987, pages 116–128.

[8] I. H. Sloan and J. N. Lyness, "The Representation of Lattice Quadrature Rules as Multiple Sums," *Math. of Comp.*, **52**, January 1989, pages 81–94.

[9] I. H. Sloan and J. N. Lyness, "Lattice Rules: Projection Regularity and Unique Representations," *Math. of Comp.*, **54**, No. 190, April 1990, pages 649–660.

[10] I. H. Sloan and L. Walsh, "A Computer Search of Rank-2 Lattice Rules for Multidimensional Quadrature," *Math. of Comp.*, **54**, No. 189, January 1990, pages 281–302.

[11] R. T. Worlet, "On Integration Lattices," *BIT*, **31**, 1991, pages 529–539.

[12] P. Zinterhof, "Gratis Lattice Points for Multidimensional Integration," *Computing*, **38**, 1987, pages 347–353.

71. Monte Carlo Method

Applicable to Definite integrals, especially multidimensional integrals.

Yields

A numerical approximation derived from random numbers.

Idea

Random numbers may be used to approximate the value of a definite integral.

Procedure 1

Suppose we wish to approximate numerically the value of the definite integral

$$I = \int_B g(x)\, dx, \tag{71.1}$$

where B is some bounded region. Since B is bounded, it may be enclosed in some rectangular parallelepiped R. Let $1_B(x)$ represent the indicator function of B, that is

$$1_B(x) = \begin{cases} 1 & \text{if } x \in B, \\ 0 & \text{if } x \notin B. \end{cases}$$

Then the integral I may be written in the form

$$I = \int_R \Big(g(x)1_B(x)\Big)\, dx = \frac{1}{V(R)} \int_R \Big(g(x)1_B(x)V(R)\Big)\, dx \tag{71.2}$$

where $V(R)$ represents the volume of the region R. Equation (71.2) may be interpreted as the expectation of the function $h(X) = g(X)1_B(X)V(R)$ of the random variable X, which is uniformly distributed in the parallelepiped R (i.e., it has the density function $1/V(R)$).

The expectation of $h(X)$ can be obtained by simulating random deviates from X, determining h at these points, and then taking the average of the h values. Hence, simulation of the random variable X will lead to an approximate numerical value of the integral I. If N trials are taken, then the following estimate is obtained:

$$I \simeq \widehat{I} = \frac{1}{N} \sum_{i=1}^{N} h(x_i) = \frac{V(R)}{N} \sum_{i=1}^{N} g(x_i)1_B(x_i) \tag{71.3}$$

where each x_i is uniformly distributed in R.

Another way to think about (71.3) is that $g(\xi_i)$, where ξ_i is chosen uniformly in B, is an independent random variable with expectation I. Averaging several of these estimates together, which is what (71.3) does, results in an unbiased estimator of I.

Procedure 2

Importance sampling is the term given to sampling from a non-uniform distribution so as to minimize the variance of the estimate for I. Consider writing (71.1) as

$$I = E_U[g(x)] \tag{71.4}$$

where $E_U[\cdot]$ denotes the expectation taken with respect to the uniform distribution on B. In other words, I is the mean of $g(x)$ with respect to the uniform distribution. Associated with this mean is a variance, defined by

$$\sigma_U^2 := E_U\left[(g(x) - I)^2\right] = E_U\left[g^2\right] - I^2. \tag{71.5}$$

Approximations to I obtained by sampling from the uniform distribution will have errors that scale with σ_U.

If $f(x)$ represents a different density function to sample from, then we may write

$$I = \int_B \left(\frac{g(x)}{f(x)}\right) f(x)\, dx = E_f\left[\frac{g(x)}{f(x)}\right]$$

where $E_f[\cdot]$ denotes the expectation taken with respect to the density $f(x)$. In other words, I is the mean of $g(x)/f(x)$ with respect to the distribution $f(x)$. Associated with this mean is a variance; defined by

$$\sigma_f^2 := E_f\left(\left\{\frac{g(x)}{f(x)} - I\right\}^2\right) = E_f\left[\frac{g^2}{f^2}\right] - I^2 = \int_B \frac{g^2(x)}{f(x)} dx - I^2$$

Approximations to I obtained by sampling from $f(x)$ will have errors that scale with σ_f.

A minimum variance estimator may be obtained by finding the $f(x)$ such that σ_f^2 is minimal. Using the calculus of variations the density function for the minimal estimator is determined to be

$$f_{\text{opt}}(x) = C|g(x)| = \frac{|g(x)|}{\int_B |g(x)|\, dx} \tag{71.6}$$

where the constant C has been chosen so that $f_{\text{opt}}(x)$ is appropriately normalized. (Since $f_{\text{opt}}(x)$ is a density function, it must integrate to unity.) Clearly, finding $f_{\text{opt}}(x)$ is as difficult as determining the original integral I! However, (71.6) indicates that $f_{\text{opt}}(x)$ should have the same general behavior as $|g(x)|$. As Example 2 shows, sometimes an approximate $f(x) \approx f_{\text{opt}}(x)$ can be chosen.

Procedure 3

Another type of Monte Carlo method is the hit-or-miss Monte Carlo method (see Hammersley and Handscomb [6]). It is very inefficient but is very easy to understand; it was the first application of Monte Carlo methods. Suppose that $0 \le f(x) \le 1$ when $0 \le x \le 1$. If we define

$$g(x,y) = \begin{cases} 0 & \text{if } f(x) < y, \\ 1 & \text{if } f(x) > y, \end{cases}$$

then we may write $I = \int_0^1 f(x)\,dx = \int_0^1 \int_0^1 g(x,y)\,dy\,dx$. This integral may be estimated by

$$I \approx \widehat{I} = \frac{1}{n} \sum_{i=1}^{n} g(\xi_{2i-1}, \xi_{2i}) = \frac{n^*}{n} \tag{71.7}$$

where the $\{\xi_i\}$ are chosen independently and uniformly from the interval $[0,1]$. The summation in (71.7) counts the number of points in the unit square which are below the curve $y = f(x)$ (this defines n^*), and divides by the total number of sample points (i.e., n). We emphasize again that the hit-or-miss method is computationally very inefficient.

Example 1

We choose to approximate the integral $I = \int_0^1 3x^2\,dx$, whose value is 1. To implement the method in (71.3),

$$I \simeq \widehat{I} = \frac{1}{N} \sum_{i=1}^{N} 3x_i^2, \qquad \text{for } x_i \text{ uniformly distributed on } [0,1]$$

the FORTRAN program in Program 71 was constructed. The program takes the results of many trials and averages these values together. Note that the program uses a routine called RANDOM, whose source code is not shown, which returns a random value uniformly distributed on the interval from zero to one.

The result of the program is as follows:

```
AFTER   100 TRIALS, THE AVERAGE IS  1.006
AFTER   200 TRIALS, THE AVERAGE IS  1.084
AFTER   300 TRIALS, THE AVERAGE IS  1.046
AFTER   400 TRIALS, THE AVERAGE IS  1.033
AFTER   500 TRIALS, THE AVERAGE IS  0.996
AFTER   600 TRIALS, THE AVERAGE IS  1.028
AFTER   700 TRIALS, THE AVERAGE IS  1.035
AFTER   800 TRIALS, THE AVERAGE IS  1.029
AFTER   900 TRIALS, THE AVERAGE IS  1.032
AFTER  1000 TRIALS, THE AVERAGE IS  1.038
```

We can also approximate I by using hit-or-miss Monte Carlo. (First, we scale the integrand by a factor of 3, to be $f_s(x) = x^2$, so that it is in the range $[0,1]$.) Now random deviates x_i and y_i (both obtained uniformly from

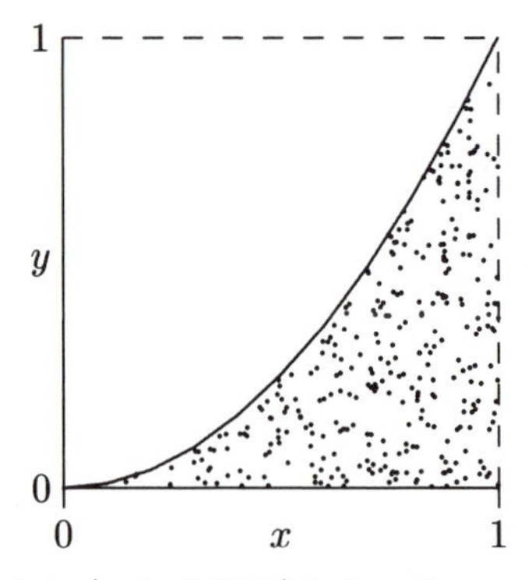

Figure 71. The 323 points (out of 1000) below the curve $y = x^2$.

the interval [0, 1]) are obtained. For each pair of values, n is incremented by one. If, for that pair of values, $y_i \leq f_s(x_i) = x_i^2$, then n^* is also incremented by one. Use of (71.7) then results in an estimate for I.

Performing this algorithm 1000 times, we obtained 323 instances when the y_i was less than y_i^2 (Figure 71 shows the locations of these points). Hence, the estimate of I becomes

$$\widehat{\widehat{I}} = 3 \cdot \frac{1}{N} \sum_{i=1}^{N} 1_B(x_i) = \frac{3}{1000} \cdot 323 = 0.969.$$

Example 2

Consider the integral

$$J = \int_0^1 \cos\left(\frac{\pi x}{2}\right) dx = \frac{2}{\pi}.$$

If we let ξ_i represent a sample from the uniform distribution from [0, 1] then J may be approximated by J_U

$$J \simeq J_U = \frac{1}{N} \sum_{i=1}^{N} \cos \frac{\pi \xi_i}{2}.$$

The variance of this estimator for J is

$$\sigma_U^2 = \int_0^1 \cos^2\left(\frac{\pi x}{2}\right) dx - J^2 = \frac{1}{2} - \frac{4}{\pi^2} \simeq .0947\ldots.$$

Now we want to obtain a density function that more closely approximates the integrand. Since $\cos\left(\frac{\pi x}{2}\right) = 1 - \frac{\pi^2}{8} x^2 + O(x^4)$ for small values of x, we choose a $f(x)$ that has a similar form. We take

$$f(x) = C(1 - x^2) = \tfrac{3}{2}(1 - x^2). \tag{71.8}$$

(The factor $\frac{3}{2}$ arises from the normalization $\int_0^1 (1-x^2)\,dx = \frac{2}{3}$.) Using this new density function we find (see the Notes for how to generate deviates from this distribution)

$$J = \int_0^1 \left(\frac{2}{3} \frac{\cos\left(\frac{\pi x}{2}\right)}{1-x^2} \right) f(x)\,dx. \tag{71.9}$$

If we let ζ_i represent a random variable coming from a distribution that has the density $f(x)$, then (71.9) may be sampled to yield an approximation to J:

$$J \simeq J_f = \frac{1}{N} \sum_{i=1}^{N} \frac{2}{3} \frac{\cos \frac{\pi \zeta_i}{2}}{1-\zeta_i^2}.$$

The variance of this second estimator for J is

$$\sigma_f^2 = \int_0^1 \left(\frac{2}{3} \frac{\cos\left(\frac{\pi x}{2}\right)}{1-x^2} \right)^2 f(x)\,dx - J^2 \simeq .00099\ldots.$$

Since σ_U is approximately 10 times larger than σ_f, the errors in using J_f to approximate J will be about 10 times smaller than using J_U to approximate J, for the same number of trials. Of course, in practical cases it will not generally be possible to exactly determine the variances σ_U and σ_f. However, estimates can be obtained for the variances by approximating the defining integrals.

Program 71

```
          SUM=0
          DO 10 J=1,1000
          X=RANDOM(T)
          VAL=3.*X**2
          SUM=SUM+VAL
          IF( MOD(J,100) .NE. 0 ) GOTO 10
          AVERAG=SUM/FLOAT(J)
          WRITE(6,5) J,AVERAG
5         FORMAT(' AFTER',I6, ' TRIALS, THE AVERAGE IS',F7.3)
10        CONTINUE
          END
```

Notes

[1] This method is of particular importance when multi-dimensional integrals are to be approximated numerically. For multi-dimensional integrals, Monte Carlo techniques may be the only techniques that will obtain an estimate in a reasonable amount of computer time. This is because the error in a Monte Carlo computation scales with $\sigma/\sqrt{N}$, where N is the number of samples of the integrand (trials), independent of dimension. For traditional methods, the number of samples of the integrand varies exponentially with the dimension (i.e., scales as α^N for some α).

[2] While the classical Monte Carlo method converges with order $1/\sqrt{N}$, where N is the number of samples, the quasi-Monte Carlo method can achieve an order of $(\log N)^\alpha/N$ for some $\alpha > 0$. See Niederreiter [10]–[11] and Wozniakowski's method on page 333.

[3] For some integrals, the variance in (71.5) may not exist. For example, with $I = \int_0^1 dx/\sqrt{x}$, the variance is computed to be $\sigma_U^2 = \int_0^1 dx/x - I^2$, and the first term is infinite. Use of importance sampling can result in a new integral that has a finite variance. See Kalos and Whitlock [7].

[4] Masry and Cambanis [8] discuss how the trapezoidal rule can be used in the Monte Carlo computation of the integral $I = \int_0^1 f(x)\,dx$. Choose n random deviates independently and uniformly on the interval $[0, 1]$. Numerically order these deviates to form the sequence $t_{n,1} < t_{n,2} < \cdots < t_{n,n}$, and then add the points $t_{n,0} := 0$ and $t_{n,n+1} := 1$. The sequence of $\{t_{n,i}\}$, used in the trapezoidal rule, produces an estimate of I:

$$I \approx I_n = \frac{1}{2}\sum_{i=0}^{n}\Big[f(t_{n,i}) + f(t_{n,i+1})\Big](t_{n,i+1} - t_{n,i}).$$

If f has a continuous second derivative on the interval $[0, 1]$, then it can be shown that

$$E\left[I - I_n\right]^2 = \frac{\left[f'(1) - f'(0)\right]^2 + o(1)}{4(n+1)(n+2)(n+3)(n+4)}.$$

Hence, the error varies as $O\left(n^{-4}\right)$ for large n.

[5] The integral $I = \int_0^1 g(x)\,dx$ may be written as $I = \int_0^1 \frac{1}{2}\left(g(x) + g(1-x)\right)\,dx$. Hence, the estimator

$$\widehat{I} = \frac{1}{N}\sum_{i=1}^{N}\frac{1}{2}\Big(g(x_i) + g(1 - x_i)\Big), \tag{71.10}$$

where the x_i are chosen from the uniform distribution, can be used to approximate I. When $g(x)$ is linear, this approximator gives the exact answer. (See Siegel and O'Brien [15] for techniques that are exact for other polynomials.) In cases where the function is nearly linear, the variance can be substantially reduced. This is known as the method of *antithetic variates.*

For example, consider the integral $I = \int_0^1 e^x\,dx = e - 1$. Using a straightforward Monte Carlo evaluation the variance is found to be $\sigma^2 = \int_0^1 \left[e^x - (e-1)\right]^2\,dx = (3-e)(e-1)/2 \simeq 0.242\ldots$. Using (71.10) reduces the variance to 0.0039, a substantial reduction.

[6] Error estimates are available for many different approximation schemes (see Cambanis and Masry [2]). Consider the integral $I(g) = \int_0^1 g(t)\,dt$, and let each U_* below represent an independent random variable, uniformly distributed over the interval $[0,1]$.

For the approximation $I(g) \approx I_1(g) = \frac{1}{n}\sum_{i=1}^n g(U_i)$, the mean-square error is given by: $E\left[(I(g) - I_1(g))^2\right] = \left[I(g^2) - I^2(g)\right]/n$.

In the *stratified sampling scheme* the interval $[0,1]$ is partitioned into n subintervals of equal length and the point $U_{n,i}$ is chosen uniformly in the i-th interval. For the approximation $I(g) \approx I_2(g) = \frac{1}{n}\sum_{i=1}^n g(U_{n,i})$, the mean-square error is given by: $\lim_{n\to\infty} n^3 E\left[(I(g) - I_2(g))^2\right] = \frac{1}{12}\int_0^1 \left[g'(t)\right]^2\,dt$. (See also Press and Farrar [13].)

In the *stratified and symmetrized scheme* the interval $[0,1]$ is partitioned into n subintervals of equal length and the point $U_{n,i}$ is chosen uniformly in the i-th interval. Let $U'_{n,i}$ represent the symmetrically opposite point to $U_{n,i}$ in the i-th interval. If g has a continuous second derivative, then for the approximation $I(g) \approx I_3(g) = \frac{1}{2n}\sum_{i=1}^n \left[g(U_{n,i}) + g(U'_{n,i})\right]$, the mean-square error is given by: $\lim_{n\to\infty} (2n)^5 E\left[(I(g) - I_3(g))^2\right] = \frac{2}{45}\int_0^1 \left[g''(t)\right]^2\,dt$.

[7] There are many other variance reduction techniques that are sometimes used in Monte Carlo calculations. These include the use of (see Hammersley and Handscomb [6]): *control variates*, *regression methods*, *orthonormal functions*, and *group sampling*.

[8] Suppose that $I = \int f(x)\,dx$ is approximated by a Monte Carlo computation. Suppose also that the integrals of some "reference functions" (functions which can be analytically integrated) are also approximated using the same set of Monte Carlo points. Then the accuracy of the estimate of I can be improved by using the estimated integrals of the reference functions. See Eberhard and Schneider [5] for details.

[9] Ogata [12] investigates the two test integrals: $\int_0^1\int_0^1\cdots\int_0^1 e^{x_1}e^{x_2}\ldots e^{x_k}\,d\mathbf{x}$ and $\int_{-\infty}^{\infty}\int_{-\infty}^{\infty}\cdots\int_{-\infty}^{\infty} e^{-\mathbf{x}B\mathbf{x}^{\mathrm{T}}}\,d\mathbf{x}$ where B is a specific Toeplitz matrix.

[10] Details on how to simulate a random variable from different distributions may be found in Devroye [3]. For example, random deviates from the density in (71.8), $f(x) = \frac{3}{2}(1-x^2)$ for x in the range $[0,1]$, may be obtained as follows:

- Generate ξ_1 and ξ_2 independently and uniformly on the interval $[0,1]$.
- If $\xi_2 \le \frac{\xi_1(3-\xi_1)}{2}$, then $x = 1 - \xi_1$; otherwise, $x = \frac{1}{2}\left(\sqrt{9 - 8\xi_2} - 1\right)$.

References

[1] G. Bhanot, "The Metropolis Algorithm," *Rep. Progr. Phys.*, **51**, No. 3, 1988, pages 429–457.

[2] S. Cambanis and E. Masry, "Trapezoidal Stratified Monte Carlo Integration," *SIAM J. Numer. Anal.*, **29**, No. 1, February 1992, pages 284–301.

[3] L. Devroye, *Non-Uniform Random Variate Generation*, Springer–Verlag, New York, 1986.

[4] P. J. Davis and P. Rabinowitz, *Methods of Numerical Integration*, Second Edition, Academic Press, Orlando, Florida, 1984, pages 389–415.

[5] P. H. Eberhard and O. P. Schneider, "Reference Functions to Decrease Errors in Monte Carlo Integrals," *Comput. Physics Comm.*, **67**, 1992, pages 363–377.

[6] J. M. Hammersley and D. C. Handscomb, *Monte Carlo Methods*, John Wiley & Sons, New York, 1965.

[7] M. H. Kalos and P. A. Whitlock, *Monte Carlo Methods. Volume I: Basics*, John Wiley & Sons, New York, 1986, Chapter 4, pages 89–116.

[8] E. Masry and S. Cambanis, "Trapezoidal Monte Carlo Integration," *SIAM J. Numer. Anal.*, **27**, No. 1, February 1990, pages 225–246.

[9] B. J. T. Morgan, *Elements of Simulation*, The University Press, Cambridge, United Kingdom, 1984.

[10] H. Niederreiter, "Quasi-Monte Carlo Methods and Pseudo-random Numbers," *Bull. Amer. Math. Soc.*, **84**, No. 6, 1978, pages 957–1041.

[11] H. Niederreiter, "Quasi-Monte Carlo Methods for Multidimensional Numerical Integration," *Numerical Integration, III (Oberwolfach, 1987)*, Internat. Schriftenreihe Numer. Math., #85, Birkhäuser, Basel, 1988, pages 157–171.

[12] Y. Ogata, "A Monte Carlo Method for High Dimensional Integration," *Numer. Math.*, **55**, No. 2, 1989, pages 137–157.

[13] W. H. Press and G. R. Farrar, "Recursive Stratified Sampling for Multidimensional Monte Carlo Integration," *Comput. in Physics*, March/April 1990, pages 191–195.

[14] Y. A. Shreider (ed.), *Method of Statistical Testing*, American Elsevier Publishing Company, New York, 1964.

[15] A. F. Siegel and F. O'Brien, "Unbiased Monte Carlo Integration Methods with Exactness for Low Order Polynomials," *SIAM J. Sci. Stat. Comput.*, **6**, No. 1, January 1985, pages 169–181.

72. Number Theoretic Methods

Applicable to Definite integrals, especially multidimensional integrals.

Yields

Quadrature rules.

Idea

Using points from a deterministic sequence might be useful when numerically approximating an integral.

Procedure

There are two procedures in this section, both for the simple numerical integration rule

$$\int f(\mathbf{x})\,d\mathbf{x} \approx \frac{1}{N}\sum_{i=1}^{N} f(\mathbf{x}_i). \tag{72.1}$$

The question is: how are the nodes $\{\mathbf{x}_i\}$ to be determined? In the first method, a different set of nodes is chosen for each value of N; for the second method, the first N nodes from an infinitely long sequence are chosen.

For both methods we will use the notation $\{z\}$ to denote the fractional part of z. For example, $\{\pi\} = 0.14159\ldots$.

Procedure 1

In the general d-dimensional case of interest, we would like to use a formula of the form

$$\int_0^1 \cdots \int_0^1 f(\mathbf{x})\,d\mathbf{x} \approx \frac{1}{N}\sum_{k_d=1}^{N}\cdots\sum_{k_1=1}^{N} f\left(\left\{\frac{\theta_1 k_1}{N}\right\},\ldots,\left\{\frac{\theta_d k_d}{N}\right\}\right) \tag{72.2}$$

for some integer values $\{\theta_i\}$, which are relatively prime to N. What are "good" values of $\{\theta_i\}$?

The usual definition for a good set of values base the decision on an error bound obtained by using (72.2) in (72.1). With this definition, there are several theorems (see Stroud [5]) on how to determine "optimal" values for the $\{\theta_i\}$, when N is a prime number or the product of two prime numbers. For some values of N, there is a constant a_N such that the values $\theta_1 = 1$ and $\theta_i = a_N^i$ yield an "optimal" vector.

Stroud [5] includes several tables of values. We have, for example, the following recommended values:

- for $d = 3$:
  - for $N = 101$, $\{\theta_i\} = \{1, 40, 85\}$
  - for $N = 1069$, $\{\theta_i\} = \{1, 136, 323\}$
  - for $N = 10007$, $\{\theta_i\} = \{1, 544, 5733\}$
  - for $N = 100063 = 47 \cdot 2129$, $\{\theta_i\} = \{1, 53584, 37334\}$
- for $d = 4$:
  - for $N = 307$, $\{\theta_i\} = \{1, 42, 229, 101\}$
  - for $N = 1069$, $\{\theta_i\} = \{1, 71, 765, 865\}$
  - for $N = 10007$, $\{\theta_i\} = \{1, 1784, 430, 6588\}$
  - for $N = 100063 = 47 \cdot 2129$, $\{\theta_i\} = \{1, 92313, 24700, 95582\}$
- for $d = 5$:
  - for $N = 1069$, $\{\theta_i\} = \{1, 63, 762, 970, 177\}$
  - for $N = 10007$, $\{\theta_i\} = \{1, 198, 9183, 6967, 8507\}$
  - for $N = 100063 = 47 \cdot 2129$, $\{\theta_i\} = \{1, 90036, 77477, 27253, 6222\}$

Procedure 2

A deterministic sequence of points $\{x_i, x_2, \ldots\}$ in the interval $[a, b]$ is said to be equidistributed, or uniformly distributed on the interval $[a, b]$, if

$$\lim_{n\to\infty} \frac{b-a}{n} \sum_{i=1}^{n} f(x_i) = \int_a^b f(x)\,dx$$

for all bounded Riemann integrable functions $f(x)$.

It is not difficult to construct uniformly distributed sequences. For example, if ζ is an irrational number, then $x_n = \{n\zeta\}$ is a uniformly distributed sequence. These sequences, when used in a simple quadrature formula, have very good error estimates. For example, Davis and Rabinowitz [2] (page 400) have the estimate

> **Theorem:** Let $f(x)$ be periodic in $[0,1]$ and be of class $C^3[0,1]$ so that we have $f(0) = f(1)$, $f'(0) = f'(1)$, and $f''(0) = f''(1)$. Let ζ be a quadratic irrational number. Then
>
> $$\left| \frac{1}{N} \sum_{i=1}^{N} f(\{n\zeta\}) - \int_0^1 f(x)\,dx \right| \leq \frac{c}{N}$$
>
> for some constant c.

Multi-dimensional uniformly distributed sequences can also be constructed. Suppose that $\zeta_1, \zeta_2, \ldots, \zeta_d$ are irrational numbers that are linearly independent over the rational numbers. (That is, $1 + a_1\zeta_1 + a_2\zeta_2 + \ldots, +a_d\zeta_d \neq 0$ for any rational numbers $\{a_i\}$.) Then the d-dimensional nodes

$$P_n = \Big(\{\zeta_1\}, \{\zeta_2\}, \ldots, \{\zeta_d\} \Big)$$

are equidistributed over the hypercube $0 \leq x_i \leq 1$, $i = 1, 2, \ldots, d$. (That is, $\lim_{N\to\infty} \frac{1}{N} \sum_{i=1}^{n} f(P_i) = \int_0^1 \cdots \int_0^1 f(\mathbf{x})\,d\mathbf{x}$ for any bounded Riemann-integrable function $f(\mathbf{x})$.)

Example

Using the values $N = 101$ and $\{\theta_i\} = \{1, 40, 85\}$ the sequence of $101^3 \simeq 10^6$ nodes is: $\big\{ \left(\frac{1}{101}, \frac{40}{101}, \frac{85}{101}\right), \left(\frac{1}{101}, \frac{40}{101}, \frac{69}{101}\right), \left(\frac{1}{101}, \frac{40}{101}, \frac{53}{101}\right), \left(\frac{1}{101}, \frac{40}{101}, \frac{37}{101}\right), \ldots, \left(\frac{1}{101}, \frac{80}{101}, \frac{85}{101}\right), \left(\frac{1}{101}, \frac{19}{101}, \frac{85}{101}\right), \left(\frac{1}{101}, \frac{59}{101}, \frac{85}{101}\right), \ldots, \left(\frac{2}{101}, \frac{40}{101}, \frac{85}{101}\right), \left(\frac{2}{101}, \frac{40}{101}, \frac{85}{101}\right), \ldots, (0,0,0) \big\}$.

Notes

[1] Using the first method, the number of nodes at which the integrals are to be evaluated grows very quickly. For example, for the (relatively) modest values of $N = 100063$ and $d = 3$, there are 10^{15} nodes required!

[2] Davis and Rabinowitz [2] give an example of the numerical evaluation of a 4-dimensional integral in which the irrational numbers chosen were: $\sqrt{2}$, $\sqrt{3}$, $\frac{1}{3}\sqrt{6}$, and $\sqrt{10}$.

[3] There is an common criterion by which a "best rule" may be determined (see Sloan and Walsh [7]). Let m be a fixed even positive integer (i.e., $m = 2$) and then define the function $f_m(\mathbf{x}) = \phi_m(x_1)\phi_m(x_2)\dots\phi_m(x_n)$, on $\mathbf{R}^n$ where

$$\phi_m(x) = 1 - (-1)^{m/2}\frac{(2\pi)^m B_m(x)}{m!}.$$

Here, $B_m(x)$ is a Bernoulli polynomial of degree m.

Among all quadrature rules Q_i on the n-dimensional unit hypercube, the "best" one is the one that minimizes $Q_i[f_m] - \int f_m = Q_i[f_m] - 1$.

[4] In Wozniakowski's method (page 333), a different infinite sequence of values is used.

[5] The number theoretic methods are generalized in the *lattice rules*, see page 300.

References

[1] R. Cranley and T. N. L. Patterson, "Randomization of Number Theoretic Methods for Multiple Integration," *SIAM J. Numer. Anal.*, **13**, No. 6, December 1976, pages 904–914.

[2] P. J. Davis and P. Rabinowitz, *Methods of Numerical Integration*, Second Edition, Academic Press, Orlando, Florida, 1984, Section 5.9.3, pages 396–410.

[3] L. K. Hua and Y. Wang, *Application of Number Theory to Numerical Analysis*, Springer–Verlag, New York, 1981.

[4] N. M. Korobov, "The Approximate Computation of Multiple Integrals," *Dokl. Akad. Nauk SSSR*, **124**, 1959, pages 1207–1210. (Russian)

[5] A. H. Stroud, *Approximate Calculation of Multiple Integrals*, Prentice–Hall Inc., Englewood Cliffs, NJ, 1971, Section 6.3, pages 198–208.

73. Parallel Computer Methods

Applicable to Definite integrals.

Yields

Ways in which a parallel computer may be used.

Idea

Parallel computers can sometimes speed up the numerical computation of an integral.

Procedure

There are many procedures that can be used, depending on the type of parallel computer under consideration. The most common use of a parallel computer is to partition an integration interval into many sub-intervals, and have the integration on each sub-interval performed in parallel.

Notes

[1] Genz [2] studied the implementation of a globally adaptive algorithm for single integrals on a SIMD computer.

[2] Rice [6] considers algorithms for single one-dimensional integrals on a MIMD computer.

[3] Burrage [1] considers a single one-dimensional integral being evaluated on a linear chain of transputers of arbitrary length. (A transputer is a local memory MIMD computer.) For some classes of problems, numerical evidence suggests that linear speed-ups are achievable with any number of transputers.

[4] Monte Carlo methods (see page 304) are generally the easiest to implement on parallel machines. While these methods are robust, they are generally the least efficient.

References

[1] K. Burrage, "An Adaptive Numerical Integration Code for a Chain of Transputers," *Parallel Computing*, **16**, 1990, pages 305–312.

[2] A. C. Genz, "Numerical Multiple Integration on Parallel Computers," *Comput. Physics Comm.*, **26**, 1982, pages 349–352.

[3] A. Genz, "Parallel Adaptive Algorithms For Multiple Integrals," in *Mathematics for Large Scale Computing*, Lecture Notes in Pure and Appl. Math., #120, Marcel Dekker, New York, 1989, pages 35–47.

[4] J. M. Lemme and J. R. Rice, "Speedup in Parallel Algorithms for Adaptive Quadrature," *J. ACM*, **26**, 1979, pages 65–71.

[5] J. R. Rice, "Parallel Algorithms for Adaptive Quadrature II. Metalgorithm Correctness," *Acta Inf.*, **5**, 1973, pages 273–285.

[6] J. R. Rice, "Adaptive Quadrature: Convergence of Parallel and Sequential Algorithms," *Bull. Amer. Math. Soc.*, **80**, 1974, pages 1250–1254.

[7] J. R. Rice, "Parallel Algorithms for Adaptive Quadrature III. Program Correctness," *ACM Trans. Math. Software*, **2**, 1976, pages 1–30.

74. Polyhedral Symmetry Rules

Applicable to Multidimensional integrals.

Yields

A numerical quadrature scheme.

Idea

Some quadrature rules are invariant under a symmetry group operating on the nodes.

Procedure

Consider the construction of a quadrature rule of the form

$$\iint_B w(\mathbf{x}) f(\mathbf{x})\, d\mathbf{x} \approx \sum_{k=1}^{N} w_k f(\mathbf{x}_k), \tag{74.1}$$

where $w(\mathbf{x})$ is a specific weight function. Generally, it is required that a quadrature rule be exact for all polynomials below some degree. Forcing (74.1) to be exact for low degree polynomials results in a large system of nonlinear algebraic equations that must be solved for the nodes $\{\mathbf{x}_k\}$, the weights $\{w_k\}$, or both.

To reduce the number of equations, we can impose full polyhedral symmetry on the formula. This may result in a formula that is not of the highest order, but it sometimes allows a formula to be quickly developed.

Example

Consider devising a 4-node quadrature rule on the square:

$$\int_{-1}^{1}\int_{-1}^{1} f(x,y)\, dx\, dy \approx \sum_{i=1}^{4} w_i f(x_i, y_i). \tag{74.2}$$

To completely specify the quadrature rule we require 12 values, $\{w_i, x_i, y_i \mid i = 1, \dots, 4\}$. To force (74.2) to be exact for all polynomials of degree 0, 1, and 2, the unknowns must satisfy

$$\begin{aligned}
f(x,y) &= 1: & w_1 + w_2 + w_3 + w_4 &= 4 \\
f(x,y) &= x: & w_1x_1 + w_2x_2 + w_3x_3 + w_4x_4 &= 0 \\
f(x,y) &= y: & w_1y_1 + w_2y_2 + w_3y_3 + w_4y_4 &= 0 \\
f(x,y) &= x^2: & w_1x_1^2 + w_2x_2^2 + w_3x_3^2 + w_4x_4^2 &= \tfrac{4}{3} \\
f(x,y) &= y^2: & w_1y_1^2 + w_2y_2^2 + w_3y_3^2 + w_4y_4^2 &= \tfrac{4}{3} \\
f(x,y) &= xy: & w_1x_1y_1 + w_2x_2y_2 + w_3x_3y_3 + w_4x_4y_4 &= 0.
\end{aligned} \tag{74.3}$$

Solving these nonlinear algebraic equations is a non-trivial task. Even determining the number of solutions to these equations is difficult.

Suppose, however, that we require the quadrature formula to be symmetric under rotation. That is, whenever the node (x_i, y_i) is in (74.2), then the nodes $(-x_i, y_i)$, $(x_i, -y_i)$, and $(-x_i, -y_i)$ should also be in (74.2). This constraint adds the following restrictions to (74.3): $w_1 = w_2 = w_3 = w_4$, $x_1 = x_2 = -x_3 = -x_4$, and $y_1 = -y_2 = -y_3 = y_4$. With these constraints, the equations in (74.3) become:

$$
\begin{aligned}
f(x,y) &= 1: & 4w_1 &= 4 \\
f(x,y) &= x: & 0 &= 0 \\
f(x,y) &= y: & 0 &= 0 \\
f(x,y) &= x^2: & 4w_1x_1^2 &= \tfrac{4}{3} \\
f(x,y) &= y^2: & 4w_1y_1^2 &= \tfrac{4}{3} \\
f(x,y) &= xy: & 0 &= 0.
\end{aligned}
\tag{74.4}
$$

The equations in (74.4) can be readily solved to determine that $w_1 = 1$ and $x_1 = y_1 = 1/\sqrt{3}$.

Notes

[1] The quadrature rule found in the example is also exact for polynomials of degree 3 (i.e., the rules are exact for the functions $f(x,y) = x^3$, $f(x,y) = x^2y$, $f(x,y) = xy^2$, and $f(x,y) = y^3$).

[2] The groups under which the nodes are mapped into themselves are the reflection groups of polyhedra. In three dimensions there only exist three such groups:

(A) the extended tetrahedral group of order 24 (the orbits of this group can have 24, 12, 6, 4, or 1 point(s));
(B) the extended octahedral group of order 48 (the orbits of this group can have 48, 24, 12, 8, 6, or 1 point(s));
(C) the extended icosahedral group of order 120 (the orbits of this group can have 120, 60, 30, 20, 12, or 1 point(s)).

[3] Cools and Haegemans [3] consider quadrature formulas that are invariant with respect to a transformation group and prove sufficient conditions for such formulas to have positive weights.

References

[1] M. Beckers and A. Haegemans, "The Construction of Three-Dimensional Invariant Cubature Formulae," *J. Comput. Appl. Math.*, **35**, 1991, pages 109–118.

[2] R. Cools and A. Haegemans, "Construction of Fully Symmetric Cubature Formula of Degree $4k-3$ for Fully Symmetric Planar Regions," *J. ACM*, **17**, No. 1–2, 1987, pages 173–180.

[3] R. Cools and A. Haegemans, "Why Do So Many Cubature Formulae Have So Many Positive Weights?," *BIT*, 1988, **28**, No. 4, pages 792–802.

[4] D. A. Dunavant, "Efficient Symmetrical Cubature Rules for Complete Polynomials of High Degree over the Unit Cube," *Int. J. Num. Methods Eng.*, **23**, 1986, pages 397–407.

[5] T. O. Espelid, "On the Construction of Good Fully Symmetric Integration Rules," *SIAM J. Numer. Anal.*, **24**, No. 4, 1987, pages 855–881.

[6] A. C. Genz and A. A. Malik, "An Imbedded Family of Fully Symmetric Numerical Integration Rules," *SIAM J. Numer. Anal.*, **20**, 1983, pages 580–588.

[7] P. Keast and J. N. Lyness, "On the Structure of Fully Symmetric Multidimensional Quadrature Rules," *SIAM J. Numer. Anal.*, **16**, 1979, pages 11–29.
[8] S. L. Sobolev, "Cubature Formulas on the Sphere Invariant Under Finite Groups of Rotations," *Soviet Math. Dokl.*, **3**, 1962, pages 1307–1310.
[9] T. Sørevik and T. O. Espelid, "Fully Symmetric Integration Rules for the 4-Cube," *BIT*, **29**, No. 1, 1989, pages 148–153.

75. Polynomial Interpolation

Applicable to Definite integrals.

Yields

Integration rules on a finite interval using uniformly spaced nodes.

Idea

When values of a function at a discrete set of points are known, an interpolating polynomial can be passed through those points. The integral of the original function will approximate the integral of the interpolating polynomial.

Procedure

Given the interval $[a,b]$, discretize it into n segments of equal length by inserting the $n+1$ nodes: $\{x_i \mid x_i = a+ih, i=0,1,\ldots,n\}$ where $h=(b-a)/n$. If the function $f(x)$ is known at the $n+1$ nodes (i.e., $f_i = f(x_i)$), then the interpolatory polynomial $P_n(x)$, of degree n or less, that goes through all $n+1$ pairs (x_i, f_i) is given by Lagrange's interpolation formula

$$P_n(x) = \sum_{i=0}^{n} f_i L_i(x), \qquad L_i(x) = \prod_{\substack{k=0 \\ k\neq i}}^{n} \frac{x-x_k}{x_i-x_k}$$

(since $L_i(x_j)=\delta_{ij}$, where δ_{ij} is the Kronecker delta). Writing $x=a+th$, and using $x_i=a+ih$, this can be written as $L_i(x) = K_i(t) = \prod_{\substack{k=0 \\ k\neq i}}^{n} \frac{t-k}{i-k}$.

Integrating $P_n(x)$ from $x=a$ to $x=b$ leads to

$$\begin{aligned} \int_a^b P_n(x)\,dx &= \sum_{i=0}^{n} f_i \int_a^b L_i(x)\,dx \\ &= h\sum_{i=0}^{n} f_i \int_0^n K_i(t)\,dt \qquad (75.1) \\ &= h\sum_{i=0}^{n} w_i f_i \end{aligned}$$

Table 75.1. Newton–Cotes rules obtained from polynomial interpolation.

n	w_i							Error	Name
1	$\frac{1}{2}$	$\frac{1}{2}$						$h^3\frac{1}{12}f^{(2)}(\xi)$	trapezoidal rule
2	$\frac{1}{3}$	$\frac{4}{3}$	$\frac{1}{3}$					$h^5\frac{1}{90}f^{(4)}(\xi)$	Simpson's rule
3	$\frac{3}{8}$	$\frac{9}{8}$	$\frac{9}{8}$	$\frac{3}{8}$				$h^5\frac{3}{80}f^{(4)}(\xi)$	Simpson's 3/8–rule
4	$\frac{14}{45}$	$\frac{64}{45}$	$\frac{24}{45}$	$\frac{64}{45}$	$\frac{14}{45}$			$h^7\frac{8}{945}f^{(6)}(\xi)$	Milne's rule*
5	$\frac{95}{288}$	$\frac{375}{288}$	$\frac{250}{288}$	$\frac{250}{288}$	$\frac{375}{288}$	$\frac{95}{288}$		$h^7\frac{275}{12096}f^{(6)}(\xi)$	—
6	$\frac{41}{140}$	$\frac{216}{140}$	$\frac{27}{140}$	$\frac{272}{140}$	$\frac{27}{140}$	$\frac{216}{140}$	$\frac{41}{840}$	$h^9\frac{9}{1400}f^{(8)}(\xi)$	Weddle's rule

where $w_i = \int_0^n K_i(t)\,dt$. Since the $\{w_i\}$ do not depend on the function $f(x)$, they can be pre-computed. For example, for $n = 2$ we find:

$$\begin{aligned}
w_0 &= \int_0^2 \left(\frac{t-1}{0-1}\right)\left(\frac{t-2}{0-2}\right)\,dt = \frac{1}{2}\int_0^2 \left(t^2 - 3t + 2\right)\,dt = \frac{1}{3}\\
w_1 &= \int_0^2 \left(\frac{t-0}{1-0}\right)\left(\frac{t-2}{1-2}\right)\,dt = -\int_0^2 \left(t^2 - 2t\right)\,dt = \frac{4}{3}\\
w_2 &= \int_0^2 \left(\frac{t-0}{2-0}\right)\left(\frac{t-1}{2-1}\right)\,dt = \frac{1}{2}\int_0^2 \left(t^2 - t\right)\,dt = \frac{1}{3}.
\end{aligned}$$

This gives rise to the integration rule:

$$\int_a^b f(x)\,dx \approx \int_a^b P_2(x)\,dx = \frac{h}{3}\left(f_0 + 4f_1 + f_2\right)$$

where $h = (b-a)/2$ and $f_k = f(a+kh)$. This is known as Simpson's rule.

The error in using the integration rules in (75.1) can be shown to be given by

$$\int_a^b P_n(x)\,dx - \int_a^b f(x)\,dx = h^{p_n+1}E_n f^{(p_n)}(\xi)$$

for some $\xi \in (a,b)$ where p_n and E_n are functions of n and not of $f(x)$.

The approximations given in (75.1) are known as the Newton–Cotes rules. Some tabulated values of the $\{w_i\}$, as well as the corresponding error term, are presented in Table 75.1. As indicated in that table, some of the Newton–Cotes rules also have other names.

* Also known as Boole's rule.

Table 75.2. Open Newton–Cotes rules obtained from polynomial interpolation.

n	w_i					Error	Name
2	2					$h^3 \frac{1}{3} f^{(2)}(\xi)$	midpoint rule
3	$\frac{3}{2}$	$\frac{3}{2}$				$h^3 \frac{3}{4} f^{(2)}(\xi)$	–
4	$\frac{8}{3}$	$-\frac{4}{3}$	$\frac{8}{3}$			$h^5 \frac{14}{45} f^{(4)}(\xi)$	–
5	$\frac{55}{24}$	$\frac{5}{24}$	$\frac{5}{24}$	$\frac{55}{24}$		$h^5 \frac{95}{144} f^{(4)}(\xi)$	–
6	$\frac{33}{10}$	$-\frac{21}{5}$	$\frac{39}{5}$	$-\frac{21}{5}$	$\frac{33}{10}$	$h^7 \frac{41}{140} f^{(6)}(\xi)$	–

Example

Consider the integral $I = \dfrac{1}{\sin 3} \displaystyle\int_0^3 \cos x \, dx = 1$. Define I_n to be the result of using the n-node rule from Table 75.1. Then we obtain the following approximations to I:

$$
\begin{aligned}
I &\approx I_2 = 0.1063722664 \\
I &\approx I_3 = 1.0379687263 \\
I &\approx I_4 = 1.0163528101 \\
I &\approx I_5 = 0.9994771775 \\
I &\approx I_6 = 0.9997082294 \\
I &\approx I_7 = 1.0000059300.
\end{aligned}
$$

Notes

[1] Newton–Cotes rules for large values of n are not often used since some of the weights $\{w_i\}$ become negative and numerical cancellation occurs in the computation.

[2] In the above we discretized the interval $[a, b]$ into $n+1$ nodes that included the endpoints a and b. Hence, the above formulas are sometimes called closed Newton–Cotes rules.

If we only consider the interior nodes, $\{x_i \mid x_i = a + ih, i = 1, \ldots, n-1\}$ (where, as before, $h = (b-a)/n$), and then approximate the given integral by the integral of the interpolating polynomial, then we will have derived on open formula. These formulas are sometimes called open Newton–Cotes rules. The first few such formulas are in Table 75.2.

[3] Instead of just interpolating the value of $f(x)$ at the nodes $\{x_i\}$, the values of $f(x)$ and $f'(x)$ may be used. For example, if values for the derivatives at the endpoints of the interval are given, then the approximate formula

$$
\int_a^b f(x) \, dx \approx Z_2(h) = \frac{h}{2} \left(f(a) + f(b) \right) + \frac{h^2}{12} \left(f'(a) - f'(b) \right)
$$

may be used (here, $h = (b - a)$). It can be shown that the error is given by

$$Z_2(h) - \int_a^b f(x)\, dx = \frac{h^5}{720} f^{(4)}(\xi)$$

with $\xi \in (a, b)$. See page 287.

[4] Gaussian quadrature rules (see page 289) are also interpolatory, but the nodes are not equidistant from one another. Instead, the node locations are chosen to make the rule have as high a degree as possible.

[5] Sometimes a quadrature formula is desired that integrates trigonometric polynomials, not ordinary polynomials, exactly. A trigonometric polynomial of degree m is a linear combination of the functions $\{1, \cos x, \sin x, \cos^2 x, \cos x \sin x, \sin^2 x, \ldots, \cos^m x, \cos^{m-1} x \sin x, \ldots \sin^m x\}$. Equivalently, a trigonometric polynomial of degree m is a linear combination of the functions $\{1, \cos x, \sin x, \cos 2x, \sin 2x, \ldots \cos mx, \sin mx\}$. The approximation

$$\int_0^{2\pi} f(x)\, dx \approx \sum_{k=1}^{n} h f\left(\beta + (k-1)h\right)$$

where $h = 2\pi/n$ and β is any real number satisfying $0 \leq \beta < h$, is exact for all trigonometric polynomials of degree $n - 1$ or less (see Mysovskikh [5]).

Vanden Berghe *et al.* [7] consider quadrature rules that exactly integrate ordinary polynomials and trigonometric polynomials.

[6] Given data values defined on a set of nodes, one polynomial can be fit to all of the data values, as shown above. Alternatively, the region of integration may be broken into smaller regions, with an interpolatory polynomial fit to the data values in each sub-region. Köhler [3] considers the case when the interpolatory polynomial on a sub-region uses data values from outside that sub-region.

[7] There are many interpolatory formulas, other than polynomials, that can be used to interpolate data. For example, De Meyer *et al.* [4] interpolate a set of values using the function $f(x) = e^{kx} \sum_{i=0}^{n} a_i x^i$. This interpolating function can then be integrated to obtain, for example, their modified trapezoidal rule:

$$\int_{x_0}^{x_0+h} f(x)\, dx \approx h \left\{ \frac{e^{\theta} - 1 - \theta}{\theta^2} f(x_0) + \frac{e^{-\theta} - 1 + \theta}{\theta^2} f(x_0 + h) \right\},$$

where $\theta = hk$ and k is an arbitrary parameter. This parameter is chosen in practice, by minimizing the error term. For the above rule, the leading order error term has the form $E = \dfrac{h^3}{\theta^2}\left[1 - \dfrac{4\sinh^2(\theta/2)}{\theta^2}\right](D_x - k)^2 f(\eta)$, where η is in the range $x_0 < \eta < x_0 + h$.

References

[1] P. J. Davis and P. Rabinowitz, *Methods of Numerical Integration*, Second Edition, Academic Press, Orlando, Florida, 1984, pages 74–81.

[2] F. B. Hiderbrand, *Introduction to Numerical Analysis*, McGraw–Hill Book Company, New York, 1974.

[3] P. Köhler, "On a Generalization of Compound Newton–Cotes Quadrature Formulas," *BIT*, **31**, 1991, pages 540–544.

[4] H. De Meyer, G. Vanden Berghe, and J. Vanthournout, "Numerical Quadrature Based on an Exponential Type of Interpolation," *Int. J. Comp. Math.*, **38**, 1991, pages 193–209.

[5] I. P. Mysovskikh, "On Cubature Formulas that Are Exact for Trigonometric Polynomials," *Soviet Math. Dokl.*, **36**, No. 2, 1988, pages 229–232.

[6] J. Stoer and R. Bulirsch, *Introduction to Numerical Analysis*, translated by R. Bartels, W. Gautschi, and C. Witzgall, Springer–Verlag, New York, 1976, pages 118–123.

[7] G. Vanden Berghe, H. De Meyer, and J. Vanthournout, "On a Class of Modified Newton–Cotes Quadrature Formulae Based Upon Mixed-type Interpolation," *J. Comput. Appl. Math.*, **31**, 1991, pages 331–349.

76. Product Rules

Applicable to Multidimensional integrals.

Yields

A numerical quadrature scheme.

Idea

Suppose two numerical quadrature schemes are known, one for r-dimensional Euclidean space and one for s-dimensional Euclidean space. The "product" of these two rules can be used to formulate a numerical quadrature scheme for $(r+s)$-dimensional Euclidean space.

Procedure

Let R (S) be a region in r-dimensional (s-dimensional) Euclidean space. Suppose we have the n-node and m-node quadrature rules

$$\begin{aligned} \int_R f(\mathbf{x})\,d\mathbf{x} &\simeq \sum_{j=1}^{n} w_j f(\mathbf{x}_j), \\ \int_S g(\mathbf{y})\,d\mathbf{y} &\simeq \sum_{k=1}^{m} v_k g(\mathbf{y}_k). \end{aligned} \tag{76.1.a–b}$$

Then an $(n+m)$-node quadrature rule for the region $B = R \times S$, in an $(r+s)$-dimensional space, is given by

$$\int_B h(\mathbf{x},\mathbf{y})\,d\mathbf{x}\,d\mathbf{y} \simeq \sum_{j=1}^{n}\sum_{k=1}^{m} w_j v_k h(\mathbf{x}_j,\mathbf{y}_k). \tag{76.2}$$

Example

On the interval $[a, b]$, Simpson's rule with three nodes approximates an integral by

$$\int_a^b f(x)\, dx \simeq \frac{h}{3}\left(f_0 + 4f_1 + f_2\right) \tag{76.3}$$

where $f_n = f(a + nh)$ and $h = (b - a)/2$. Likewise, on the interval $[c, d]$, Simpson's rule with five nodes approximates an integral by

$$\int_c^d g(y)\, dy \simeq \frac{k}{3}\left(g_0 + 4g_1 + 2g_2 + 4g_3 + g_4\right) \tag{76.4}$$

where $g_m = g(c + mk)$ and $k = (d - c)/4$.

Taking the product of the rules in (76.3) and (76.4) results in the following approximation of a two-dimensional integral

$$\begin{aligned}\int_a^b \int_c^d h(x, y)\, dx\, dy \simeq \frac{hk}{9}(&h_{0,0} + 4h_{0,1} + 2h_{0,2} + 4h_{0,3} + h_{0,4} \\ &+ 4h_{1,0} + 16h_{1,1} + 8h_{1,2} + 16h_{1,3} + 4h_{1,4} \\ &+ h_{2,0} + 4h_{2,1} + 2h_{2,2} + 4h_{2,3} + h_{2,4})\end{aligned}$$

where $h_{n,m} = h(a + nh, c + mk)$.

Notes

[1] If (76.1.a) exactly integrates $f(\mathbf{x})$, and if (76.1.b) exactly integrates $g(\mathbf{y})$, and $h(\mathbf{x}, \mathbf{y}) = f(\mathbf{x})g(\mathbf{y})$, then (76.2) will exactly integrate $h(\mathbf{x}, \mathbf{y})$.

[2] This technique can be used for general Cartesian product regions, not just parallelpipeds; for instance, circular cylinders, circular cylindrical shell, and triangular prisms.

[3] Stroud [3] analyzes product rules by use of transformations. Suppose the region of integration is S, and the integrals of interest have the weight function $w(\mathbf{x})$. If the quadrature rule is to be exact for polynomials, then $I = \iint\limits_S \cdots \int w(\mathbf{x})\, x_1^{\alpha_1} x_2^{\alpha_2} \ldots x_n^{\alpha_n}\, d\mathbf{x}$ must be integrated exactly for some set of $\{\boldsymbol{\alpha}_i\}$. If there exists a transformation of the form $\mathbf{x} = \mathbf{x}(\mathbf{u})$ that turns I into the product $I = \left(\int w_1(u_1) g_1(u_1)\, du_1\right) \ldots \left(\int w_n(u_n) g_n(u_n)\, du_n\right)$, and if suitable formulas are known for these single integrals, then one has obtained a product rule.

[4] Using product rules, the number of nodes at which the integrand must be evaluated grows exponentially with the dimension of the integration. If a one-dimensional quadrature rule that uses 19 nodes is the basis for a 7-dimensional quadrature rule, then $19^7 \simeq 10^9$ integrand evaluations are required.

[5] The rules devised by this technique are often not the most efficient in terms of number of integrand evaluations.

[6] Often, a more computationally efficient quadrature rule for a multidimensional integral can be found. For example, Acharya and Mohapatra [1] give the two-dimensional quadrature rule for analytic functions:

$$\int_{z_0-h}^{z_0+h}\int_{z_0-h'}^{z_0+h'} f(z,z')\,dz\,dz' \approx hh'\Big[-256f_{00}+25\,(f_{11}+f_{13}+f_{31}+f_{33}) + 40\,(f_{20}+f_{02}+f_{40}+f_{04})\Big] \tag{76.5}$$

where $f_{\alpha\beta} = f(z_\alpha, z'_\beta)$, $z_\alpha = z_0 + hki^{\alpha-1}$, and $z'_\alpha = z_0 + h'ki^{\alpha-1}$. When $k = 1/\sqrt{15}$, the rule in (76.5) has degree of precision 5.

[7] A quadrature rule for the n-cube C_n can sometimes be used as the basis for a quadrature rule for C_m (with $m > n$). The rule for C_m is then called an extended rule. Product methods, described in this section, are only one way in which a rule can be extended. For other methods, see Stroud [3] (Chapter 4).

References

[1] B. P. Acharya and T. Mohapatra, "Approximations of Double Integrals of Analytic Functions of Two Complex Variables," *Computing*, **37**, 1986, pages 357–364.

[2] P. J. Davis and P. Rabinowitz, *Methods of Numerical Integration*, Second Edition, Academic Press, Orlando, Florida, 1984, pages 354–363.

[3] A. H. Stroud, *Approximate Calculation of Multiple Integrals*, Prentice–Hall Inc., Englewood Cliffs, NJ, 1971, Chapter 2, pages 23–47.

77. Recurrence Relations

Applicable to Integrals for which a recurrence relation can be found.

Yields

An asymptotic approximation, or a numerical computation scheme.

Idea

If a recurrence relation can be found for an integral, then it may be used to determine an asymptotic approximation, or it can form the basis of a numerical computation.

Procedure

Often, an integral can be written in terms of a recurrence relation and some initial (or boundary) condition(s). (This is frequently accomplished by integration by parts.) This recurrence relation can be used to obtain asymptotic information about the integral.

Or, the recurrence relation can be iteratively applied to determine numerical values for the integral. These numerical computations should only be performed after an asymptotic analysis has been performed, to prevent roundoff errors from ruining the numerical accuracy.

Consider the three term recurrence relation

$$y_{n+1} + a_n y_n + b_n y_{n-1} = 0, \qquad n = 1, 2, \ldots, \tag{77.1}$$

where $\{a_n\}$ and $\{b_n\}$ are given sequences of real or complex numbers, and $b_n \neq 0$. From Van der Laan and Temme [6] we have the following theorem:

> Let a_n and b_n have the asymptotic behavior
>
> $$a_n \sim an^\alpha, \quad b_n \sim bn^\beta \quad \text{as } n \to \infty \tag{77.2}$$
>
> with $ab \neq 0$ and both of α and β real. Let t_1 and t_2 be roots of the characteristic polynomial $t^2 + at + b = 0$, with $|t_1| \geq |t_2|$. Then there are three cases:
>
> (1) If $\alpha > \frac{1}{2}\beta$ then (77.1) has two linearly independent solutions u_n and v_n for which
>
> $$\frac{u_{n+1}}{u_n} \sim -an^\alpha, \qquad \frac{v_{n+1}}{v_n} \sim -\frac{b}{a}n^{\beta-\alpha} \quad \text{as } n \to \infty. \tag{77.3}$$
>
> (2) If $\alpha = \frac{1}{2}\beta$ then (77.1) has two linearly independent solutions u_n and v_n for which
>
> $$\frac{u_{n+1}}{u_n} \sim t_1 n^\alpha, \qquad \frac{v_{n+1}}{v_n} \sim t_2 n^\alpha \quad \text{as } n \to \infty \tag{77.4}$$
>
> provided that $|t_1| > |t_2|$. If $|t_1| = |t_2|$ then
>
> $$\lim_{n\to\infty} \sup \left(|y_n| \, (n!)^{-\alpha}\right)^{1/n} = |t_1| \tag{77.5}$$
>
> for all non-trivial solutions of (77.1).
>
> (3) If $\alpha < \frac{1}{2}\beta$ then
>
> $$\lim_{n\to\infty} \sup \left(|y_n| \, (n!)^{-\beta/2}\right)^{1/n} = |b|^{1/2} \tag{77.6}$$
>
> for all non-trivial solutions of (77.1).

The solution of the recurrence relation in (77.1) can be written in the form $y_n = Au_n + Bv_n$, where A and B are constants.

Table 77.1. A recursion computation of the Bessel functions $\{Y_n\}$ using (77.7). This computation is stable.

$\{y_n\}$	$Y_n(5)$
$y_1 = 0.1478631434$	$Y_1(5) = 0.1478631434$
$y_2 = 0.3676628826$	$Y_2(5) = 0.3676628826$
$y_5 = -0.4536948225$	$Y_5(5) = -0.4536948225$
$y_{10} = -25.129110$	$Y_{10}(5) = -25.129110$
$y_{12} = -382.9821416$	$Y_{12}(5) = -382.9821416$
$y_{14} = -8693.938814$	$Y_{14}(5) = -8693.938814$
$y_{16} = -272949.0350$	$Y_{16}(5) = -272949.0350$
$y_{20} = -593396529.7$	$Y_{20}(5) = -593396529.7$
$y_{30} = -4.028568418 \times 10^{18}$	$Y_{30}(5) = -4.028568418 \times 10^{18}$

Example

The Bessel functions $Y_n(z)$, when n is an integer, are defined by (see page 174):

$$Y_n(z) = \frac{1}{\pi}\int_0^{\pi} \sin(z\sin\theta - n\theta)\,d\theta - \frac{1}{\pi}\int_0^{\infty}\left[e^{nt} + (-1)^n e^{-nt}\right] e^{-z\sinh t}\,dt.$$

From this, or otherwise (see Abramowitz and Stegun [1] 9.1.27.a), it can be shown that these Bessel functions satisfy the recurrence relation (using $y_n = Y_n(z)$)

$$y_{n+1} - \frac{2n}{z}y_n + y_{n-1} = 0. \tag{77.7}$$

The above theorem can be used on this recurrence relation with $a = -2/z$, $\alpha = 1$, $b = 1$, and $\beta = 0$. We find that case (1) applies and results in $u_{n+1}/u_n \sim 2n/z$ and $v_{n+1}/v_n \sim z/2n$. This implies that

$$Y_n(z) \sim A\left(\frac{2n}{z}\right)^n + B\left(\frac{z}{2n}\right)^n,$$

as $n \to \infty$, for some values of A and B. Note that if $A \neq 0$, then the first term dominates the asymptotic expansion.

It can be shown that $Y_n(z) \sim n!\,(2/z)^n$ as $n \to \infty$ (see Abramowitz and Stegun [1], 9.1.8). This asymptotic expansion agrees with the results of the theorem; here we have $A \neq 0$. If (77.7) is used to compute $\{Y_n(z)\}$, then roundoff errors will result in $\{v_n\}$ terms. Since the evolution of these terms is much smaller than the $\{Y_n(z)\}$ terms, this will be stable computation. Table 77.1 shows the computation of $\{Y_n(5)\}$ using (77.7). The values of $Y_1(5)$ and $Y_2(5)$ were used to initialize the recurrence relation. For large value of n, the computation is accurate to all decimal places.

Table 77.2. A recursion computation of the Bessel functions $\{J_n\}$ using (77.7). This computation is not stable.

$\{y_n\}$	$J_n(5)$
$y_1 = -0.3275791376$	$J_1(5) = -0.3275791376$
$y_2 = 0.04656511628$	$J_2(5) = 0.04656511628$
$y_5 = 0.2611405461$	$J_5(5) = 0.2611405461$
$y_{10} = 0.00146780258$	$J_{10}(5) = 0.00146780265$
$y_{12} = 0.0000762771$	$J_{12}(5) = 0.0000762781$
$y_{14} = 0.000002778$	$J_{14}(5) = 0.000002801$
$y_{16} = -0.00000065$	$J_{16}(5) = -0.000000077$
$y_{20} = -0.0016$	$J_{20}(5) = 2.7 \times 10^{-11}$
$y_{30} = -10^7$	$J_{30}(5) = 2.7 \times 10^{-21}$

There are other Bessel functions that satisfies the recursion in (77.7), the $\{J_n(z)\}$. It can be shown that $J_n(z) \sim \dfrac{1}{n!}\left(\dfrac{z}{2}\right)^n$ as $n \to \infty$ (see Abramowitz and Stegun [1], 9.1.7), which indicates that $J_n(z) \sim B\left(z/2n\right)^n$. If (77.7) is used to compute $\{J_n(z)\}$, then roundoff errors will result in $\{u_n\}$ terms. Since the evolution of these terms is much greater than the $\{J_n(z)\}$ terms, this will be an unstable computation. Table 77.2 shows the computation of $\{J_n(5)\}$ using (77.7). The values of $J_1(5)$ and $J_2(5)$ were used to initialize the recurrence relation. For a few values of n, the computation is accurate. For n above about 10, however, the computational values are not meaningful.

Notes

[1] Van der Laan and Temme [6] indicate the results of applying the above theorem to Bessel functions (as we have), confluent hypergeometric functions (two different recursions), incomplete beta functions, Legendre functions (recursion with respect to order and with respect to degree), Jacobi polynomials. and repeated integrals of the error function. All cases of the theorem are illustrated.

[2] The exponential integrals $E_n(z) = \int_1^\infty t^{-n} e^{-tz}\,dt$ have the recurrence relation $nE_{n+1}(z) = e^{-z} - zE_n(z), n = 1, 2, \ldots$. This relation is studied in Gautschi [3].

[3] For the integrals I_n, J_n, and K_n

$$I_n(c) := \int_0^\infty \frac{te^{-ct}}{(1+t^2)\sqrt{t}}\left(\frac{t^2}{1+t^2}\right)^n dt,$$

$$J_n(c) := \int_0^\infty \frac{e^{-ct}}{(1+t^2)\sqrt{t}}\left(\frac{t^2}{1+t^2}\right)^n dt,$$

$$K_n(c) := \int_0^\infty \frac{e^{-ct}}{\sqrt{t}}\left(\frac{t^2}{1+t^2}\right)^n dt,$$

with $c \geq 0$ and $n \geq 0$, Acton [2] finds the recurrence relations:

$$\begin{aligned} I_{n-1} &= \frac{4nI_n + 2cK_n}{4n-1}, \\ J_{n-1} &= \frac{4nJ_n + 2cI_{n-1}}{4n-3}, \\ K_{n-1} &= K_n + J_{n-1}. \end{aligned}$$

References

[1] M. Abramowitz and I. A. Stegun, *Handbook of Mathematical Functions*, National Bureau of Standards, Washington, DC, 1964.

[2] F. S. Acton, "Recurrence Relations for the Fresnel Integral $\int_0^\infty \frac{\exp(-ct)\,dt}{\sqrt{t}(1+t^2)}$ and Similar Integrals," *Comm. ACM*, **17**, No. 8, 1974, pages 480–481.

[3] W. Gautschi, "Recursive Computation of Certain Integrals," *J. ACM*, **8**, 1961, pages 21–40.

[4] W. Gautschi, "Computational Aspects of Three-terms Recurrence Relations," *SIAM Review*, **9**, 1967, pages 24–82.

[5] W. Gautschi, "Recursive Computation of the Repeated Integrals of the Error Function," *Math. of Comp.*, **15**, 1961, pages 227–232.

[6] C. G. Van der Laan and N. M. Temme, *Calculation of Special Functions: The Gamma Function, the Exponential Integrals and Error-like Functions*, Centrum voor Wiskunde en Informatica, Amsterdam, 1984.

[7] E. W.-K. Ng, "Recursive Formulae for the Computation of Certain Integrals of Bessel Functions," *J. Math. and Physics*, **46**, 1967, pages 223–224.

[8] J. Wimp, *Computations with Recurrence Relations*, Pitman Publishing Co., Marshfield, MA, 1984.

78. Symbolic Methods

Applicable to Definite integrals.

Yields

A numerical quadrature scheme.

Idea

Using symbolic operators, quadrature rules can be devised.

Procedure

Define the following operators:

- The forward-differencing operator Δ:
  $\Delta f(x_0) = f(x_0 + h) - f(x_0)$
  $\Delta^2 f(x_0) = \Delta[\Delta f(x_0)] = \Delta f(x_0 + h) - \Delta f(x_0)$
- The backward-differencing operator ∇:
  $\nabla f(x_0) = f(x_0) - f(x_0 - h)$
  $\nabla^2 f(x_0) = \nabla[\nabla f(x_0)] = \nabla f(x_0) - \nabla f(x_0 - h)$
- The stepping operator E:
  $Ef(x_0) = f(x_0 + h)$
  $E^2 f(x_0) = E[Ef(x_0)] = f(x_0 + 2h)$
  $E^n f(x_0) = f(x_0 + nh)$.

The obvious relationships among these operators are $\Delta = E - 1$ and $\nabla = 1 - E^{-1}$. Since the above operators are linear operators, the usual laws of algebra can be applied. Hence, several other relationships between the operators can be developed, such as: $E\nabla = \Delta$, $E^n\nabla^n = \Delta^n$, and $\Delta^n f_0 = \nabla^n f_n$ (where $f_n = f(x_n) = f(x_0 + nh)$).

Since $y_n = E^n y_0$ and $x = nh$, we have

$$y_n' = \frac{d}{dx}(E^n y_0) = \frac{1}{h}\frac{d}{dn}(E^n y_0) = \frac{1}{h}(\log E)(E^n y_0),$$

or $D = \dfrac{1}{h}\log E$, where D denotes the differentiation operator. This is equivalent to $E = e^{hD}$. The integration operation, which is the inverse operator to D, can be represented as $\int = D^{-1} = h/\log E$.

Example

Using the above operators, we can derive many quadrature rules. Start with $f_s = f(x_s) = E^s f_0$, multiply by $dx = h\,ds$, and then integrate from x_0 to x_1 (i.e., from $s = 0$ to $s = 1$) to obtain

$$\int_{x_0}^{x_1} f(x)\,dx = h\int_0^1 E^s f_0\,ds = \left(\frac{h}{\log E}E^s f_0\right)\Bigg|_{s=0}^{s=1} = \frac{h(E-1)}{\log E}f_0.$$

Using $E = 1 + \Delta$, the expression $\log(1+\Delta)$ may be formally expanded in a power series to obtain $\log(1+\Delta) = \Delta - \frac{1}{2}\Delta^2 + \frac{1}{3}\Delta^3 - \frac{1}{4}\Delta^4 + \ldots$. Dividing $\Delta = E - 1$ by this last expression results in

$$\begin{aligned}\int_{x_0}^{x_1} f(x)\,dx &= \frac{h\Delta}{\Delta - \frac{1}{2}\Delta^2 + \frac{1}{3}\Delta^3 - \frac{1}{4}\Delta^4 + \ldots}f_0 \\ &= h\left(f_0 + \tfrac{1}{2}\Delta f_0 - \tfrac{1}{12}\Delta^2 f_0 + \tfrac{1}{24}\Delta^3 f_0 - \ldots\right) \\ &= h\left(f_0 + \tfrac{1}{2}(f_1 - f_0) - \tfrac{1}{12}(f_2 - 2f_1 + f_0) + \ldots\right).\end{aligned} \tag{78.1}$$

If n terms of this formula are used, then a polynomial of degree n is being fit to the points $x_0, x_1, \ldots, x_n$; this interpolating polynomial is then integrated between x_0 and x_1. If only the first two terms are used, then the trapezoidal rule is obtained in the form

$$\int_{x_0}^{x_1} f(x)\,dx \approx f_0 + \frac{1}{2}(f_1 - f_0) = \frac{1}{2}(f_0 + f_1).$$

Notes

[1] Analogous to the result in the example, we can obtain

$$\int_{x_0}^{x_2} f(x)\,dx = \left(\frac{hE^s}{\log E} f_0\right)\Bigg|_{s=0}^{s=2} = \frac{h(E^2-1)}{\log E} f_0$$
$$= h\left(2f_0 + 2\Delta f_0 + \tfrac{1}{3}\Delta^2 f_0 + \ldots\right).$$

[2] If we define the central-differencing operator δ by $\delta f(x_0) = f(x_0 + \frac{1}{2}h) - f(x_0 - \frac{1}{2}h)$, then the following integration rules can be found (see Beyer [1]):

$$\int_{-1/2}^{1/2} f(x)\,dx = \left(1 + \tfrac{1}{24}\delta^2 - \tfrac{17}{5760}\delta^4 + \frac{367}{967680}\delta^6 - \ldots\right) f_0$$
$$\int_{-1}^{1} f(x)\,dx = 2\left(1 + \tfrac{1}{6}\delta^2 - \tfrac{1}{180}\delta^4 + \frac{1}{1512}\delta^6 - \ldots\right) f_0$$
$$\int_{-2}^{2} f(x)\,dx = 4\left(1 + \tfrac{2}{3}\delta^2 + \tfrac{7}{90}\delta^4 - \frac{2}{945}\delta^6 + \ldots\right) f_0.$$

[3] The operators defined above can also be used in some clever manipulations of integrals. For example, Ullah [4] evaluates $I = \displaystyle\int_{-1}^{1} \frac{P_l^m(x)P_k^n(x)}{(1-x^2)^{p+1}}\,dx$, (with some restrictions on l, m, k, n, and p), in closed form, by first writing it in the form:

$$I = \int_{-1}^{1} (1-x^2)^{-n/2-p-1} P_l^m(x) e^{(1-x^2)D_\lambda} \lambda^{n/2} P_k^n\left(\sqrt{1-\lambda}\right)\Bigg|_{\lambda=0},$$

where $D_\lambda = \partial/\partial\lambda$.

References

[1] W. H. Beyer (ed.), *CRC Standard Mathematical Tables and Formulae*, 29th Edition, CRC Press, Boca Raton, Florida, 1991.

[2] C. F. Gerald and P. O. Wheatley, *Applied Numerical Analysis*, Addison–Wesley Publishing Co., Reading, MA, 1984, pages 256–257.

[3] F. B. Hiderbrand, *Introduction to Numerical Analysis*, McGraw–Hill Book Company, New York, 1974.

[4] N. Ullah, "Evaluation of an Integral Involving Associated Legendre Polynomials and Inverse Powers of $(1-x^2)$," *J. Math. Physics*, **4**, No. 4, April 1984, pages 872–873.

79. Tschebyscheff Rules

Applicable to One-dimensional definite integrals.

Yields

A numerical quadrature rule.

Idea

In Tschebyscheff rules the weight associated with each node is constant.

Procedure

Tschebyscheff rules are quadrature rules of the form

$$\int_{-1}^{1} f(x)\,dx \approx \sum_{k=1}^{N} w f(x_k). \tag{79.1}$$

Note that the weight function on the right-hand side is constant for all nodes in the summation. This formula can be made to be exact for all polynomials $f(x) = x^n$ for $n = 0, 1, \ldots, N$. Setting up the necessary equations, and then solving them, we find that $w = 2/N$, and the $\{x_k\}$ are the roots of the polynomial part of

$$P_N = u^N \exp\left(-\frac{s_1}{u} - \frac{s_2}{2u^2} - \frac{s_3}{3u^3} - \cdots\right)$$

where $s_i = \dfrac{N}{2}\displaystyle\int_{-1}^{1} u^i\,du = \begin{cases} 0 & \text{for odd } i, \\ \dfrac{N}{i+1} & \text{for even } i. \end{cases}$ The polynomial part, Π_N, has degree N.

Example

For $N = 2$ we find

$$\begin{aligned} P_2 &= u^2 \exp\left(-\frac{2}{6u^2} - \frac{2}{12u^4} - \cdots\right) \\ &= u^2\left(1 - \frac{2}{6u^2} + O\left(\frac{1}{u^4}\right)\right)\left(1 - \frac{2}{12u^4} + O\left(\frac{1}{u^8}\right)\right)\left(1 + O\left(\frac{1}{u^6}\right)\right) \\ &= \left(u^2 - \frac{1}{3}\right) + O\left(\frac{1}{u^2}\right). \end{aligned}$$

Therefore, $\Pi_2 = x^2 - 1/3$ and so $x_k = \pm 1/\sqrt{3} \approx \pm 0.5773$.

For $N = 3$ we find the polynomial part of P_3 to be $\Pi_3 = x^3 - x/2$. Hence, for $N = 3$, the nodes are located at $x_k = \{0, \pm 1/\sqrt{2}\}$.

Notes

[1] The reason that Tschebyscheff rules are of interest is that sometimes the integrand values, the $f(x_i)$, are observed with errors. In this case the variance of the quadrature rule, $\int_a^b f(x)\,dx \approx \sum_{i=1}^n w_i f(x_i)$, is given by $\sigma^2 = \sum_{i=1}^n w_i^2$. The value of σ^2 is minimized when all the weights are equal. (Presuming that the integration rule is going to integrate constants exactly, which results in the constraint $\sum_i w_i = 1$.)

[2] The equations $\prod_N(x) = 0$ have complex roots for $N = 8$ and for $N \geq 10$, so that there do not exist useful Tschebyscheff rules in these cases.

[3] Tschebyscheff rules have been extended to quadrature rules with weight functions, $\int_a^b w(x)f(x)\,dx \approx \sum_{i=1}^n wf(x_i)$, see Förster [2].

[4] Guerra and Vincenti [6] consider quadrature rules of the form

$$\int_{-1}^{1} w(x)f(x)\,dx = \sum_{h=0}^{r} A_{2h} \sum_{j=1}^{n} f^{(2h)}(x_j) + R(f).$$

References

[1] S. Iyanaga and Y. Kawada, *Encyclopedic Dictionary of Mathematics*, MIT Press, Cambridge, MA, 1980, page 929.

[2] K.-J. Förster, "On Chebyshev Quadrature for a Special Class of Weight Functions," *BIT*, **26**, No. 3, 1986, pages 327–332.

[3] K.-J. Förster, "On Weight Functions Admitting Chebyshev Quadrature," *Math. of Comp.*, **49**, No. 179, 1987, pages 251–258.

[4] K.-J. Förster, "On Chebyshev Quadrature and Variance of Quadrature Formulas," *BIT*, **28**, No. 2, 1988, pages 360–363.

[5] K.-J. Förster and G. P. Ostermeyer, "On Weighted Chebyshev-type Quadrature Formulas," *Math. of Comp.*, **46**, No. 174, 1986, pages 591–599, S21–S27.

[6] S. Guerra and G. Vincenti, "The Chebyshev Problem for Quadrature Formulas with Derivatives of the Integrand," **22**, No. 3, 1985, pages 335–349.

[7] D. K. Kahaner, "On Equal and Almost Equal Weight Quadrature Formulas," *SIAM J. Numer. Anal.*, **6**, 1968, pages 551–556.

[8] W. Squire, *Integration for Engineers and Scientists*, American Elsevier Publishing Company, New York, 1970, pages 132–135.

80. Wozniakowski's Method

Applicable to Multidimensional integrals on a hypercube.

Yields

A numerical approximation scheme which uses the smallest number of nodes for a specified average error.

Idea

Wozniakowski has devised a way to choose nodes optimally to approximate a multidimensional integral numerically.

Procedure

Wozniakowski's technique applies to an integral over the d-dimensional unit hypercube

$$I = \underbrace{\int_0^1 \int_0^1 \cdots \int_0^1}_{d} f(\mathbf{x})\, d\mathbf{x}$$

where $\mathbf{x} = (x_1, x_2, \ldots, x_d)$. Given a (small) value of ε, we would like to approximate this integral by the formula $I \approx \frac{1}{n} \sum_{k=1}^{n} f(\mathbf{x}_k)$, and have the "average error" be less than ε. This requires determining the number of nodes n and determining the location of the nodes $\{\mathbf{x}_k\}$.

If we restrict ourselves to integrands that are real and continuous (some class must be specified for an average error to make sense), then Wozniakowski finds that as $\varepsilon \to 0$, the number of nodes needed is

$$n = O\left(\frac{1}{\varepsilon} |\log \varepsilon|^{(d-1)/2}\right). \tag{80.1}$$

Note that the number of nodes required to obtain the same accuracy using Monte Carlo techniques ($n_{\text{MonteCarlo}}$, see page 304) or using a uniform grid (n_{uniform}, see page 323) are

$$n_{\text{MonteCarlo}} = O\left(\frac{1}{\varepsilon^2}\right), \qquad n_{\text{uniform}} = O\left(\frac{1}{\varepsilon^d}\right).$$

Hence, the number of integrand evaluations in (80.1) is far less than the number of evaluations needed by these other two methods.

Wozniakowski does not give formulae on how to determine the optimal nodes $\{\mathbf{x}_k\}$ exactly, but he does give formulae for determining nodes $\{\widehat{\mathbf{x}}_k\}$ and $\left\{\widehat{\widehat{\mathbf{x}}}_k\right\}$ that are "nearly as good." That is, more of these nodes are required for the same average accuracy, but the number of nodes required only increases a little (the exponent $(d-1)/2$ in (80.1) changes to $(d-1)$ or d). Even with this many nodes, it represents a substantial improvement over using Monte Carlo methods or a uniform grid.

The procedure for determining the node locations is straightforward. For each value of k, start by writing k in base 2, then in base 3, then base 5, etc., using the first $d-1$ primes as the bases. As an example, we choose $d = 6$ and $k = 42$. Then we find:

$$42 = (101010)_2 = (1120)_3 = (132)_5 = (60)_7 = (39)_{11}.$$

Now, for each of these primes p, calculate the "radical inverse" for k, $\phi_p(k)$, which is obtained by reversing the digits of the base-p representation of k, then dividing the result by p^i, where i is the number of digits in that representation. We have:

$$\begin{aligned} \phi_2(42) &= \frac{(010101)_2}{2^6} = \frac{21}{64}, \qquad \phi_3(42) = \frac{(0211)_3}{3^4} = \frac{22}{81}, \\ \phi_5(42) &= \frac{66}{125}, \qquad \phi_7(42) = \frac{6}{49}, \qquad \phi_{11}(42) = \frac{102}{121}. \end{aligned} \tag{80.2}$$

Finally, the $\{\mathbf{x}_k\}$ are determined by

$$\mathbf{x}_k = (1, 1, ..., 1) - \left(\frac{k+t}{n}, \phi_2(k), \phi_3(k), \ldots, \phi_{p_{d-1}}(k)\right)$$

where t is some constant which Wozniakowski does not evaluate. Since this constant is unknown, we could delete it and use instead

$$\widehat{\mathbf{x}}_k = (1, 1, ..., 1) - \left(\frac{k}{n}, \phi_2(k), \phi_3(k), \ldots, \phi_{p_{d-1}}(k)\right). \tag{80.3}$$

If we do this, then the number of nodes needed varies as $\frac{1}{\varepsilon}|\log \varepsilon|^q$ with $q = d - 1$ rather than the optimal $q = (d - 1)/2$. Observe that, for $d = 6$, this results in (using the values in (80.2))

$$\widehat{\mathbf{x}}_{42} = \left(\tfrac{n-42}{n}, \tfrac{43}{64}, \tfrac{59}{81}, \tfrac{59}{125}, \tfrac{43}{49}, \tfrac{19}{121}\right).$$

Note that we will have to recalculate the first component of each $\mathbf{x}_k$ if we decide to increase n. We can avoid this recalculation by using the first d primes (rather than the first $d - 1$) and instead of the $\mathbf{x}_k$ or the $\widehat{\mathbf{x}}_k$ above, choose the nodes to be

$$\widehat{\widehat{\mathbf{x}}}_k = (1, 1, ..., 1) - (\phi_2(k), \phi_3(k), ..., \phi_{p_d}(k)).$$

This results in the error estimate $\frac{1}{\varepsilon}|\log \varepsilon|^q$ with $q = d$. Observe that, for $d = 6$, this produces the 42nd node

$$\widehat{\widehat{\mathbf{x}}}_{42} = \left(\tfrac{43}{64}, \tfrac{59}{81}, \tfrac{59}{125}, \tfrac{43}{49}, \tfrac{19}{121}, \tfrac{127}{169}\right).$$

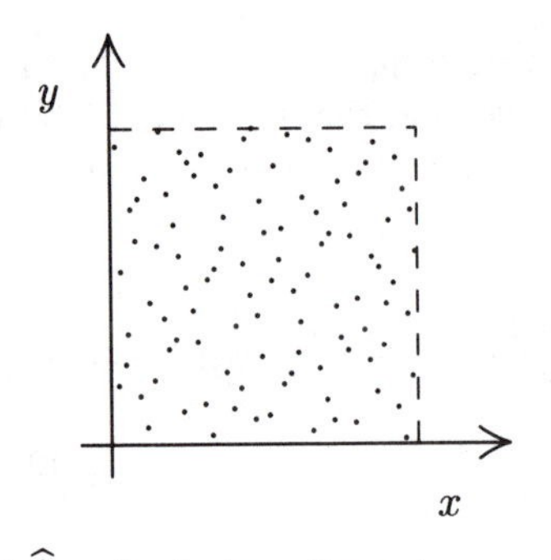

Figure 80. The first 100 $\widehat{\widehat{\mathbf{x}}}$ nodes in two dimensions.

Example

If 100 nodes are be used in an integration routine, then the (nearly) "optimal nodes," the $\{\widehat{\widehat{\mathbf{x}}}_k\}$ nodes, are shown Figure 80.

Notes

[1] The technique presented in this section comes with two obvious caveats:

(A) There is no estimate of the overall multiplicative constants needed to find the requisite n from the desired error ε, nor do we know how small ε must be before we get into the asymptotic regime where the expressions for n are valid.

(B) This technique is based on the average error, not the worst-case error, so we could be unlucky and do very badly for the integrand we are really interested in.

[2] In one dimension, the number of nodes needed to obtain a minimal average error has a more concise answer. For the class of r times continuously differentiable functions equipped with a specific type of probability distribution, the number of nodes required is $O\left(\left(\sqrt{\frac{|B_{2r+1}|}{(2r+2)!}\frac{1}{\varepsilon}}\right)^{1/(r+1)}\right)$, where B_{2r+2} is a Bernoulli number. See Traub *et al.* [4] for details.

[3] Writing the nodes in (80.3) as $\widehat{\mathbf{x}}_k = (1, 1, ..., 1) - \mathbf{z}_k$, we recognize the $\{\mathbf{z}_k\}$ to be *Hammersley points* (see Davis and Rabinowitz [2]). Removing the first component from $\{\mathbf{z}_k\}$ results in the *Halton points*. Berblinger and Schlier [1] used the *Halton points* as a "quasi-random" sequence of nodes in a Monte Carlo type computation.

[4] In this section we have chosen nodes to minimize the average error. Many results are known for the number of nodes needed to minimize the maximum error. For example, considering the Sobolev class of real functions defined on the d-dimensional unit hypercube whose r-th distribution derivatives exist and are bounded by one in the L_p norm, the number of nodes required is $O\left(\varepsilon^{-d/r}\right)$ when $pr > d$. See Novak [3] for details.

References

[1] M. Berblinger and C. Schlier, "Monte Carlo Integration with Quasi-random Numbers: Some Experience," *Comput. Physics Comm.*, **66**, 1991, pages 157–166.

[2] P. J. Davis and P. Rabinowitz, *Methods of Numerical Integration*, Second Edition, Academic Press, Orlando, Florida, 1984, pages 350–354.

[3] E. Novak, *Deterministic and Stochastic Error Bounds in Numerical Analysis*, Lecture Notes in Mathematics #1349, Springer–Verlag, New York, 1988.

[4] J. F. Traub, G. W. Wasilkowski, and H. Woźniakowski, *Information-based Complexity*, Academic Press, New York, 1988.

[5] H. Wozniakowski, "Average Case Complexity of Multivariate Integration," *Bull. Amer. Math. Soc.*, January 1991, **24**, No. 1, pages 185–194.

81. Tables: Numerical Methods

Applicable to Definite integrals.

Yields

Quadrature rules.

Idea

Many tables of numerical methods for integrals have been created.

Procedure

As indicated in this book, there are many ways in which to derive a scheme that will evaluate an integral numerically. Many books and papers have tabulated schemes for numerically evaluating integrals. While these are not adaptive schemes, they are useful because they may be directly entered into a computer.

In this section we merely reference where collections of quadrature rules may be found. Some one-dimensional and two-dimensional quadrature rules may be found starting on page 340.

Example 1

The book by Stroud [8] is perhaps the most comprehensive book on quadrature rules. It contains rules for the following regions:

[1] the n-dimensional cube: C_n
[2] the n-dimensional cubical shell: C_n^{shell}
[3] the n-dimensional sphere: S_n
[4] the n-dimensional spherical shell: S_n^{shell}
[5] the surface of S_n: U_n
[6] the n-dimensional octahedron: G_n
[7] the n-dimensional simplex: T_n
[8] entire n-dimensional space with weight function $\exp\left(-x_1^2 - \ldots - x_n^2\right)$

[9] entire n-dimensional space with weight function $\exp\left(-\sqrt{x_1^2+\ldots+x_n^2}\right)$
[10] the two-dimensional hexagon: H_2
[11] the two-dimensional ellipse with weight function $\left[(x-c)^2+y^2\right]^{-1/2}$ $\left[(x+c)^2+y^2\right]^{-1/2}$: E_{LP}
[12] the parabolic region bounded by $x=a^2-y^2/4a^2$ and $x=y^2/4b^2-b^2$: PAR
[13] the parabolic region bounded by $y=b-bx^2/a^2$ and the x-axis.: PAR_2
[14] the parabolic region bounded by $y=b-bx^2/a^2$ and $y=bx^2/a^2-b$: PAR_3
[15] a three dimensional pyramid: CN:C_2
[16] a three-dimensional cone: CN:S_2
[17] a three-dimensional torus with circular cross section: $\text{TOR}_3 : S_2$
[18] a three-dimensional torus with square cross section: $\text{TOR}_3 : C_2$

where Stroud's notation for some of the regions has been used.

For each of these regions, quadrature rules of varying degree and with a varying number of nodes are given. For example, Stroud reports the following quadrature rules for the n-dimensional cube (the numbering of the methods is his):

1-1 Degree 1, 1 Node (Centroid formula)

Node	Weight
$(0,0,\ldots,0)$	V

where V represents the volume of C_n, that is $V=2^n$.

1-2 Degree 1, 2^n Nodes (Product trapezoidal rule)

Nodes	Weights
$(\pm1,\pm1,\ldots,\pm1)$	$V/2^n$

where the symbol $(\pm s,\pm s,\ldots,\pm s)$ denotes a set of 2^n nodes, the n signs $\pm$ are assumed independent.

2-2 Degree 2, $2n+1$ Nodes

Nodes	Weights
$(2r,2r,\ldots,2r,2r)$	V
$(1,r,\ldots,r,r)_S$	$-rV$
$(-1,r,\ldots,r,r)_S$	rV

where $r=\sqrt{3}/6$ and a subscript of S indicates a symmetric set (i.e., all permutations) of nodes. For example: $(1,r,\ldots,r,r)_S$ denotes the n nodes: $\{(1,r,r,\ldots,r), (r,1,r,\ldots,r), (r,r,1,\ldots,r), \ldots, (r,r,r,\ldots,1)\}$.

3-4 Degree 3, 2^n Nodes (Product Gauss rule)

Nodes	Weights
$(\pm r,\pm r,\ldots,\pm r)$	$V/2^n$

where $r=1/\sqrt{3}$.

5-9 Degree 5, 3^n Nodes (Product Gauss formula)

Nodes $(r_{i_1}, r_{i_2}, \ldots, r_{i_n})$ Weights $(A_{i_1} A_{i_2} \cdots A_{i_n})$

where each of the subscripts $i_1, i_2, \ldots, i_n$ ranges independently over the integer $1, 2, 3$ and $r_1 = -\sqrt{3/5}$, $r_2 = 0$, $r_3 = \sqrt{3/5}$, $A_1 = 5/9$, $A_2 = 8/9$, $A_3 = 5/9$.

Example 2

The book by Stroud and Secrest [9] contains tables of Gaussian quadrature rules for the following types of integrals:

[1] $\int_{-1}^{1} f(x)\,dx \approx \sum_{i=1}^{N} A_i f(x_i)$

[2] $\int_{-1}^{1} (1-x^2)^{\alpha} f(x)\,dx \approx \sum_{i=1}^{N} A_i f(x_i)$

[3] $\int_{-1}^{1} (1+x)^{\beta} f(x)\,dx \approx \sum_{i=1}^{N} A_i f(x_i)$

[4] $\int_{-1}^{1} |x|^{\alpha} f(x)\,dx \approx \sum_{i=1}^{N} A_i f(x_i)$

[5] $\int_{-\infty}^{\infty} e^{-x^2} f(x)\,dx \approx \sum_{i=1}^{N} A_i f(x_i)$

[6] $\int_{0}^{\infty} e^{-x} f(x)\,dx \approx \sum_{i=1}^{N} A_i f(x_i)$

[7] $\int_{-\infty}^{\infty} |x|^{\alpha} e^{-x^2} f(x)\,dx \approx \sum_{i=1}^{N} A_i f(x_i)$

[8] $\int_{-\infty}^{\infty} |x|^{\alpha} e^{-|x|} f(x)\,dx \approx \sum_{i=1}^{N} A_i f(x_i)$

[9] $\int_{0}^{1} \log(1/x)\, f(x)\,dx \approx \sum_{i=1}^{N} A_i f(x_i)$

[10] $\frac{1}{2\pi i} \int_{c-i\infty}^{c+i\infty} p^{-1} e^{p} F(p)\,dp \approx \sum_{i=1}^{N} A_i f(x_i)$

[11] $\int_{-1}^{1} f(x)\,dx \approx A f(-1) + \sum_{i=1}^{N} A_i f(x_i) + A f(+1)$

[12] $\int_{-1}^{1} f(x)\,dx \approx A f(-1) + \sum_{i=1}^{N} A_i f(x_i)$

[13] $\int_{-1}^{1} f(x)\,dx \approx \sum_{i=1}^{N} A_i f(x_i) + \sum_{k=0}^{M} B_{2k} f^{(2k)}(0)$

[14] $\int_{-\infty}^{\infty} e^{-x^2} f(x)\,dx \approx \sum_{i=1}^{N} A_i f(x_i) + \sum_{k=0}^{M} B_{2k} f^{(2k)}(0)$.

Example 3

The book by Krylov and Pal'tsev [5] contains tables of quadrature rules for the following types of integrals:

[1] $$\int_0^1 x^{\alpha} \log\frac{e}{x} f(x)\,dx \approx \sum_{i=1}^{n} A_i f(x_i)$$

[2] $$\int_0^1 x^{\beta} \log\frac{e}{x} \log\frac{e}{1-x} f(x)\,dx \approx \sum_{i=1}^{n} A_i f(x_i)$$

[3] $$\int_0^1 \log\left(\frac{1}{x}\right) f(x)\,dx \approx \sum_{i=1}^{n} A_i f(x_i)$$

[4] $$\int_0^{\infty} x^{\beta} e^{-x} \log\left(1+\frac{1}{x}\right) f(x)\,dx \approx \sum_{i=1}^{n} A_i f(x_i).$$

Notes

[1] Stroud [8] includes in his tables all quadrature rules known to him that have, in his opinion, some major importance. It is an extensive list of quadrature rules. The ones that he deems to be particularly useful are specifically indicated. (Of course, there are infinitely many product rules for some regions; only representative samples of these rules are given.)

[2] Appendix 4 of Davis and Rabinowitz [2] contains a comprehensive bibliography of tabulated quadrature rules.

References

[1] M. Abramowitz and I. A. Stegun, *Handbook of Mathematical Functions*, National Bureau of Standards, Washington, DC, 1964.

[2] P. J. Davis and P. Rabinowitz, *Methods of Numerical Integration*, Second Edition, Academic Press, Orlando, Florida, 1984.

[3] A. Fletcher, J. C. P. Miller, L. Rosenhead, and L. J. Comrie, *An Index of Mathematical Tables*, Addison–Wesley Publishing Co., Reading, MA, 1962.

[4] A. S. Kronrod, *Nodes and Weights of Quadrature Formulas*, Consultants Bureau, New York, 1965.

[5] V. I. Krylov and A. A. Pal'tsev, *Tables for Numerical Integration of Functions with Logarithmic and Power Singularities*, Israel Program for Scientific Translations, Jerusalem, 1971.

[6] V. I. Krylov and N. S. Skoblya, *Handbook of Numerical Inversion of Laplace Transforms*, Israel Program for Scientific Translations, Jerusalem, 1969.

[7] C. H. Love, *Abscissas and Weights for Gaussian Quadrature for $N = 2$ to 100, and $N =$ 125, 150, 175, 200*, National Bureau of Standards Monograph 98, December 1966, National Bureau of Standards, Washington, DC.

[8] A. H. Stroud, *Approximate Calculation of Multiple Integrals*, Prentice–Hall Inc., Englewood Cliffs, NJ, 1971.

[9] A. H. Stroud and D. Secrest, *Gaussian Quadrature Formulas*, Prentice–Hall Inc., Englewood Cliffs, NJ, 1966.

82. Tables: Formulas for Integrals

Applicable to One-dimensional and two-dimensional definite integrals.

Yields

A numerical scheme.

Idea

There exist standard quadrature rules for numerically approximating integrals over intervals and different geometric shapes.

Procedure

There exist standard quadrature rules for numerically integrating different types of integrands. In this section are quadrature rules for numerically integrating one-dimensional and two-dimensional integrals, these formulas are organized by geometric shape. Most of the two-dimensional quadrature rules are from Abramowitz and Stegun [1].

Notes

[1] For Gaussian quadrature rules, see the tables on pages 290 and 298.

[2] For Newton–Cotes rules, see the table on page 320. For open Newton–Cotes rules, see the table on page 321.

One Dimensional Integration Rules

In the following, ξ is some number between x_0 and x_n, and $f_j = f(x_j) = f(x_0 + jh)$ where $h = (x_n - x_0)/n$.

(A) Trapezoidal Rule

$$\int_{x_0}^{x_n} f(x)\,dx = h\left(\frac{1}{2}f_0 + f_1 + f_2 + \cdots f_{n-1} + \frac{1}{2}f_n\right) - \frac{nh^3}{12}f^{(2)}(\xi).$$

(B) Modified Trapezoidal Rule

$$\begin{aligned}\int_{x_0}^{x_n} f(x)\,dx =& h\left(\frac{1}{2}f_0 + f_1 + \ldots + f_{n-1} + \frac{1}{2}f_m\right)\\ &+ \frac{h}{24}\left(-f_{-1} + f_1 + f_{n-1} - f_{n-1}\right) + \frac{11n}{720}h^5 f^{(4)}(\xi).\end{aligned}$$

(C) Simpson's rule

$$\int_{x_0}^{x_2} f(x)\,dx = \frac{h}{3}\left(f_0 + 4f_1 + f_2\right) - \frac{h^5}{90}f^{(4)}(\xi).$$

(D) Extended Simpson's rule

$$\begin{aligned}\int_{x_0}^{x_{2n}} f(x)\,dx = \frac{h}{3}\Big[& f_0 + 4\left(f_1 + f_3 + \ldots + f_{2n-1}\right)\\ &+ 2\left(f_2 + f_4 + \ldots + f_{2n-2}\right)\Big] - \frac{nh^5}{90}f^{(4)}(\xi).\end{aligned}$$

(E) Euler–Maclaurin summation formula (also known as the composite trapezoidal rule)

$$\begin{aligned}\int_{x_0}^{x_n} f(x)\,dx &= h\left[\frac{1}{2}f_0 + f_1 + f_2 + \ldots + f_{n-1} + \frac{1}{2}f_n\right] \\ &\quad - \frac{B_2}{2!}h^2\left(f_n' - f_0'\right) - \ldots - \frac{B_{2k}}{(2k)!}h^{2k}\left(f_n^{(2k-1)} - f_0^{(2k-1)}\right) + R_{2k}\end{aligned}$$

where $R_{2k} = \dfrac{\theta n B_{2k+2} h^{2k+3}}{(2k+2)!} \max_{x_0 \leq x \leq x_n} \left|f^{(2k+2)}(x)\right|$, with $-1 \leq \theta \leq 1$. Here, the B_k's are Bernoulli numbers (see [1], 25.4.7).

(F) Five-point rule for analytic functions (f must be analytic)

$$\begin{aligned}\int_{z_0-h}^{z_0+h} f(z)\,dz = \frac{h}{15}\Big[&24f(z_0) + 4f(z_0+h) + 4f(z_0-h) \\ &- 4f(z_0+ih) + 4f(z_0-ih)\Big] + R\end{aligned}$$

where $|R| \leq \dfrac{|h|^7}{1890} \max_{z \in S} \left|f^{(6)}(z)\right|$ and S is the square with vertices at $\{z_0 + i^k h \mid k = 0, 1, 2, 3\}$.

Integration Formulae for Different Geometric Shapes

Circumference of a circle If Γ represents the circumference of the circle $x^2 + y^2 = h^2$, then we have the approximate integration rules (see [1], equation 25.4.60)

$$\frac{1}{2\pi h}\int_{\Gamma} f(x,y)\,ds = \frac{1}{2m}\sum_{n=1}^{2m} f\left(h\cos\frac{\pi n}{m}, h\sin\frac{\pi n}{m}\right) + O\left(h^{2m-2}\right)$$

where $m \geq 1$. The following figure indicates the location of the 12 points when $m = 6$.

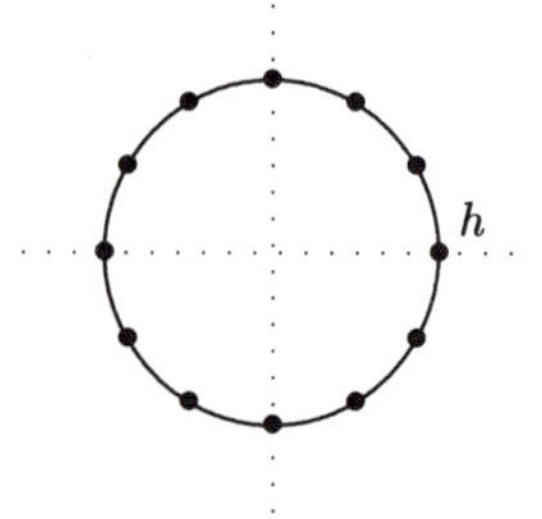

Circular region If C represents the circle with radius h $(x^2+y^2 \leq h^2)$, then we have the approximate integration rules (see [1], equation 25.4.61)

$$\frac{1}{\pi h^2} \iint_C f(x,y)\,dx\,dy = \sum_{i=1}^{n} w_i f(x_i, y_i) + R:$$

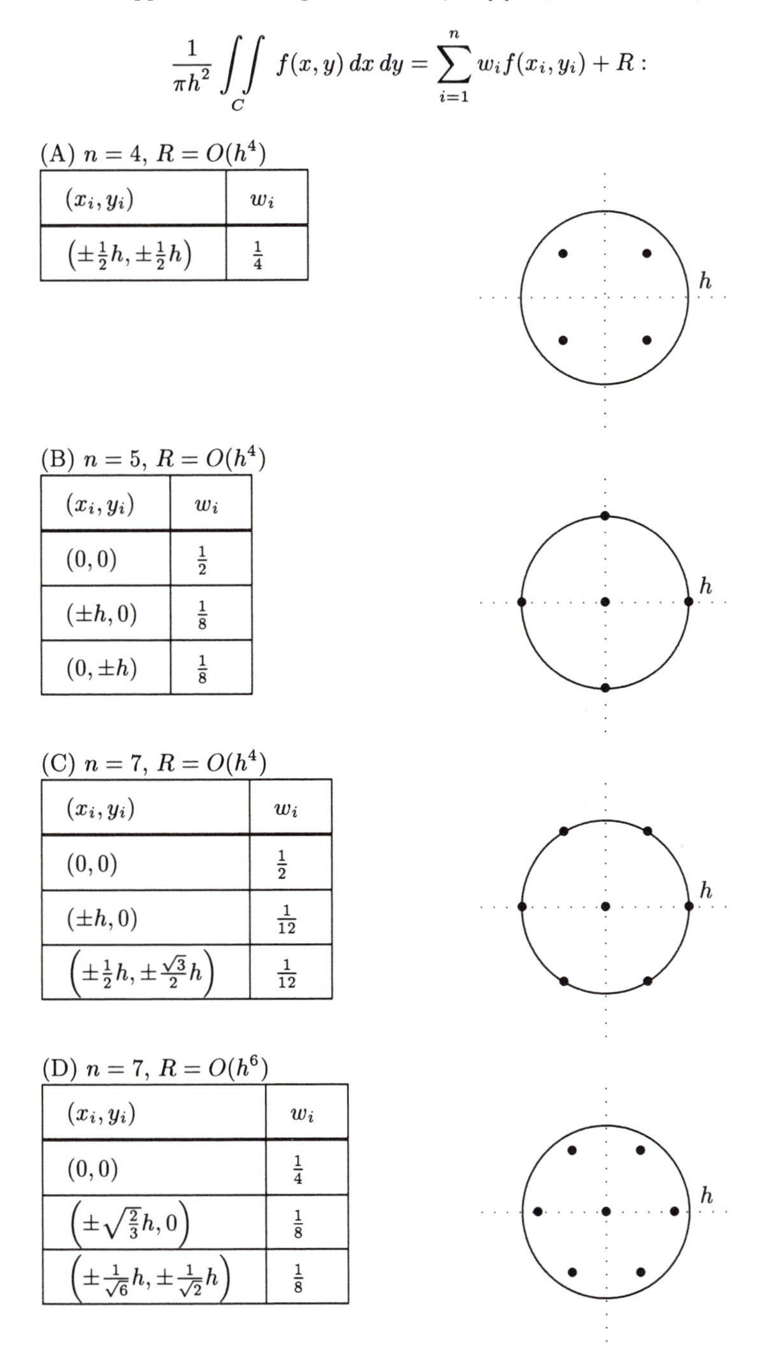

(A) $n=4$, $R=O(h^4)$

(x_i, y_i)	w_i
$\left(\pm\frac{1}{2}h, \pm\frac{1}{2}h\right)$	$\frac{1}{4}$

(B) $n=5$, $R=O(h^4)$

(x_i, y_i)	w_i
$(0,0)$	$\frac{1}{2}$
$(\pm h, 0)$	$\frac{1}{8}$
$(0, \pm h)$	$\frac{1}{8}$

(C) $n=7$, $R=O(h^4)$

(x_i, y_i)	w_i
$(0,0)$	$\frac{1}{2}$
$(\pm h, 0)$	$\frac{1}{12}$
$\left(\pm\frac{1}{2}h, \pm\frac{\sqrt{3}}{2}h\right)$	$\frac{1}{12}$

(D) $n=7$, $R=O(h^6)$

(x_i, y_i)	w_i
$(0,0)$	$\frac{1}{4}$
$\left(\pm\sqrt{\frac{2}{3}}h, 0\right)$	$\frac{1}{8}$
$\left(\pm\frac{1}{\sqrt{6}}h, \pm\frac{1}{\sqrt{2}}h\right)$	$\frac{1}{8}$

(E) $n = 9$, $R = O(h^6)$

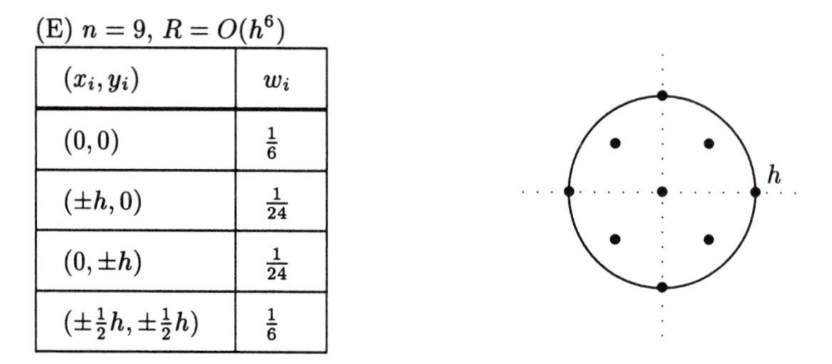

(x_i, y_i)	w_i
$(0, 0)$	$\frac{1}{6}$
$(\pm h, 0)$	$\frac{1}{24}$
$(0, \pm h)$	$\frac{1}{24}$
$(\pm\frac{1}{2}h, \pm\frac{1}{2}h)$	$\frac{1}{6}$

Squares If S represents a square with each side of length $2h$ ($|x| \leq h$, $|y| \leq h$), then we have the approximate integration rules (see [1], equation 25.4.62 and Stroud [3])

$$\frac{1}{4h^2} \iint\limits_S f(x, y)\, dx\, dy = \sum_{i=1}^{n} w_i f(x_i, y_i) + R:$$

(A) $n = 3$, $R = O(h^3)$

(x_i, y_i)	w_i
$\left(\sqrt{\frac{2}{3}}h, 0\right)$	$\frac{1}{3}$
$\left(-\frac{1}{\sqrt{6}}h, \frac{1}{\sqrt{2}}h\right)$	$\frac{1}{3}$
$\left(-\frac{1}{\sqrt{6}}h, -\frac{1}{\sqrt{2}}h\right)$	$\frac{1}{3}$

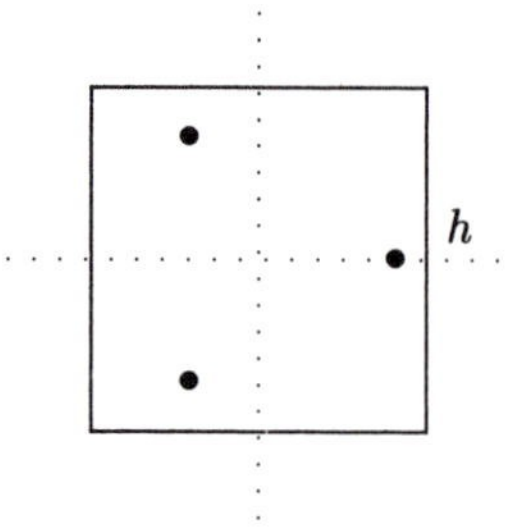

(B) $n = 3$, $R = O(h^3)$

(x_i, y_i)	w_i
(h, h)	$\frac{1}{7}$
$(\frac{1}{3}h, -\frac{2}{3}h)$	$\frac{3}{8}$
$(-\frac{5}{9}h, \frac{2}{9}h)$	$\frac{27}{56}$

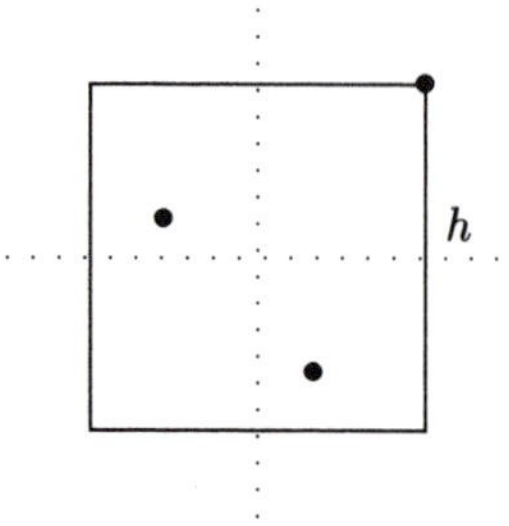

(C) $n = 3$, $R = O(h^3)$

(x_i, y_i)	w_i
$(h, \frac{4}{3}h)$	$\frac{3}{28}$
$(\frac{1}{3}h, 0)$	$\frac{3}{4}$
$(h, -h)$	$\frac{1}{7}$

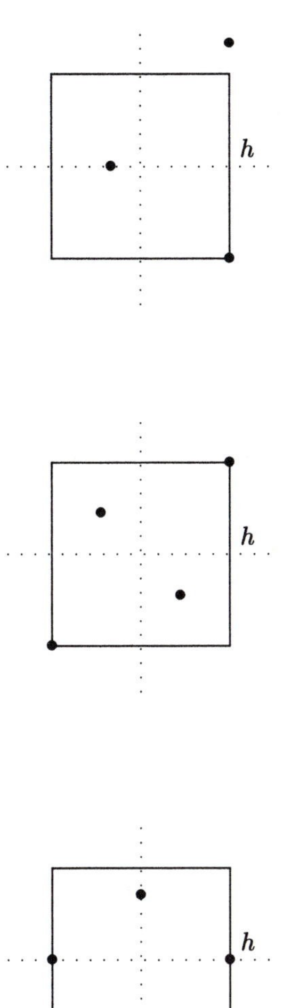

(D) $n = 4$, $R = O(h^4)$

(x_i, y_i)	w_i
(h, h)	$\frac{1}{12}$
$(-h, -h)$	$\frac{1}{12}$
$\left(-\frac{1}{\sqrt{5}}h, \frac{1}{\sqrt{5}}h\right)$	$\frac{5}{12}$
$\left(\frac{1}{\sqrt{5}}h, -\frac{1}{\sqrt{5}}h\right)$	$\frac{5}{12}$

(E) $n = 4$, $R = O(h^4)$

(x_i, y_i)	w_i
$(\pm h, 0)$	$\frac{1}{6}$
$\left(0, \pm\frac{1}{\sqrt{2}}h\right)$	$\frac{1}{3}$

(F) $n = 4$, $R = O(h^4)$

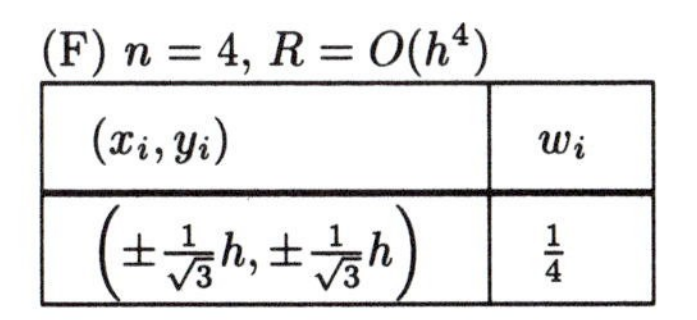

(x_i, y_i)	w_i
$\left(\pm\frac{1}{\sqrt{3}}h, \pm\frac{1}{\sqrt{3}}h\right)$	$\frac{1}{4}$

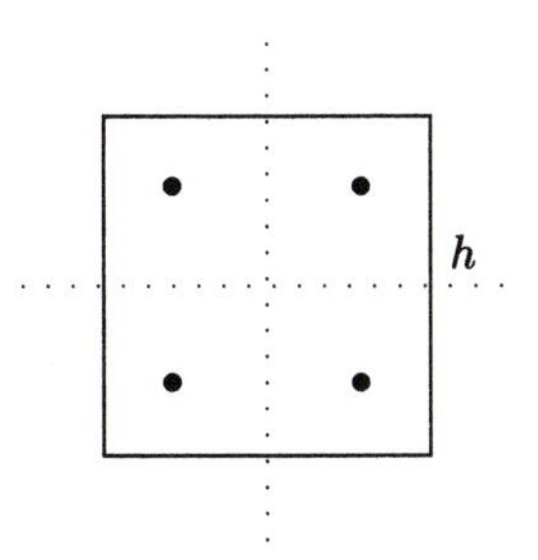

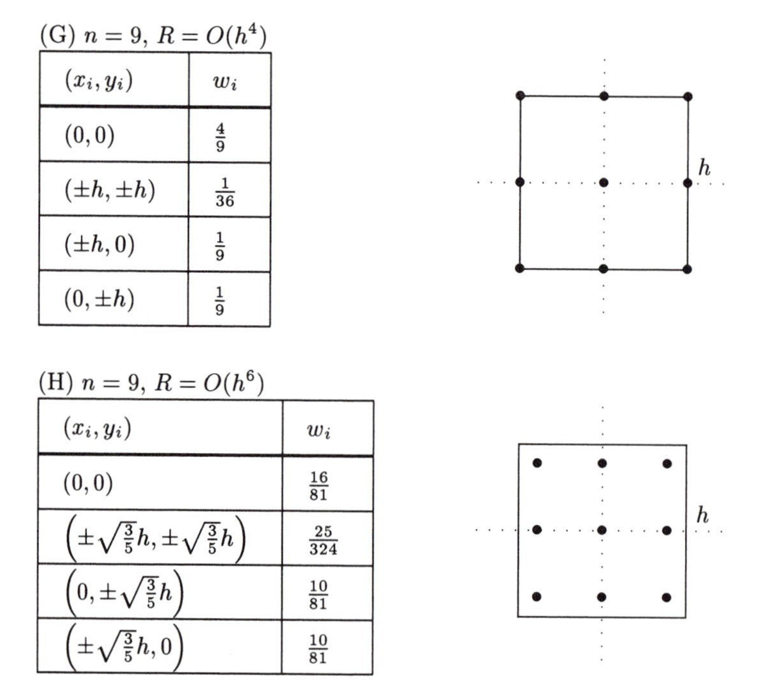

(G) $n = 9$, $R = O(h^4)$

(x_i, y_i)	w_i
$(0, 0)$	$\frac{4}{9}$
$(\pm h, \pm h)$	$\frac{1}{36}$
$(\pm h, 0)$	$\frac{1}{9}$
$(0, \pm h)$	$\frac{1}{9}$

(H) $n = 9$, $R = O(h^6)$

(x_i, y_i)	w_i
$(0, 0)$	$\frac{16}{81}$
$\left(\pm\sqrt{\frac{3}{5}}h, \pm\sqrt{\frac{3}{5}}h\right)$	$\frac{25}{324}$
$\left(0, \pm\sqrt{\frac{3}{5}}h\right)$	$\frac{10}{81}$
$\left(\pm\sqrt{\frac{3}{5}}h, 0\right)$	$\frac{10}{81}$

Equilateral triangle If T represents an equilateral triangle, then we have the approximate integration rules (see [1], equation 25.4.63)

$$\frac{1}{\frac{3}{4}\sqrt{3}h^2} \iint_T f(x, y)\, dx\, dy = \sum_{i=1}^{n} w_i f(x_i, y_i) + R :$$

(A) $n = 4$, $R = O(h^3)$

(x_i, y_i)	w_i
$(0, 0)$	$\frac{3}{4}$
$(h, 0)$	$\frac{1}{12}$
$\left(-\frac{1}{2}h, \pm\frac{\sqrt{3}}{2}h\right)$	$\frac{1}{12}$

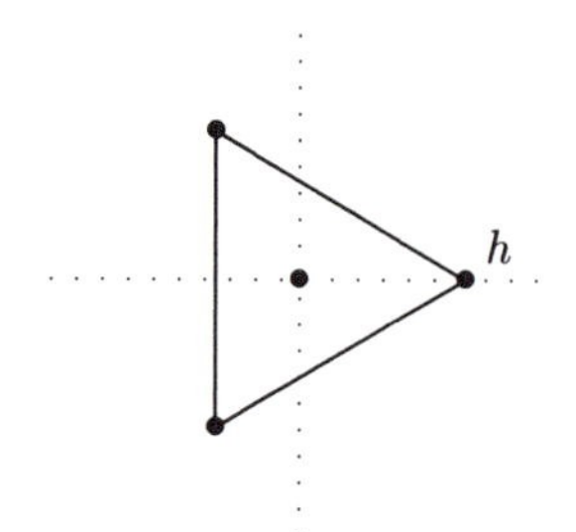

(B) $n = 7$, $R = O(h^4)$

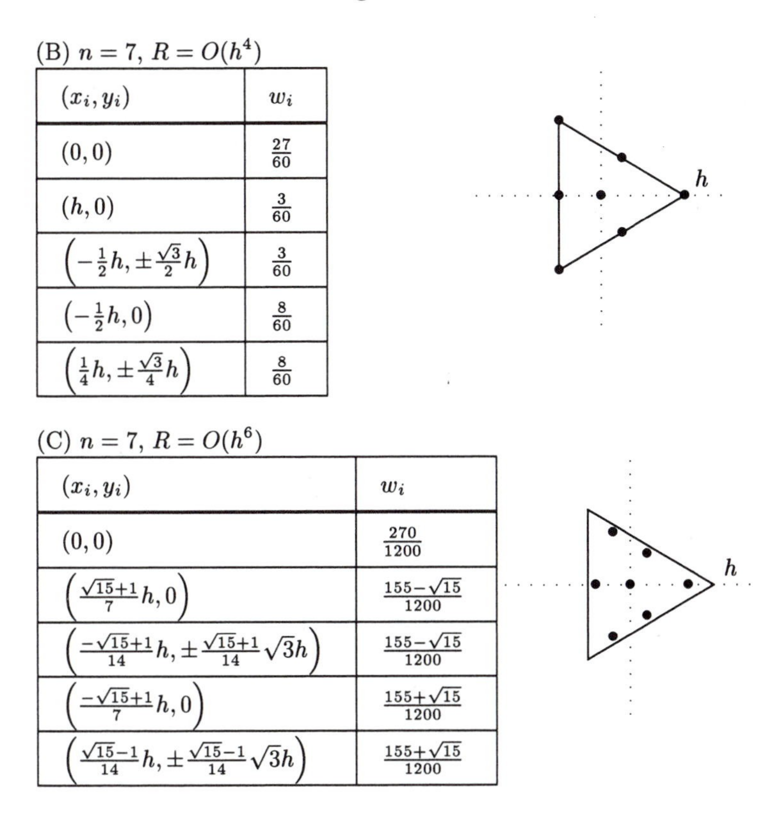

(x_i, y_i)	w_i
$(0, 0)$	$\frac{27}{60}$
$(h, 0)$	$\frac{3}{60}$
$\left(-\frac{1}{2}h, \pm\frac{\sqrt{3}}{2}h\right)$	$\frac{3}{60}$
$\left(-\frac{1}{2}h, 0\right)$	$\frac{8}{60}$
$\left(\frac{1}{4}h, \pm\frac{\sqrt{3}}{4}h\right)$	$\frac{8}{60}$

(C) $n = 7$, $R = O(h^6)$

(x_i, y_i)	w_i
$(0, 0)$	$\frac{270}{1200}$
$\left(\frac{\sqrt{15}+1}{7}h, 0\right)$	$\frac{155-\sqrt{15}}{1200}$
$\left(\frac{-\sqrt{15}+1}{14}h, \pm\frac{\sqrt{15}+1}{14}\sqrt{3}h\right)$	$\frac{155-\sqrt{15}}{1200}$
$\left(\frac{-\sqrt{15}+1}{7}h, 0\right)$	$\frac{155+\sqrt{15}}{1200}$
$\left(\frac{\sqrt{15}-1}{14}h, \pm\frac{\sqrt{15}-1}{14}\sqrt{3}h\right)$	$\frac{155+\sqrt{15}}{1200}$

Regular hexagon If H represents a regular hexagon, then we have the approximate integration rules (see [1], equation 25.4.64)

$$\frac{1}{\frac{3}{2}\sqrt{3}h^2} \iint_H f(x, y)\, dx\, dy = \sum_{i=1}^{n} w_i f(x_i, y_i) + R :$$

(A) $n = 7$, $R = O(h^4)$

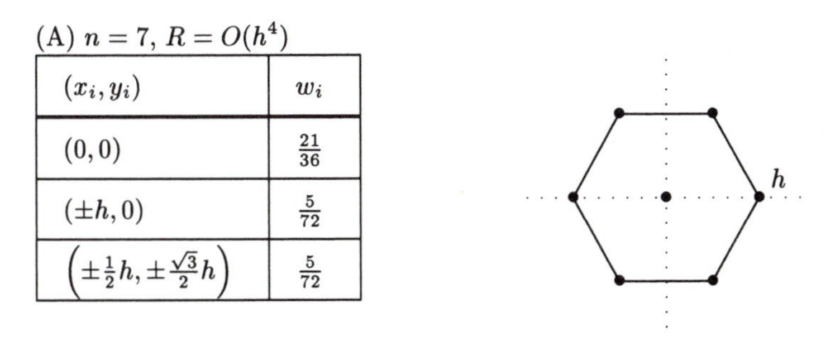

(x_i, y_i)	w_i
$(0, 0)$	$\frac{21}{36}$
$(\pm h, 0)$	$\frac{5}{72}$
$\left(\pm\frac{1}{2}h, \pm\frac{\sqrt{3}}{2}h\right)$	$\frac{5}{72}$

(B) $n = 7$, $R = O(h^6)$

(x_i, y_i)	w_i
$(0,0)$	$\frac{258}{1008}$
$(\pm\frac{\sqrt{14}}{5}h, 0)$	$\frac{125}{1008}$
$\left(\pm\frac{\sqrt{14}}{10}h, \pm\frac{\sqrt{42}}{10}h\right)$	$\frac{125}{1008}$

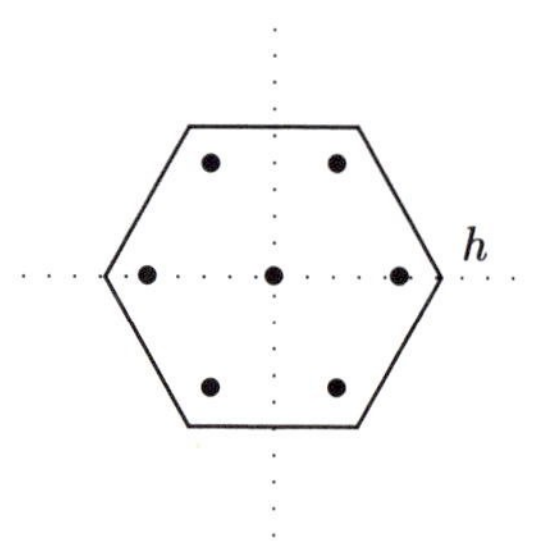

References

[1] M. Abramowitz and I. A. Stegun, *Handbook of Mathematical Functions*, National Bureau of Standards, Washington, DC, 1964.

[2] F. B. Hiderbrand, *Introduction to Numerical Analysis*, McGraw–Hill Book Company, New York, 1974.

[3] A. H. Stroud, *Approximate Calculation of Multiple Integrals*, Prentice–Hall Inc., Englewood Cliffs, NJ, 1971, Chapter 3, pages 53–126.

83. Tables: Numerically Evaluated Integrals

Applicable to Specific definite integrals.

Yields

An exact evaluation.

Idea

The numerical evaluation of many integrals has been tabulated.

Procedure

When computers were not as common as they are now, it was useful to have tabulated values of certain definite integrals. Tables were published containing these values.

These tables are less useful now that computers are readily available to perform numerical integrations as needed. They are still of occasional use, though, for testing new computational routines.

Example 1

(A) For tables of Anger functions, see Abramowitz and Stegun [2] or Bernard and Ishimaru [6].

(B) For tables of Bessel functions, see Abramowitz and Stegun [2] or Haberman and Harley [9].

(C) For tables of the cosine integral, see Abramowitz and Stegun [2], or reference [15].

(D) For tables of elliptic functions, see Abramowitz and Stegun [2], Belyakov *et al.* [4], Fettis and Caslin [8], or Selfridge and Maxfield [13].

(E) For tables of the Fresnel integral, see Abramowitz and Stegun [2], Martz [10], or Pearcey [11].

(F) For tables of the Gamma function, see Abramov [1] or Pearson [12].

(G) For tables of the sine integral, see Abramowitz and Stegun [2], or reference [15].

(H) For a table of the transport integral $\int_0^x \frac{e^z z^n \, dz}{(e^z - 1)^2}$, see Rogers and Powell [14].

(I) For a table of the function $\int_0^x \frac{\gamma(y, \xi)}{\xi} \, d\xi$, see Anker and Gafarian [3].

Example 2

(A) For a computation method for computing the polygamma function, see DiMarzio [7].

Notes

[1] Abramowitz and Stegun [2] also have tables of Clausen's integral, Debye function, dilogarithm, exponential integral, Sievert integral, and Struve functions.

[2] Many of the tables referenced in this section are now superfluous as the numerical values of the integrals can be readily computed. For example, Mathematica [16] has special commands for computing elliptic integrals, Fresnel integrals, dilogarithms, and many other functions. Also, for those functions for which a single command does not exist, the numerical integration routine can be used.

References

[1] A. A. Abramov, *Tables of* Ln $\Gamma[z]$ *for Complex Arguments*, translated by D. G. Fry, Pergamon Press, New York, 1960.

[2] M. Abramowitz and I. A. Stegun, *Handbook of Mathematical Functions*, National Bureau of Standards, Washington, DC, 1964.

[3] C. J. Anker, Jr. and A. V. Gafarian, *The Function* $J(x, y) = \int_0^x \frac{\gamma(y, \xi)}{\xi} \, d\xi$; *Some Properties and a Table*, System Development Corporation, 2500 Colorado Ave, Sanata Monica, CA, April 1962.

[4] V. M. Belyakov, P. I. Kravtsova, and M. G. Rappoport, *Tables of Elliptic Integrals*, translated by P. Basu, The MacMillan Company, New York, 1965.

[5] O. S. Berlyand, R. I. Gavrilova, and A. P. Prudnikov, *Tables of Integral Error Functions and Hermite Polynomials*, translated by P. Basu, The MacMillan Company, New York, 1962.

[6] G. D. Bernard and A. Ishimaru, *Tables of the Anger and Lommel–Weber Functions*, University of Washington Press, Seattle, 1962.

[7] F. DiMarzio, "An Improved Procedure for the Accurate Evaluation of Polygamma Functions with Integer and Half-Integer Argument," *Comput. Physics Comm.*, **39**, 1986, pages 343–345.

[8] H. E. Fettis and J. C. Caslin, *A Table of the Complete Elliptic Integral of the First Kind for Complex Values of the Modulus. Part I*, ARL 69-0172, 1969, *A Table of the Complete Elliptic Integral of the First Kind for Complex Values of the Modulus. II*, ARL 69-0173, 1969, *A Table of the Complete Elliptic Integral of the First Kind for Complex Values of the Modulus: III. Auxiliary Tables*, ARL 70-0081, 1970, United States Air Force, Wright–Patterson Air Force Base, OH.

[9] W. L. Haberman and E. E. Harley, *Numerical Evaluation of Integrals Containing Modified Bessel Functions*, Hydromechanics Laboratory, Research and Development Report 1580, March 1964, Department of the Navy, Washington, DC.

[10] C. W. Martz, "Tables of the Complex Fresnel Integral," NASA SP-3010, NASA, Washington, DC, 1964.

[11] T. Pearcey, "Table of the Fresnel Integral to Six Decimal Places," Cambridge University Press, New York, 1956.

[12] K. Pearson, *Tables of the Incomplete Γ-Function*, Cambridge University Press, 1934.

[13] R. G. Selfridge and J. E. Maxfield, *A Table of the Incomplete Elliptic Integral of the Third Kind*, Dover Publications, Inc., New York, 1958.

[14] W. M. Rogers and R. L. Powell, *Tables of Transport Integrals* $J_n(x) \equiv \int_0^x \frac{e^z z^n \, dz}{(e^z - 1)^2}$, National Bureau of Standards Circular 595, July 3, 1958, National Bureau of Standards, Washington, DC.

[15] Staff of the Computation Laboratory, *Tables of Generalized Sine- and Cosine-Integral Functions: Part I*, Harvard University Press, Cambridge, MA, 1949.

[16] S. Wolfram, *Mathematica: A System of Doing Mathematics by Computer*, Second Edition, Addison–Wesley Publishing Co., Reading, MA, 1991.

Mathematical Nomenclature

$C^p[a,b]$ The class of functions that are continuous and have p continuous derivatives, on the interval $[a,b]$.

E the expectation operator, see page 30.

E the stepping operator, see page 329.

$\mathcal{F}[u(x)]$ the Fourier transform of the function $u(x)$, defined by $\dfrac{1}{\sqrt{2\pi}}\displaystyle\int_{-\infty}^{\infty} u(x)e^{ix\omega}\,dx$. The inverse transform, denoted by $\mathcal{F}^{-1}[U(\omega)]$, is defined by $\dfrac{1}{\sqrt{2\pi}}\displaystyle\int_{-\infty}^{\infty} U(\omega)e^{-ix\omega}\,d\omega$.

$H(x)$ the Heaviside or step function, defined by
$H(x) := \int_{-\infty}^{x} \delta(x)\,dx = \begin{cases} 0 & \text{if } x < 0 \\ 1/2 & \text{if } x = 0 \\ 1 & \text{if } x > 0 \end{cases}$.

$\mathcal{H}[u(x)]$ the Hilbert transform of the function $u(x)$, defined by $\displaystyle\int_{-1}^{1} \frac{u(x)}{x-t}\,dx$.

$\mathcal{L}[u(x)]$ the Laplace transform of the function $u(x)$, defined by $\int_0^\infty u(x)e^{-sx}\,dx$. The inverse transform, denoted by $\mathcal{L}^{-1}[U(s)]$, is defined by the Bromwich integral $\int_{\mathcal{C}} U(s)e^{sx}\,ds$, where $\mathcal{C}$ is Bromwich contour (a contour that is to the right of all of the singularities of $U(s)$, and is closed in the left half plane).

$L_p[a,b]$ The class of functions $\{u(x)\}$ that satisfy $\int_{-\infty}^{\infty} |u(x)|^p\,dx < \infty$.

$\mathcal{M}[f(t)]$ the Mellin transform of the function $f(t)$, defined by $\int_0^\infty t^{z-1}f(t)\,dt$.

O We say that $f(x) = O(g(x))$ as $x \to x_0$ if there exists a positive constant C and a neighborhood U if x_0 such that $|f(x)| \le C|g(x)|$ for all x in U.

o We say that $f(x) = o(g(x))$ as $x \to x_0$ if, given any $\mu > 0$ there exists a neighborhood U of x_0 such that $|f(x)| < \mu|g(x)|$ for all x in U.

$\mathcal{P}$ the Weierstrass function (an elliptic integral). See page 156.

sgn the signum function, this returns the sign of its argument. If $x > 0$, then $\operatorname{sgn} x$ is $+1$. If $x < 0$, then $\operatorname{sgn} x$ is -1. If $x = 0$, then $\operatorname{sgn} x$ is indeterminate.

δ the central-differencing operator, see page 329.

δ_{ij} the Kronecker delta, it has the value 1 if $i = j$ and the value 0 if $i \neq j$.

$\delta(x)$ the delta function, it has the properties that $\delta(x) = 0$ for $x \neq 0$, but $\int_{-\infty}^{\infty} \delta(x)\, dx = 1$.

Δ the forward-differencing operator, see page 329.

ε generally a small number, usually assumed to be much less than one in magnitude.

∇ the backward-differencing operator, see page 329.

∂S If S is a region or volume, then ∂S denotes its boundary.

$||\cdots||$ the norm of the argument.

$⨎$ This is a Hadamard finite part integral, see page 73.

$\mathcal{I}\!\!\int$ This is an Ito integral, see page 187.

$\oint$ This is a line integral around a closed curve, see page 50 or page 168.

$⨍$ This is a principle value integral, see page 92.

$\mathcal{S}\!\!\int$ This is a Stratonovich integral, see page 189.

$[\,]^-$ The negative part of a function:

$$[f(x)]^- := -\sup\{-f(x), 0\} = \tfrac{1}{2}\left(|f(x)| - f(x)\right) = \begin{cases} 0 & \text{if } f(x) \geq 0, \\ f(x) & \text{if } f(x) < 0. \end{cases}$$

$[\,]^+$ The positive part of a function:

$$[f(x)]^+ := \sup\{f(x), 0\} = \tfrac{1}{2}\left(|f(x)| + f(x)\right) = \begin{cases} f(x) & \text{if } f(x) \geq 0, \\ 0 & \text{if } f(x) < 0, \end{cases}$$

Index

A

B

C

D

E

F

I

N

O

P

Q

R

S

U

V

W

Y

Z